Elementary Statistics
for Geographers

Elementary Statistics
for Geographers

GERALD M. BARBER

THE GUILFORD PRESS
New York   London

© 1988 The Guilford Press

A Division of Guilford Publications, Inc.

72 Spring Street, New York, N.Y. 10012

Printed in the United States of America

Last digit is print number: 9 8 7 6 5 4 3 2

Library of Congress Cataloging-in-Publication Data

Barber, Gerald M.
   Elementary statistics for geographers.

   Bibliography: p.
   Includes index.
   1. Geography—Statistical methods.   I. Title.
G70.3.B37   1988   910'.21   87-8801
ISBN 0-89862-777-X

Preface

The recent trend to teach the principles of statistical analysis in the social sciences has led to a remarkable increase in the number of introductory statistics courses being taught outside the traditional mathematics curriculum. Many of these courses are taken by students whose mathematical backgrounds are limited. With such a clientele, it has been especially common to teach introductory statistical methods without presenting many of the fundamental ideas of statistical analysis. The implicit argument is that these principles cannot be taught without a mathematically rigorous presentation. I believe that the acceptance of this notion has led to the development of courses in geography that treat statistical methods in cookbook manner and tend to emphasize breadth rather than depth of coverage. Students are taught how to do it, but not why. The basic logic of statistical procedures is usually glossed over, and typical students, while being exposed to many different aspects of statistical analysis, often fail to understand the most basic ideas of the field. When these students are subsequently urged to utilize these techniques or interpret the results of research that incorporates statistical methods, they are at a loss to do so. However, even without an extensive treatment of the underlying mathematics, many of the principles of statistical analysis can be effectively presented to geography students.

For this reason this book gives a much more extensive treatment to certain topics commonly avoided in introductory statistical methods courses. Many of the ideas of statistical inference are based on probability, and it is foolish to suppose that geographers can somehow understand the logic of statistical inference without it. Most superior statistical textbooks in the social sciences (particularly in economics) recognize this fact. It is my belief that this material, if competently presented, is not beyond the ability of the typical geography student. It is equally important that the standard at which statistical methods are taught to geography students be raised, particularly if they are to compete with other social science students. However, clearly not all geography students require an understanding of statistics at the depth presented in this text. For many students, though, it is essential.

This text is divided into three main parts. Part I contains two chapters covering descriptive statistics. Chapter 2 includes material from both conventional univariate descriptive statistics and newer techniques of exploratory data analysis. Chapter 3 discusses methods of summarizing geographical data expressed in areal, point, or directional form. Part II comprises seven chapters of material covering inferential statistics for one variable. Chapters 4 and 5 present sufficient probability theory for the student to understand and appreciate the logic of statistical inference. Emphasis is placed on the notion of a random variable and its role in the process of statistical inference. Chapter 6 discusses sampling, in both geographical and nongeographical contexts.

The core of statistical inference procedures is explained in Chapters 7 through 10. Some emphasis is placed on the technique of PROB-VALUE (or p-value) testing, which is shown to be closely related to conventional hypothesis testing. Since many statistical software packages report tests of significance in this way, this knowledge should help students in relating the theory of statistical inference to the typical computer output from statistical software.

Part III examines statistical relationships between two variables. In Chapter 11, measures of the strength of the linear relationship between two variables are presented. Chapters 12 and 13 present descriptive and inferential aspects of linear regression analysis. In all three parts, examples of the use of the Statistical Analysis System (SAS) are provided. I believe that all courses at this level should contain an introduction to a statistical software package, whether it be SAS, SPSS (Statistical Procedures for the Social Sciences), MINITAB, or some other suitable system. Although I tend to favor instruction on microcomputers, the use of any type of computing environment is highly recommended.

Because of the variety of statistical and quantitative methods courses in the discipline, this text has been designed to serve either a one-semester, two-quarter sequence or a year-long introductory course. In a one-semester (or even a two-quarter) sequence, the core material includes Chapters 1 through 8 and covers the topics commonly labeled classical statistical inference. In addition, up to three other chapters of the remaining five might be selected. If the goal of the course is to provide a good background in statistical inference, Chapters 9 and 10 might be added. These two chapters cover two-sample inferential tests and nonparametric statistics. A more common alternative is probably to add the last three chapters of the text covering correlation and regression analysis. A third alternative is to include the chapter on nonparametric procedures (Chapter 10) and the chapters introducing correlation and regression analysis (Chapters 11 and 12). Again, it is desirable to include instruction in statistical analysis using the computer. Some sections of certain chapters can be considered optional. For a course lasting an entire academic year, a two-semester sequence, or a three-quarter sequence, virtually the entire book can be used, along with suitable instruction in computer software. Again, certain sections of some chapters might be omitted. The possible formats for courses of varying length are summarized in the accompanying diagrams.

At the ends of the chapters are sections called References, Further Reading, and Problems. The References include a complete citation of any article or book referred to in the chapter, and Further Reading contains a selection of readings which students may find useful on particular topics. A short description of the contents of the citations is also included. The goal of this section is not to be comprehensive or to point out the most advanced material available. Rather, the intention is to provide reference to a few widely available items that cover the relevant topic in a different,

One-semester course (or two-quarter sequence).

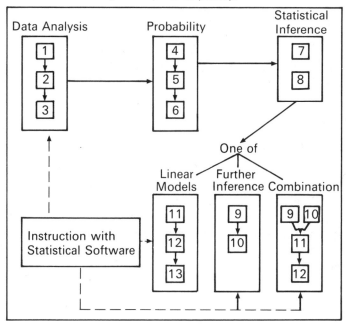

Year-long course, two-semester sequence (or three-quarter sequence).

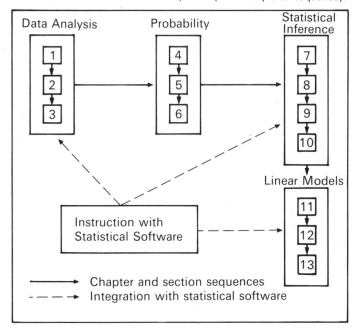

but not necessarily more mathematically rigorous, manner than that provided in the text. Students sometimes find it useful to consult a textbook which simply presents the material in a slightly different way. However, this is not always the case. Sometimes, students consulting other textbooks are befuddled by a slightly different notation. Nevertheless, I think it important to offer a few alternative presentations, applications, or extensions of the material presented in most chapters. Instructors are encouraged to supplement (or even replace) this list with other applications suitable to their particular courses. I often find it irritating to see examples of the use of statistics drawn from odd places—for example, pebble orientation in the Upper Crawsdale region of the Lower Pennines—and my impression is that they are of dubious value to students.

The Problems given at the end of the chapter serve two purposes. First, I like to simply list the terms or concepts introduced in the chapter. Students are urged to write short definitions or explanations of these terms to reinforce their understanding of new material. Many of these concepts are highlighted within the text, especially in the first few chapters. Important formulas are likewise highlighted. These conventions may minimize the problems a student faces when confronted by the new and seemingly endless jargon of statistical analysis. Not all the Problems at the end of each chapter describe geographical applications of material contained in the chapter. The emphasis has been on easy-to-understand examples, often of an everyday problem. Sometimes for simplicity the problem is purely numerical. Most problems are simple applications of the techniques presented in that chapter. At times, students are asked to examine data that have been analyzed in a different way in another chapter.

Appendices are included in several chapters and also serve a number of purposes. A few chapters contain reviews of simple mathematical material (e.g., the geometry of a straight line in Chapter 12). Other appendices contain mathematical proofs of theorems presented in the chapter. These are not considered essential but are offered for completeness. A few students may have sufficient background to understand this material and are urged to do so. Finally, applications of the concepts in a chapter are sometimes illustrated with a program and computer output of SAS. Most often, the data used in these programs are the same data used in the development of a key example in the chapter.

No preface would be complete without acknowledging the help received in developing the text. Students of the Department of Geography at the University of Victoria, through their reactions to my presentations of most material included here, helped me improve the way in which I introduce difficult concepts. I have also had considerable help from Janet Crane of The Guilford Press, several reviewers whom she has called upon to comment upon various drafts of this text, and a meticulous copyeditor supervised by Judith Grauman. Finally, I must acknowledge the considerable debt to my family, who have, at times, paid the price for my decision to write this book. I dedicate this book to them: Moe, Eric, Gordie, and Mike.

Contents

Elementary Statistics
for Geographers

1

Statistics and Geography

Hardly a day goes by without almost everyone being affected by the applications of statistics. Commonly the word *statistics* refers to any collection of numerical data. *Vital statistics* such as birth and death rates, *economic indicators* such as unemployment rates and money supply figures, and *social statistics* such as juvenile delinquency rates are all statistics used in this sense. A brief examination of a major daily newspaper in almost any city reveals several stories dealing with numerical data. Statistical data of this type have been used for centuries by various governments to improve administration. In antiquity, statistics were compiled to determine the number of citizens liable to both taxation and military service. The role of statistics expanded rapidly following the industrial revolution. Today, statistical data are collected by many individuals, government agencies, public or nonprofit organizations, and businesses. These data are included in a variety of information systems. Many people owe their current employment to the explosion in the use of statistical data. With the rapid developments in computer technology, increasingly sophisticated systems have been developed to handle the enormous task of data management.

The word *statistics* has another, more specialized meaning: It is the methodology for collecting, presenting, and analyzing data. This methodology can be used as a basis for investigation in such diverse academic fields as education, physics and engineering, medicine, the biological sciences, and the social sciences including geography. Even the traditionally nonquantitative disciplines in the humanities are finding increasing uses for statistical methodology. No matter which discipline utilizes this methodology, statistical analysis begins with the collection of data. Analyses of these data are then undertaken for one or more of the following purposes:

1. To help *summarize* the findings of some inquiry, for example, a study of travel behavior of elderly or handicapped citizens or the estimation of timber reforestation requirements
2. To obtain a *better understanding* of the phenomenon under study, primarily as an aid in *generalization* or *theory validation*, for example, to validate a theory of urban land rent
3. To make a *forecast* of some variable, for example, interest rates, voter behavior, or house prices

4. To *evaluate* the performance of some program, for example, a particular form of industrial subsidy or an innovative medical or educational program
5. To help *select* a course of action among a set of possible alternatives or to *plan* some system, for example, school locations

The fact that elements of statistical methodology are applicable to such a wide variety of problems attests to its versatility.

It is convenient to divide statistical analysis into two parts. *Descriptive statistics* deals with the organization and summary of data. The purpose of descriptive statistics is to replace an often very large set of numbers contained in some data set by a smaller number of summary measures. Whenever this replacement is made, there is inevitably some loss of information. It is impossible to *retain* all the information in a data set using a smaller set of numbers. One goal of descriptive statistics is to minimize the effects of this information loss. Understanding which statistical measure should be used as a summary index in a particular case is an important goal in descriptive statistics. If we understand the derivation and use of descriptive statistics and are aware of its limitations, we can help to avoid the propagation of misleading results. Much of the distrust of statistical method felt by some individuals derives from their dissatisfaction with studies in which the statistical methodology may have been inappropriately applied or interpreted. Just as a photographer can use a lens to distort the scene, so can a statistician distort the information in a data set through a poor choice of summary statistics. Understanding what descriptive statistics can tell use, *as well as what it cannot*, is a key concern in statistical analysis.

The second portion of statistical methodology is termed *inferential statistics*. In this form of statistical analysis, descriptive statistics is linked with probability theory so that an investigator can generalize the results of a study of a few individuals to some larger group. To clarify this process, it is necessary to introduce a few simple definitions. The set of persons, regions, areas, or objects in which an investigator has an interest is known as the *population* for the study.

DEFINITION: STATISTICAL POPULATION

A statistical population is the total set of elements (objects, persons, regions, neighborhoods, rivers, etc.) under study in a particular problem.

For instance, for a geographer studying farm practices in a particular region, the relevant population consists of all farms in the region on a certain date or within a certain period. As a second example, the population for a study of voter behavior in a city would include all potential voters; these people are usually contained in an eligible-voters list.

In many instances, the statistical population under consideration is finite; that is, the total number of elements in the population can be listed. The eligible-voters list and the assessment rolls of a city or country are examples of *finite* populations. At other times, the population may be *hypothetical*. For example, a steel manufacturer wishing to test the quality of output may select a batch of 100 castings over a few weeks of production. The population under study is really the future set of castings to be produced by the manufacturer using this equipment. Of course, this population does not exist and has an infinitely large number of elements. Statistical analysis is relevant to both finite and hypothetical populations.

Usually, we are interested in one or more characteristics of the population.

DEFINITION: POPULATION CHARACTERISTIC
A population characteristic is any measurable attribute of an element in the population.

A fluvial geomorphologist studying stream flow in a certain watershed may be interested in a number of different measurable properties of these streams. Stream velocity, discharge, sediment load, and many other characteristic channel data may be collected during some study. Since a population characteristic usually takes on different values for different elements of the population, it is usually termed a *variable*. The fact that the population characteristic does take on different values is what makes the process of statistical inference necessary. If a population characteristic does not vary within the population, it is of little interest to the investigator from an inferential point of view.

DEFINITION: VARIABLE
A variable is a population characteristic which takes on different values for the elements comprising the population.

There are two ways of collecting information about a population. The first is to determine the value of the variable(s) of interest for *each and every* element of the population. This is known as a *population census*, or *population enumeration*. It is clearly feasible only for finite populations.

DEFINITION: POPULATION CENSUS
A population census is a complete tabulation of the relevant population characteristic for all elements in the population.

It is usually both necessary and desirable to determine the values of a variable for only a subset of the individuals in a population.

DEFINITION: SAMPLE
A sample is a subset of the elements in the population used to make inferences about certain characteristics of the population as a whole.

For practical considerations, namely, time and cost, it is convenient to sample rather than enumerate the entire population. Of course, there is one distinct disadvantage to sampling. Restricting the study to a small proportion of the population makes it impossible to be as accurate about population characteristics as is possible with a complete census. There is a risk of making errors.

DEFINITION: SAMPLING ERROR
Sampling error is the difference between the value of a population characteristic and the value of that characteristic inferred from a sample.

To illustrate sampling error, consider the population characteristic of the average selling price of homes in a given metropolitan area in a certain year. If each and every house sold is examined, it is found that the average selling price is $60,000. However, if only 25 homes per month and the average selling price of the 300 homes in the sample (12 months × 25 homes) are sampled, the average selling price in the sample may be $65,000. All other things being equal, we could say that the difference of $65,000 − $60,000 = $5000 is due to sampling error.

What do we mean by *all other things being equal*? Our error of $5000 may be partly due to factors other than sampling. Perhaps the selling price for one home in the sample was incorrectly identified as $52,000 instead of $42,000. Many errors of this type can occur in large data sets. Information obtained from personal interviews or questionnaires can contain factual errors from respondents owing to lack of recall, ignorance, or simply the respondent's desire to be devious.

DEFINITION: NONSAMPLING, OR DATA ACQUISITION, ERRORS
Errors which arise in the acquisition, recording, and editing of statistical data are termed nonsampling, or data acquisition, errors.

So that error or difference between the sample and the population can be ascribed solely to sampling error, it is important to minimize nonsampling errors. Validation checks, careful editing, and instrument calibration are all methods of reducing the chance of nonsampling errors dominating the total error and affecting subsequent statistical analyses.

The link between the sample and the population is probability theory. Inferences about the population are based on information in the sample. The quality of these inferences depends on how well the sample reflects, or *represents*, the population. Unfortunately, short of a complete census of the population, there is no way of knowing how well a sample reflects the population. So instead of selecting a *representative* sample, we select a *random* sample. In a random sample, every individual in the population has the same chance, or probability, of being included. This procedure helps ensure unbiased findings. It is possible to obtain a quite unrepresentative random sample, but the chance of doing so is usually very remote if the sample is large enough. In fact, because the sample has been randomly chosen, we can always determine the probability that the inferences made from the sample are misleading. This is why statisticians always make probabilistic judgments, never deterministic ones. The inferences are always qualified to the extent that random sampling error may lead to incorrect judgments.

The process of statistical inference is illustrated in Figure 1-1. Members, or units, of the population are selected in the process of sampling. Together these units comprise the sample. From this sample, inferences about the population are made. In

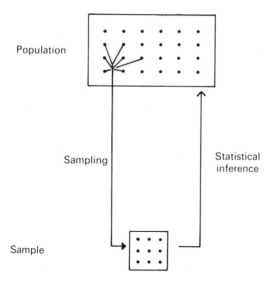

FIGURE 1-1. Process of statistical inference.

short, sampling takes us from the population to a sample; statistical inference takes us from the sample back to the population.

The aim of statistical inference is to make statements about a population characteristic based on the information in a sample. There are two ways of making inferences. One type of inference is known as *estimation*.

DEFINITION: STATISTICAL ESTIMATION

Statistical estimation involves the use of information in a sample to estimate the value of an unknown population characteristic.

The use of political polls to estimate the proportion of voters in favor of certain candidates is a well-known example of statistical estimation. Estimates are the statistician's best guess of the value of a population characteristic.

Another way of making inferences about a population characteristic is called *hypothesis testing*. In hypothesis testing, we hypothesize a value for some population characteristic and then determine the degree of support for the hypothesized value from the data in the sample.

DEFINITION: HYPOTHESIS TESTING

Hypothesis testing is a procedure of statistical inference in which we decide whether the data in a sample support a hypothesis that specifies the value (or range of values) of a certain population characteristic.

As an example, we may wish to use a political poll to find out whether a certain candidate holds an absolute majority of decided voters, that is, whether a proportion

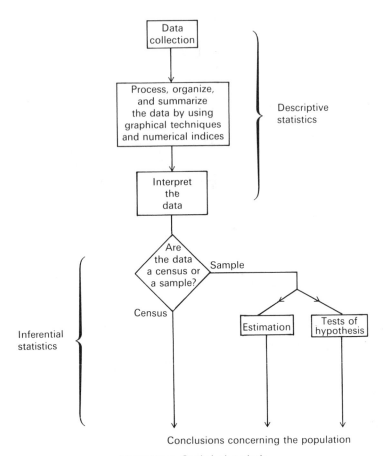

FIGURE 1-2. Statistical analysis.

of voters greater than .50 intend to vote for the candidate. We are interested not in the actual value of the population characteristic (the candidate's exact level of support), but in whether the candidate is likely to get a majority of votes. As you might guess, these two ways of making inferences are intimately related and differ more at the conceptual level. The relation between them is so intimate that, for most purposes, both can be used to answer any problem. No matter which method is used, there are two important elements of any statistical inference: the inference itself and a measure of our faith, or confidence, in it.

A useful synopsis of statistical analysis, including both descriptive and inferential statistics, is illustrated in Figure 1-2.

1.1. Statistical Analysis and Geography

The application of statistical methods to problems in geography is relatively new. Only for about the last 30 years has statistics been an accepted part of the academic training of geographers. There are, however, earlier references to uses of descriptive statistics in literature cited by geographers. For example, several nineteenth-century researchers, including Carey (1858) and Ravenstein (1885), used statistical techniques in their studies of migration and other interactions. Elementary methods of descriptive statistics are seen commonly in the geographical literature of the early twentieth century. But for the most part the three paradigms that dominated academic geography for the first half of the twentieth century—exploration, environmental determinism and possibilism, and regional geography—found few uses for statistical methods. Techniques for statistical inference were emerging at this time, but were not applied within the geographical literature.

Exploration

This paradigm is one of the earliest used in geography. Unexplored areas of the earth continued to hold the interest of geographers well into the current century. Explorations, funded by geographical societies such as the American Geographical Society (AGS) and the Royal Geographical Society (RGS), continued the tradition of geographers of collecting, collating, and disseminating information about relatively unknown parts of the world. The research sponsored by these organizations helped lead to the establishment of academic departments of geography at several universities. But, given only a passing interest in generalization and an extreme concern for the unique, little of the data generated by this research were ever analyzed by conventional statistical techniques.

Environmental Determinism and Possibilism

In attempting to generalize and to explain the diversity of human impact on the landscape, environmental determinists and possibilists focused on the nature of the physical environment as a controlling factor. Geographers began to concentrate on the physical environment as a control of human behavior. Some proponents went so far as to contend that virtually all aspects of human behavior are caused by environmental factors. Possibilists held a less extreme view, asserting that people are not totally passive agents on the landscape. Their debate with the determinists was long and, at times, bitter. Few geographers studied human-environment relations outside this paradigm. Again, very little attention was paid to statistical methodology. The debate between the determinists and possibilists occurred at a more general, methodological level.

Regional Geography

Reacting against the naive lawmaking attempts of the determinists and possibilists were proponents of regional geography. Generalization of a different character was the goal. According to this paradigm, an *integration*, or a *synthesis*, of the characteristics of areas or regions was to be undertaken by geographers. Ultimately, this would lead to a more or less complete knowledge of the areal differentiation of the world. Statistical methodology was limited to the systematic studies of population distribution, resources, industrial activity, and agricultural patterns. Emphasis was placed on the data collection and summary components of statistical analysis. These systematic studies were seen as being subsidiary, and preliminary to, the essential task of regional synthesis. The definitive work establishing this paradigm at the forefront of geographical research was Richard Hartshorne's *The Nature of Geography*, published in 1939.

Many of the contributions in this field discussed the problems of delimiting homogeneous regions. Each of the systematic specializations produced its own regionalizations. Together these regions could be synthesized to produce a regional geography. A widely held view was that regional delimitation was a personal interpretation of the findings of many systematic studies. Despite the fact that the map was considered one of the cornerstones of this approach, seldom were the extraction and the analysis of quantitative information from maps undertaken. A notable exception was Weaver's (1954) multiattribute agricultural regionalization. But this was not considered to be mainstream regional geography.

The regional paradigm dominated geographical research in the 1940s and early 1950s. However, disappointment with the research that characterized much of regional geography became increasingly common. First, many of the generalizations were extremely naive, particularly those that defined macroregions of the world. Second, the focus was almost exclusively empirical and seemed to preclude law-making or law-seeking research. Such work was becoming commonplace in other social sciences such as economics. Also, the writing of much of regional geography seemed dreary. Endless lists of facts and figures seemed to be a far cry from the significant insights into landscape evolution and interpretation that were the initial goal. There were certainly exceptions to this characterization, particularly in historical geography.

Beginning about 1950, the dominant approach to geographical research shifted away from regional geography and regionalism. To be sure, the transition took place over the succeeding two decades and did not proceed without substantial opposition. It was fueled by the increasing dissatisfaction with the regional approach and the gradual emergence of an acceptable alternative. Probably the first indication of what was to come was the rapid development of systematic specialties in the discipline. The traditional systematic branches of physical, economic, historical, and political soon gave way to urban, marketing, resource management, recreation, agricultural, transportation, population, and social geography. These systematic specialties developed close links with related disciplines. For example, economic geographers looked to economics as a useful source of ideas for research. Indeed, increased training in these so-called parent disciplines was suggested as an appropriate means of improving the quality of geographical scholarship. Throughout the 1950s and 1960s the teaching of systematic specialties and research in these fields became much more important in geography. The subservience of these fields to regional geography was reversed during this period.

Scientific Method and Logical Positivism

The new paradigm which took root at this time focused on the use of the *scientific method*. In particular, the new paradigm sought to use the scientific method to establish truly geographical laws and theories to explain spatial patterns. To some, geography could be defined only as a *spatial science*. As it was applied in geography, the scientific method utilized the *deductive* approach to explanation favored by *positivist* philosophers.

The deductive approach is summarized in Figure 1-3. The investigator begins with a perception of the real-world structure. A pattern, such as the distance decay of some form of spatial interaction, leads the investigator to develop a model of the phenomenon from which a generalization or hypothesis can be formulated. An experiment or some other kind of test is used to see whether the hypothesis suggested by the model can be verified. Data are collected from the real world, and verification of the hypothesis (or speculative law) is undertaken. If the test proves successful, laws and then theories can be developed, heightening our understanding of the real world. If these tests prove successful in many empirical applications, then the *hypothesis* gradually comes to be accepted as a *law*. Ultimately, these laws are combined to form a *theory*. This approach obviously has many parallels to the methodology for statistics outlined in the introduction to this chapter.

The deduction-based scientific method was applied in virtually all fields of geography during the 1950s and 1960s. It was (and still is) particularly important in physical geography and urban, economic, and transportation geography within human geography. Part of the reason for this strength is undoubtedly the widespread use of the scientific method in biology, geology, and economics.

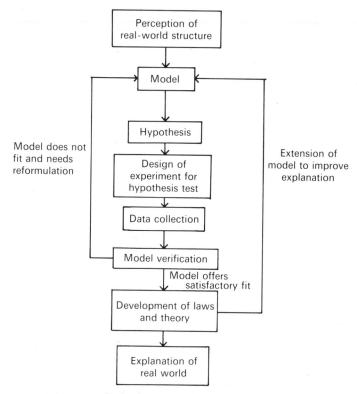

FIGURE 1-3. Deductive approach to scientific explanation.

Quantification is essential to the application of the scientific method. Mathematics and statistics play central roles in the advancement of geographic knowledge by this approach. The relatively poor training of most geographers in mathematics is probably responsible for the early emergence of a statistical methodology as the single most important procedure for making valid generalizations by using the scientific method in geography. Statistical methodology is now accepted as an important research tool by geographers. That is not to say that the methodology has been accepted throughout the discipline. Historical and cultural geographers shunned the new wave of quantitative, theoretical geography. Part of the reason was that much of the early research in this paradigm seemed long on quantification and short on theory. True positivists view quantification as only a means to an end—the development of theory through hypothesis testing. It cannot be said that this distinction was clear in the minds of all who practiced this approach to geographic generalization. The methods used by theoretical, quantitative geographers are themselves insufficient to define the field of geography. Obviously this fact was not clear to some of the practitioners of positivist approaches during the 1950s and 1960s.

Many researchers advocating the use of the scientific method also defined geography as a *spatial science*. Unfortunately, such a narrow definition seemed to preclude much of the work undertaken by cultural and historical geographers. Human geography was defined in terms of spatial structures, spatial interaction, spatial processes, or spatial organization. Distance was seen as the key variable in the field. Physical geography, which had been brought back into geography with the onset of the quantitative revolution, was once again set apart from the field. Physical processes could not be defined as solely spatial processes. Reaction against geography as a spatial science occurred for several reasons. Chief among these was the disparity between the type of model promised by advocates of the scientific method and spatial science and the theoretical developments themselves. Most of these theoretical models gave adequate descriptions of reality only at a very general level. The axioms upon which they were based seemed to provide a rather poor foundation for furthering the development of geographical theory.

By the mid-1960s, a field now known as *behavioral geography* was beginning to emerge. It had obvious and close links with psychology. Proponents of this approach did not disagree with the basic goals of logical positivism—the development of theory-based generalizations—only with how this task could be best accomplished. Behavioral geographers began to focus on individual spatial cognition and behavior, primarily from an inductive point of view. Rather than accept the unrealistic axioms of perfect knowledge and perfect rationality inherent in many models developed by this time, behavioral geographers felt that the use of more realistic assumptions about behavior might provide deeper insights into spatial structures and spatial behavior. Their induction-based approach was seen as a way of providing the necessary input into a set of richer models based on the deductive mode. Statistical methodology has a clear role in behavioral geography.

Contemporary Geography

In the last 10 to 15 years, the number of approaches manifest in the geographical literature has broadened considerably. *Positivist approaches*, based on the notion of the objectivity of scientific analysis of the world, continue to dominate geography and now occupy the mainstream. Two alternatives are becoming increasingly popular. First, there are approaches based on *humanistic philosophies*. Humanistic geographers believe that people create subjective worlds in their own minds and that their behavior can be understood only by using a methodology which can penetrate that subjectivity. By definition, geographical laws developed from positivist precepts cannot have this capability. Humanistic geography appeals most strongly to historical, cultural, and social geographers, many of whom reacted strongly against the positivist-based scientific method.

Structuralists reject both positivist and humanistic methodologies, arguing that explanations of observed spatial patterns cannot be made by a study of the pattern itself, only by the establishment of theories to explain the development of the societal conditions within which people must act. The structuralist alternative, exemplified by Marxism, emphasizes how human behavior is constrained by more general societal processes and can be understood only in those terms. For example, patterns of income segregation in contemporary cities can be understood only as a result of class conflict between the bourgeoisie, on the one hand, and the proletariat, or workers, on the other. Understanding how power and therefore resources are allocated in a society is a prerequisite to a comprehension of its spatial organization.

What is the role of statistics in contemporary geography? Certainly, it is now an important component of the research methodology in most systematic branches of geography. A substantial portion of the research in physical, urban, and economic geography, for example, employs increasingly sophisticated statistical analysis. In addition, other quantitative nonstatistical techniques are used extensively in these fields. Although many advocates of the more recent humanistic or Marxist approaches to geography reject statistical inquiry on the basis of its close affinity to the scientific method, there is an increasing use of statistical methods in empirical studies based within these paradigms. Many of the data gathered from personal interviews, mail questionnaires, and participant observation lend themselves to analysis by descriptive and inferential techniques.

For many geographers, the map is the essense of the research. Cartography is undergoing a period of rapid change in which computer-based methods continue to replace much conventional map compilation and production. This trend has forced many geographers to acquire better technical backgrounds in computer science and mathematics and has opened the door to increased use of statistical and other quantitative methods in cartography. Geographic information systems are one manifestation of this phenomenon. Large sets of data are now stored, accessed, compiled, and subjected to various cartographic display techniques using video display terminals or various plotting devices. The analysis of the spatial pattern of a single map and the comparison of set of interrelated maps are two other cartographic problems for which statistical methodology has been an important source of ideas. Many of the fundamental problems of displaying data on maps have clear parallels to the problems of summarizing data through conventional descriptive statistics.

Finally, statistical methods find numerous applications in applied geographical research. Retail location problems, transportation forecasting, and environmental impact assessment are three examples of situations in which statistical and other quantitative techniques are used in applied research. Both private consulting firms and government planning agencies at the municipal, state, and federal levels encounter problems in these areas on a day-to-day basis.

In sum, statistical analysis is commonplace in contemporary geographical research. It is now more carefully and thoughtfully applied than was the case for some of the early work and includes an ever-widening array of specific techniques. Probably statistical analysis will continue to play an important role in geographical teaching and research for some time.

1.2. Use of Statistics in Geographical Research: Two Examples

To illustrate the versatility of statistical methodology, let us consider in detail two examples of its use in geography.

EXAMPLE 1-1. Analysis of Wind Speed and Direction Statistical concepts and techniques are frequently applied in all the subfields of physical geography. Because of their close relations with the physical and life sciences, the systematic branches of physical geography, including geomorphology, hydrology, climatology or meteorology, and biogeography, felt the impact of the so-called quantitative revolution far earlier than did the systematic branches of human geography. Descriptive statistics is used to investigate the morphology of physical systems as diverse as vegetation associations, drainage basins, and climatic types. The information generated in these studies often finds practical use in applied studies of flood damage control, forest regrowth, avalanche prediction, wind power generation, and other related areas. The data sets used in these studies often contain thousands of records for different variables taken at close time intervals for long observation periods. For example, a series of temperatures may consist of hourly recordings at a meteorological station taken each day for 20 or more years! There is an obvious need for efficient graphical devices and summary descriptive measures for large data sets of this kind.

As an example, suppose a meteorologist has developed a data set containing measurements of wind speed and direction taken hourly at a series of stations for a 25-yr period. Wind speeds are obtained from an anemometer mounted 33 ft (10 m) above the ground. Speed, measured in miles per hour, is taken as the average reading over a 1-min interval closest to the hour. Directions are measured on a continuous scale graduated in degrees true and to 16 points of the compass. In addition, the data set notes the frequency of *gusts* (rapid, brief increases in wind speed) and *calm* (the absence of appreciable air motion).

The task of making sense of a long series of wind records is made substantially easier by the use of many techniques from descriptive statistics. One of the simplest questions to be asked might be, Which of these reporting stations has the highest wind speeds? We might even wish to classify the stations according to wind speeds as very windy, windy, relatively calm, and calm. A useful summary measure is the *average* wind speed by day, month, season, and year or for the entire 25-yr record of 219,000 observations (25 yr × 365 d × 24 h). Wind speeds vary systematically by hour of the day, month of the year, and at other regular frequencies. A summary table might be used to contrast the average wind speeds in two characteristic months of the year based on the 25-yr observation period:

	Average wind speed, mi/h	
Station	January	July
A	8.4	5.4
B	2.2	6.5
C	7.1	7.8
D	25.2	18.0

From this table clearly station D is the windiest, station B is the calmest and has a July maximum, station A has a January maximum, and station C has no marked monthly maximum. Other tables generated in this way might point out other phenomena which may require some kind of meteorological explanation.

In addition to the typical wind speed likely to be observed at some location, the variability or persistence of wind speed might be of interest. The variability of wind speeds at station A is illustrated in Figure 1-4. This bar chart is a percentage summary of the wind speed readings for 1 yr and consists of 8760 (24 × 365) observations. Of the observations, 10 percent represent calm wind conditions. That is, 10 percent of the time we might expect calm wind conditions at station A. Winds seem to be somewhat concentrated at speeds between 5 and 12 mi/h. Over 40 percent of the observations lie within these limits. Wind speed variations are an important consideration in studies of the dispersal of airborne pollutants. Stations having dominantly calm wind conditions may be susceptible to pollutant concentrations. Locations with persistently high winds, however, may be associated with rapid dispersal of pollutants. Of course, it may be necessary to examine other meteorological variables such as temperature and precipitation as well as topographic conditions before such conclusions could be reached.

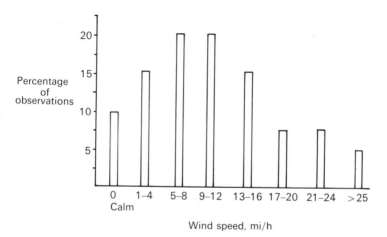

FIGURE 1-4. Variability of wind speed at station A.

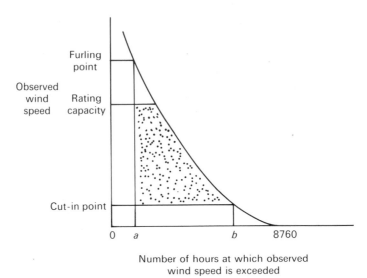

Number of hours at which observed
wind speed is exceeded

FIGURE 1-5. A power duration curve.

Sometimes, the summary of wind speeds from such an extended series of measurements needs to be compiled in a particular fashion for a specific purpose. For example, consider the problem of identifying potential sites for wind power generation. Figure 1-5 illustrates a form of a *power duration curve*, a useful graphical device for estimating the viability of a site for wind power generation. The horizontal axis measures the number of hours of the year, from 0 to 8760. The vertical axis represents the observed wind speed. The power duration curve shows the number of hours per year (or, in this case, the number of observations in the yearly wind speed series) for which the wind speed exceeds the value given on the vertical axis. Of particular importance in this curve is the percentage of time that the power generator cannot be in operation owing to either insufficient or excessive winds. If the wind speeds are below the *cut-in point*, there is insufficient wind to produce a significant amount of power. Such conditions prevail at this site for a total of $8760 - b$ h. At wind speeds above this value and up to a wind speed sufficient for maximum power generation from the wind turbines, power can be generated. Maximum power is generated at a certain wind speed, termed the *rating capacity*. Wind speeds above the rating capacity provide no increase in power. However, at a critical value of wind speed known as the *furling point*, the wind is excessive and the turbines must be shut down to avoid damage. This occurs for a h of the year. The particular form of wind speed at a location can be used to estimate power production, since power production varies with the cube of the observed wind speed. In general, the larger the shaded area in a power duration curve, the greater the possibilities of power production. Ideally, a site with wind speeds which are constantly at the rating capacity is desirable.

Summarizing the directional components of wind is another problem in descriptive statistics. Three different methods of portraying the directional components of the wind are illustrated in Figure 1-6. In (*a*), the wind directions are summarized

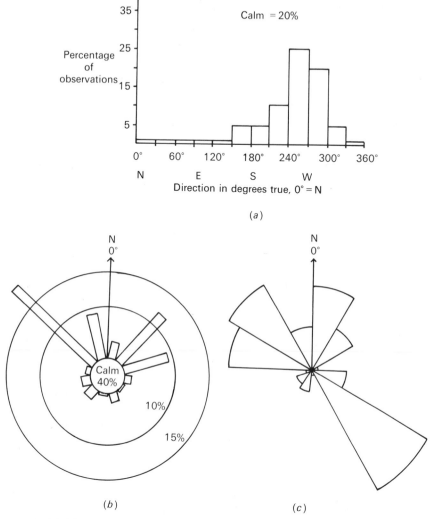

FIGURE 1-6. (*a*) Histogram of station *A*; (*b*) circular histogram of station *D*; (*c*) rose diagram of station *C*.

in a conventional *histogram*. The percentage of observed wind directions in a yearly series is plotted on the vertical axis. The wind directions are aggregated to 30-degree sectors, although a 16-point compass aggregation of $22\frac{1}{2}$-degree sectors (N, NNE, NE, ENE, E, etc.) can also be used. For station *A*, the predominance of westerly winds is apparent from the figure. The *circular histogram* of Figure 1-6(*b*) is an alternative, and possibly more easily interpretable, graphical portrayal of wind direction. In this format, the bars for the various directions are extended from a reference circle within which the percentage of calm winds is noted. The fact that winds at station *D* are dominantly northerly, especially from the northwest, is easy to see from this graph. A *rose diagram* is another easily interpretable summary graphic for directional data. A rose diagram for the observed winds at station C is illustrated in Figure 1-6(c).

As this example clearly illustrates, the types of summary statistics and graphs useful for spatial data, such as an observed series of directions, often differ from the devices used to summarize ordinary variables, such as wind speed. Part I of this text is devoted entirely to the methods of descriptive statistics. Conventional methods and relatively newer exploratory methods are discussed in Chapter 2, and procedures useful for spatial data in Chapter 3.

EXAMPLE 1-2. Residential Property Values Geographers interested in the city often center their research on the theme of *urban spatial structure*. This term refers simultaneously to both the visible pattern of individual elements of the city and the underlying interrelationships that integrate these activities to a system. The spatial structure of the residential areas of cities has been of continuing interest to geographers. One particular focus of research has been the examination of the systematic variation of house prices and/or rents across an urban area. Statistical methodology has been used extensively to answer several key questions concerning the actual variations in house prices in a city at some time. First, what major explanatory variables account for the observed variations in individual house prices? Second, what are the key spatial variables at work? Third, what is the relative importance of these spatial or locational variables vis-à-vis other factors? That is, are locational variables more or less important than the structure of the house itself and the neighborhood in which it is located? How do these results compare across cities?

The statistical approach to these problems begins with the selection of a random sample from all the homes in a city sold in some given period, for example, the previous 12 months. The most obvious variable to be collected is the actual selling price of the residential property. In addition, many other variables are noted for each house included in the sample. As shown in Table 1-1, these variables, selected from a rather large set of potential influences, are grouped as (1) structural characteristics, (2) neighborhood characteristics, and (3) location variables. Structural characteristics include any measurement of the house or lot upon which it is located. The size of the house in usable or developed square footage is one example of a structural variable. Neighborhood characteristics include measurable properties of the surrounding area. The percentage of the neighborhood in nonresidential land use is one example of a neighborhood variable. It is well known that houses located in neighborhoods with mixed land uses tend to have less value than those in exclusively or near exclusively residential areas. The location of the house within the city is another factor influencing the selling price. More accessible homes tend to be more valuable, for example. Historically, the principal variable used to measure this factor was distance to the central business district, but it is now more common to use accessibility measures that take into account the location of activities throughout the city. There is a great deal of variability in the actual set of variables used in a study of urban house prices. In some studies the list of variables may include well over 50 different variables, although a smaller subset is more common. There is extreme diversity in the number, and exact definition of, variables used in these studies.

TABLE 1-1

Factors Influencing House Price Variations in an Urban Area

Structural characteristics of the house	
Size	Number of bathrooms
Lot size	Garage
Street frontage	Central heating
Number of rooms	Air conditioning
Number of bedrooms	Architectural significance

Neighborhood characteristics
Residential density
Percentage of neighborhood in nonresidential uses
Crime rate
Level of pollution
Presence of minorities

Location variables
Distance to central business district
Distance to major artery
Freeway access
Accessibility to employment, retail, and/or recreational facilities

What can be done with these data? First, descriptive statistics can be used to summarize each of the variables gathered during the data collection. For example, what is the average house price in the sample? Real estate agents and housing market analysts often use this statistic to monitor the change in house prices in a city over time. Second, we might also wish to examine the variability of house prices. How homogeneous are these selling prices about the average house selling price? A conventional histogram, such as that used to describe wind speed variations in Figure 1-4, could be used for this purpose. Similar analyses might be undertaken for each variable collected. As geographers, we might be especially interested in the number of homes sold by neighborhood. This could be displayed on a map. Ordinary descriptive statistics and descriptive methods used to analyze maps might prove to be useful.

Third, we can examine the relationship between variables. We can see, for example, whether the house selling price tends to be associated with distance from the central business district or any other variable collected in the sample. A *scatter plot*, or *scatter diagram*, can be used for this purpose. The scatter diagram relating house selling price and distance from the central business district (CBD) of Figure 1-7 suggests a general decrease in selling prices with distance, but several houses do not follow this trend. Methods for analyzing the relationship between two variables are discussed extensively in Part III.

Fourth, studies of house price variations in cities are often concerned with identifying the *determinants* of house price variations. Unfortunately, there is no single theory of residential location and house prices which takes into account all the variables listed in Table 1-1. Nevertheless, a number of empirical studies have found consistent effects for these variables as factors influencing the price of houses in

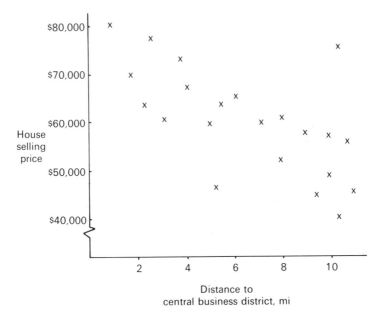

FIGURE 1-7. Scatter diagram.

North American and other western cities. It is now usual to search for the complex interactions among these variables in a structured way. One of the most commonly used techniques is known as *hedonic price estimation.* The purpose is to take a price of some aggregate product, such as a house, and divide it into implicit prices for the various attributes of the product. For a house, for example, it can be argued that part of the total price paid is to purchase square footage or a number of bedrooms, that some is paid to purchase location of the house, and that another part is paid for certain environmental variables, such as low pollution levels. Although the consumer does not actually pay these prices, they are implicit in the bids the consumer makes and thus are the ultimate selling price of the houses in a city.

In general, the method utilizes statistical techniques to determine the form of the function which links the price of a house to these attributes:

House selling price = f(structural characteristics, neighborhood
characteristics, location variables)

For example, suppose the results of the hedonic price estimation in some city indicate that the following equation describes house price variations in some city:

House selling price = $83,000 + $15,000 (house size, in thousands of square feet)
− $1000 (carbon monoxide concentration, in parts
per million)
− $5000 (distance to CBD, in miles)

By using this equation, it is possible to estimate the selling price of a 2000-ft^2 house in a neighborhood averaging 14 parts per million (ppm) of carbon monoxide and located 2 mi from the central business district as $83,000 + $15,000(2) − $1000(14) − $5000(2) = $89,000. The coefficients of this equation—15,000, 1000, and 5000—can be thought of as the implicit prices that house buyers have used in making their bids for the houses in the sample. For example, an extra 1000 ft^2 of space adds about $15,000 to the selling price of an average house.

Bivariate correlation and regression analysis is used to analyze the statistical relationship between two variables. These procedures are discussed in Part III. When one variable (such as house selling price) is simultaneously related to several other variables (such as house size, carbon monoxide concentration, and distance to the CBD), the appropriate techniques are *multiple correlation and regression analysis*. These advanced techniques are briefly introduced in Chapter 13.

The two examples presented in this section clearly illustrate the variety of situations in geographical research in which statistical methodology is relevant. There are literally hundreds, if not thousands, of instances in geographical research where the methods described in subsequent chapters of this text are useful. These applications span both human and physical geography. This range of application is further illustrated by the choice of examples for specific techniques presented throughout the remainder of this text and in the problems at the end of each chapter.

1.3. Data

Although it is suggested in Figure 1-2 that statistical analysis begins with a data set, this is not strictly true. It is not unusual for a statistician to be consulted at the earliest stages of a research investigation. As the problem becomes clearly defined and questions of appropriate data emerge, the statistician can often give invaluable advice on sources of data, the methods used to collect them, and characteristics of the data themselves. A properly executed research design will yield data that can be used to answer the questions of concern in the study. The nature of the data used should never be overlooked. As a preliminary step, let us consider a few issues relating to the sources of data, the kinds of variables amenable to statistical analysis, and several characteristics of the data such as measurement scales, precision, and accuracy.

Sources of Data

Sometimes the data required for a study are already available. When these data are available in some form in various records kept by the institution or agency undertaking the study, the data are said to be from an *internal source*.

DEFINITION: INTERNAL DATA
Data available from existing records or files of an institution undertaking a study are data from an internal source.

For example, a meteorologist employed by a weather forecasting service normally has available many key variables such as air pressure, temperature, and wind velocity from a large array of computer files which are augmented hourly, daily, or at some other predetermined frequency. Besides the ready availability of internal data, the meteorologist has the added advantage of knowing a great deal about instruments used to collect the data, the accuracy of the data, and possible errors.

When an *external source* must be used for the data required in a study, many important characteristics of the data may not be known.

DEFINITION: EXTERNAL DATA
Data obtained from an organization external to the institution undertaking the study are data from an external source.

Caution should be exercised in using any data source, but particularly those sources which are external. Since the investigator may not be thoroughly acquainted with the source, she or he may be unaware of certain important features of the data. Two different data collection procedures, or even two sets of recording instruments, may have been used to obtain the values in the data set. One set of instruments may employ different measurement procedures or exhibit a different accuracy from the other set. This may introduce significant uncontrolled nonsampling errors. If at all possible, such errors should be avoided or kept to a minimum.

Consider a set of city populations extracted from a statistical digest summarizing population growth over a 50-yr period. Such a source may not record the exact areal definitions used as a basis for the figures. Moreover, these definitions may have changed over the 50-yr study period, for example, owing to annexations or amalgamations. Only a primary source, such as the national census, would record all the relevant information. Unless such characteristics are carefully recorded in the external source, users of such data may have the false impression that no anomalies exist in the data. It is not unusual to derive results in statistical analysis that cannot be explained without a detailed knowledge of the data source.

Another useful distinction is between *primary* and *secondary* external data.

DEFINITION: PRIMARY DATA
Primary data are obtained from the organization or institution that originally collected the information.

DEFINITION: SECONDARY DATA
Secondary data are data obtained from a source other than the primary data source.

The difficulty with secondary sources is that they may contain data altered by re-recording or reediting of errors, selective data omission, rounding, aggregation, questionable merging of data sets from two or more sources, or other ad hoc procedures. For these reasons, primary data are normally preferred.

Data Acquisition

When the data required for a study cannot be obtained from an existing source, usually they are collected during the course of the study. Data acquisition methods can be classified as either *experimental* or *nonexperimental*.

DEFINITION: EXPERIMENTAL METHOD OF DATA ACQUISITION

An experimental method of data acquisition is one in which some of the factors under consideration are controlled in order to isolate their effects on the variable of interest.

Only in physical geography is this method of data collection prominent. Fluvial geomorphologists, for example, often use flumes to control such variables as stream velocity, discharge, bed characteristics, and gradient. Among the social sciences, the largest proportion of data collected from experiments is in psychology.

DEFINITION: NONEXPERIMENTAL METHOD OF DATA ACQUISITION

A nonexperimental method of data collection or statistical survey is one in which no control is exercised over the factors that may affect the population characteristic of interest.

Studies in physical geography using field data fall into this category. Many carefully planned, detailed field studies have been seriously affected by instances of unforeseen aridity, excess moisture, insufficient snow pack, and the like. It is easy to see why experiments generally provide more useful data than surveys. This seeming advantage must be weighed against the obvious disadvantage of experiments in simulating real-world processes.

There are four common survey methods. *Observation* (or field study) requires monitoring of an ongoing activity and direct recording of data. This form of data collection avoids several of the more serious problems associated with other survey techniques, including incomplete data. However, faulty or incorrectly calibrated instruments can sometimes mitigate against the advantages of observational techniques. More than one field study has been prolonged by instrument failure.

Two methods of data collection are often used to extract information from households or individuals: *personal interviews* and *telephone inteviews*. In a personal interview, a trained interviewer asks a series of questions and records responses on a specially designed form. There are obvious advantages and disadvantages to this procedure. An alternative, and often cheaper, method of securing the data from a set of households is to send a mail questionnaire. This is often termed *self-enumeration* since the individual completes the questionnaire without any assistance from the agency undertaking the study. The disadvantages of this method include non-response, partial response, and low return rates for completed questionnaires. Factors affecting the quality of data generated from mail surveys include appropriate wording, proper question order, a selection of question types, and layout and design. These topics are discussed in the References section at the end of this chapter.

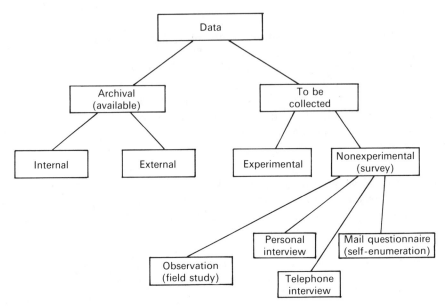

FIGURE 1-8. A typology of data sources.

A useful typology of data sources is illustrated in Figure 1-8. An important first step in a statistical inquiry is to examine the important issues relevant to the source of data. The potential usefulness of a study is directly related to the quality of the data utilized.

Characteristics of Data Sets

Statistical analysis cannot proceed until the available data have been assembled into a usable form.

DEFINITION: DATA SET

A data set is a collection of statistical information or values of the variables of interest in a study.

An example of a data set is given in Table 1-2. In this example, the *observational units* are climatic stations. Five *variables* are contained within the data set. The information on every variable from one observational unit is often termed an *observation*; it is also common to speak of the data value for a single variable as an observation since it is the observed value.

The variables in a data set can be divided into two types: quantitative and qualitative.

DEFINITION: QUANTITATIVE VARIABLE

A quantitative variable is one in which the values or outcomes are expressed numerically.

TABLE 1-2
A Geographical Data Set

			Variables			
	Climatic station	Days per year with precipitation	Annual rainfall, cm	Mean January temperature, °C	Mean July temperature, °C	Coastal or inland
	A	114	71	6	16	C
	B	42	48	12	16	C
Observational ⟶	C	54	32	−4	21	I
units	D	32	28	−8	20	I
	E	41	129	16	18	C
	F	26	18	1	22	I
	G	3	8	24	29	I

Quantitative values can be obtained either by counting or by measurement. *Discrete variables* are those that can be obtained by counting. For example, the number of children in a family, the number of cars owned, and the number of trips made in a day are all counting variables. Within the data set of Table 1-2, the number of days per year with precipitation is a counting variable. The possible values of a counting variable are the ordinary integers and zero: 0, 1, 2, ... Quantities such as rainfall, pressure, or temperature are measured and can take on any continuous value depending on the accuracy of the measurement and recording instrument. *Continuous variables* are inherently different from discrete variables.

Qualitative variables are neither measured nor counted.

DEFINITION: QUALITATIVE VARIABLE

A qualitative variable is one for which the obtainable values are nonnumerical.

Qualitative variables are also called *categorical variables* since the observational units are placed into categories. Male/female, land-use type, occupation, and plant species are all examples of qualitative variables. These variables are defined by the set of classes into which an observation can be placed. In Table 1-2, for example, climatic stations are classified as either coastal (C) or inland (I).

Numerical values are sometimes assigned to qualitative variables. For example, the yes responses to a particular question in a survey may be assigned the value 1 and the no responses a value 2. The variable sex, indicating the sex of the respondent, may be identified by males = 1 and females = 0. In both these examples, each category has been assigned an *arbitrary* numerical value. As we shall see, it is improper to perform most mathematical operations on qualitative variables expressed in this manner. Consider, for example, the operation of addition. Although this operation makes sense for a quantitative variable, it makes no sense to add the numerical values assigned to the variable sex.

Converting quantitative variables to qualitative variables is sometimes advantageous. In this case the possible numerical values of the quantitative variable are grouped into nonnumerical classes. For example, the quantitative variable called household income might be classified into three categories: low, medium, and high. Such a simplification can sometimes facilitate the interpretation of a large number of figures.

Besides being described as qualitative or quantitative, variables can also be classified according to the *scale of measurement* on which they are defined. This scale defines the amount of information the variable contains and what operations can be meaningfully undertaken and interpreted. The lowest scale of measurement is the *nominal scale*. Nominal-scale variables are those *qualitative* variables which have no implicit ordering to their categories. Even though we sometimes assign numerical values to nominal variables, they have no meaning. Consider the variable in Table 1-2 that distinguishes coastal climatic stations (C) from inland ones (I). All that we can really do is distinguish between the two types of stations. In other words, we know that stations *A*, *B*, and *E* are coastal and therefore different from stations *C*, *D*, *F*, and *G*. Also, stations *A*, *B*, and *E* are all alike according to this variable. One way of summarizing a nominal variable is to count the number of observational units (or observations) in each category. This can be summarized in a bar graph or a simple table of the following form:

Category	Count of frequency	Proportion	Percentage
C	3	0.429	42.9
I	4	0.571	57.1
	7	1.000	100.0

The use of percentages or relative proportions to summarize such data is quite common. For example, 0.429, or 42.9 percent, of the climatic stations are coastal and 0.571, or 57.1 percent, of the stations are inland.

Proportional summaries are often extremely useful in comparing responses to similar questions from two or more different surveys, or from different classes of respondents to the same question in one survey. For example, studies of migrants to cities of the third world often include questions concerning the sources of information used by individual migrants in selecting their destinations. A question of this type might be phrased in the following way: In reaching your decision to come here, you must have had some information about job possibilities, living conditions, income, etc. Which of the following gave you the *most* information? Then the question would be followed by a list of possible responses. One interesting research question which may arise in a study of this type is the differential use of the various information sources by the age or educational attainment of the respondent.

Table 1-3 summarizes the responses to this question and differentiates between two different age classes of migrants. Clearly about two-thirds of both age groups cited family or friends as the dominant information source. However, younger migrants seem to rely slightly more on family and less on friends in comparison to older migrants. Also, younger migrants seem to use newspapers more frequently, and older migrants seem to use other sources more often. These results indicate the dominant role played by kin and friends in the rural-urban migration process, but

TABLE 1-3
Percentage Distribution of Responses concerning Primary Information Source about Urban Destinations Used by Migrants

	Age of respondent, yr	
Sources of information	14–21	22–64
Newspapers	13	7
Radio	3	2
Government labor office	2	3
Family members	40	28
Friends	27	41
School teacher	4	1
Career counselor	1	1
Other sources	10	17
Total	100	100

also point out significant differences in primary sources of information depending on the age of the migrant.

If the categories of a *qualitative* variable can be put into order, then the scale of measurement of the variable is said to be *ordinal*. An example of an ordinal variable is the strength of opinion measured in the responses to a question with the following categories:

Strongly agree	Agree	Neutral	Disagree	Strongly disagree
-2	-1	0	$+1$	$+2$
1	2	3	4	5
5	4	3	2	1

Three different possible numerical assignments are given. Each is consistent with an ordinal scale for these variables. The stronger the opinion, the higher the numerical assignment. In fact, any assignment can be used as long as the values given to the categories maintain the ordering implicit in the wordings attached to the categories. The assignment of values $-2000, -10, 270, 271, 9382$ could also be used. Of course, the scales defined above have the added advantage of simplicity. The numerical difference between the values assigned to the categories has no meaning for an ordinal variable. We can neither subtract nor add the values of ordinal variables. Notice, however, that the numerical assignments used to define ordinal variables often use a constant difference or unit between successive categories. In the first scale defined above, each category differs from the next by a value of 1. This does *not* mean that a respondent who checks the box for strongly agree is 2 units higher than a respondent who checks the box for neutral. We can say only that the first respondent agrees *more* with the question than the second respondent. Only statements about order can be made from the values of an ordinal variable.

Quantitative, or numerical, variables, whether discrete or continuous, can be classified into two scales of measurement. *Interval variables* differ from ordinal variables in that the interval scale uses the concept of *unit distance*. The difference between any two numbers on this scale can always be expressed as some number of units. Both Fahrenheit and Celsius temperature scales are examples of interval-scale variables. Although it makes sense to compare differences of interval-scale variables, it is not permissible to take ratios of values. For example, we can say that 90°F is 45°F hotter than 45°F, but we cannot say that it is twice as hot. To see why, let us convert these temperatures to the Celsius scale: 90°F = 32°C and 45°F = 7°C. Note that 32°C is *not* twice as hot as 7°C.

At the highest level of measurement are *ratio-scale* variables. Any variable having the properties of an interval variable *as well as* a natural origin of zero is measured at the ratio scale. Distance measured in kilometers or miles, rainfall measured in centimeters or inches, and most other variables commonly studied by geographers are measured on the ratio scale. It is possible to compute ratios of such variables as well as to perform many other mathematical operations such as logarithms, powers, and roots. There is a logical and theoretical distinction between ratio and interval variables, but at the practical level this distinction rarely comes into play.

Of far greater interest is the specification of the level of measurement of a variable measured indirectly. For example, the scale of measurement of a variable constructed from the responses to a question or a set of questions in an interview may not be easily identified. The variable may be ordinal, interval, or even ratio. That is, respondents *may treat* the categories of scale with labels *strongly agree, agree, neutral, disagree*, and *strongly disagree* as if they were part of an interval, not ordinal, scale. A significant amount of research in psychology has examined methods of deriving scales with interval properties from test questions that are, strictly speaking, only ordinal in structure. Attitude scales and some measures of intelligence are two examples where interval properties are desirable. An investigator must sometimes decide what sort of operations on the variables collected are meaningful or whether the operations exceed the information contained in the variable. A summary of the scales of measurement, along with a list of permissible mathematical operations and examples, is given in Table 1-4.

Three other important characteristics related to the measurement of variables are precision, accuracy, and validity. The precision of a variable reflects the range of values obtained in the process of measurement. For an instrument-read variable, the precision depends on the calibration of the instrument used to take the readings. A finely calibrated thermometer may be capable of estimating temperatures within a tenth of a degree, but an ordinary thermometer might differentiate temperature only within a single degree. The first thermometer is said to be more precise than the second.

Accuracy involves the degree to which the measurement is unbiased. An improperly calibrated thermometer might always record temperatures 2° or 3° above the actual temperature. It is therefore inaccurate. The concepts of measurement precision and measurement accuracy are not synonymous. To see the difference, consider the example of recording temperatures. Four different thermometers are used to measure an air temperature known to be 20°C. Five readings are taken with

TABLE 1-4
Levels of Measurement: A Summary

Level of measurement	Permissible mathematical operations[a]	Geographical examples
Nominal	$A = B$ or $A \neq B$, counting	Presence or absence of a road linking two cities Land-use types in region D or E or F
Ordinal	$A < B$ or $A > B$ or $A = B$	Preferences for different neighborhoods in rank order Ratings of shopping center attractiveness
Interval	Addition $(A + B)$, subtraction $(A - B)$, and multiplication $(A \times B)$	Temperatures, °F or °C
Ratio	Take ratios A/B and compare them Square roots Powers Logarithms Exponentiation	Distances (imperial or metric) Density (persons per unit area) Stream discharge Shopping center square footage

[a]Permissible mathematical operations at each level of measurement include all those operations permissible for variables at lower levels of measurement.

each instrument. The recordings are illustrated in Figure 1-9. The thermometer with the readings illustrated in (*a*) is both accurate and precise. All five readings are within 0.5°. They are almost perfectly centered on the true temperature 20°C. The instrument used in (*b*) is accurate, since the recordings are also centered on 20°C, but it is imprecise. The five readings cover a range of 4°. In (*c*) the recordings of an inaccurate but precise instrument are illustrated, and in (*d*) an imprecise and inaccurate set of readings is illustrated.

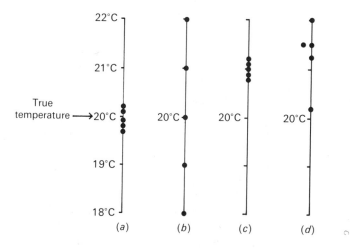

● Indicates recorded temperature.

FIGURE 1-9. Differentiating between precision and accuracy.

The concept of measurement validity is complex. In its simplest form, *validity* is the degree to which a variable measures what it is supposed to measure. For most instrument recordings and most count variables, this is seldom an issue. However, some variables are defined to measure quite general concepts, and then the issue of validity is not so clear-cut. This is true of many indices that purport to measure some concept which is not directly measurable. Attitude scales, quality-of-life indices, and measures of standard of living are examples where the question of validity always arises. The particular operational definition chosen for such theoretical concepts in any study may be inappropriate. Although a one-to-one correspondence between theoretical concepts and operational definitions would be ideal, it does not exist. Many of the concepts involved are themselves imperfectly defined. There may be ambiguities. Sometimes this problem may indicate that the concept itself is in need of a more complete specification. At other times, the particular operational definition chosen in a study reflects the fact that some of the data required were readily available and it would have been difficult to obtain better data. At still other times, the investigator might choose the operational definition which best helps his or her case; the investigator may have even ignored a more reliable data source in order to do so.

When the results of a study based on empirical data are analyzed, a good first step is always to closely examine the operational definitions chosen for the variables. Are they suitable? Are the variables unambiguously defined? Are they the best data set which could have been used? Are they sufficiently precise? Are they sufficiently accurate? Could they be responsible for any misleading inferences? The need for caution in the use of data applies to all data sources, including those which are experimental, survey, and external or internal.

1.4. Summary

Statistical analysis includes methods used to collect, organize, present, and analyze data. Descriptive statistics refers to techniques used to describe data, either numerically or graphically. Inferential statistics includes methods used to make statements about a population characteristic on the basis of information from a sample. Statistical inference includes both estimation and hypothesis testing methods.

Within geography, most applications of statistical methodology are rather recent, having become a significant part of the literature only after the "quantitative revolution" of the 1950s and 1960s. Statistical methodology is most commonly utilized by geographers advocating a scientific approach to the discipline, an approach now common in many of the systematic branches of the field. Historical and cultural geographers find fewer uses for the methodology. Recent advocates of humanistic and Marxist approaches tend to be particularly critical of the basis of the scientific method and generally reject statistical methodology. Even in these cases, however, there are some uses for statistical methodology, particularly descriptive statistics.

One of the first tasks of the statistician is to evaluate the data. One important distinction is whether the data are already available in some form or must be collected by the researcher. If the data must be collected, an experimental approach should be attempted, if possible. Whenever statistical surveys must be undertaken, there is some possibility that suitable data may not be generated. The greater the control in data collection, the better the end product (the data). Of particular importance in evaluating a set of available data are the level of measurement of the variables in the data set, their precision, their accuracy, and their validity.

REFERENCES

H. Carey, *Principles of Social Science* (Philadelphia: Lippincott, 1858).
R. Hartshorne, *The Nature of Geography* (Lancaster, Pa.: Association of American Geographers, 1939).
E. Ravenstein, "The Laws of Migration," *Journal of the Royal Statistical Society* 48 (1885) 167–235.
J. C. Weaver, "Crop Combinations in the Middle West," *Geographical Review* 44 (1954) 175–200.

FURTHER READING

Most of the material covered in this chapter is included in virtually all introductory textbooks in statistics for the social sciences. For a more in-depth treatment of the role of statistics, and quantification in general, in geography, consult Johnston (1983). Survey methods of data collection are increasingly common in geographical research. Two very readable introductions to this topic are by Dillman (1978) and Sudman and Bradburn (1983).

Don A. Dillman, *Mail and Telephone Surveys: The Total Design Method* (New York: Wiley, 1978).
R. J. Johnston, *Geography and Geographers: Anglo-American Geography since 1945*, 2d ed. (London: Edward Arnold, 1983).
S. Sudman and N. Bradburn, *Asking Questions: A Practical Guide to Questionnaire Design* (San Francisco: Jossey-Bass, 1983).

PROBLEMS

1. Test your knowledge of this chapter by explaining the meanings of the following terms:
   a. Descriptive statistics
   b. Inferential statistics
   c. Statistical population
   d. Variable
   e. Population census
   f. Sample
   g. Representative sample
   h. Random sample
   i. Sampling error
   j. Nonsampling error
   k. Statistical estimation
   l. Hypothesis test

2. Differentiate between the following pairs of terms:
   a. Internal versus external data
   b. Primary versus secondary data
   c. Experimental versus survey methods of data collection
   d. Quantitative versus qualitative variables

3. Describe the four scales of measurement. Give an example of a variable based on each scale of measurement.

4. What do we mean when we speak of (a) the precision, (b) the accuracy, and (c) the validity of a set of measurements?

5. Give an example of a research problem in geography in which (a) descriptive statistics and (b) inferential statistics might prove useful.

6. Write a brief history of the role of statistical research in geography.

7. Under what conditions might it be advantageous to undertake a population census rather than a sample?

8. Examine several recent issues of academic journals in geography and cite three articles that employ statistical techniques. Discuss the role of statistical analysis in each study.

9. Describe a research topic in geography which could be approached by using (a) a statistical experiment and (b) a survey.

10. Why do you think few experimental studies have been undertaken by human geographers?

11. Examine the census of Canada or the United States. List 10 different variables that could be obtained from the census.

12. In the reference room of your college library, locate 10 different sources of published data appropriate for statistical analysis. For each source, (a) identify the publisher, (b) note the frequency of publication, and (c) cite two specific items of information obtainable from the source.

I

DESCRIPTIVE STATISTICS

2

Univariate Descriptive Statistics

In Chapter 1 you were introduced to several types of data geographers encounter in their research. You may not have noticed that we introduced some simple procedures of descriptive statistics in our discussion of data measured at a nominal level. It was found, for example, that a large number of nominally scaled observations can be effectively summarized by recording the frequency, or count, of observations in each nominal class or category. We could even standardize the data by dividing the frequency in each class by the total number of observations and calculating proportions, or percentages. The bar chart is a useful visual summary of data of this type.

In this chapter, we concern ourselves with interval-scale variables. Whether continuous or discrete, interval-scale variables are not as easily summarized as nominal variables. There is no given set of classes to neatly summarize our counts or frequencies. Often all the values of a variable are different. Or there may be only a small number of repetitions. Consider the following example. One measure of the quality of water in a lake or a stream is the amount of dissolved oxygen (DO). As wastes are discharged into the lake or stream, the amount of DO in the water declines. This oxygen is used in the biochemical reactions by which the nutrients in the wastes are broken down. Public water supplies should have a DO content of at least 5 mg/L at any time. Table 2-1 contains a list of DO values estimated from a sample of 50 lakes in some region. Notice that the water samples from some lakes have DO levels below the public water supply standard of 5 mg/L, and others have levels above it. A closer examination of these data reveals that there are 25 different DO values in the 50 lakes. It is extremely difficult to scan this list of numbers and get a clear picture of the water quality. Since this is a relatively small data set with only $n = 50$ observations, imagine the difficulty of trying to make sense out of a sample of 500 or even 1000 observations. Our task is to try to reduce this set of numbers to some reasonable collection of descriptive terms in much the same way as we simplified the presentation of a nominal variable.

In Section 2.1 we explore various ways of organizing our data and graphical techniques which can give us an overall picture of the distribution of the variable. When we speak of the *distribution* of a variable, we are referring to the tendency for the variable to take on different values with different frequencies. Although tabular and graphical devices are very useful for obtaining some initial insights into the

TABLE 2-1
Dissolved Oxygen Values for 50 Sampled Lakes, mg/L

Observation No. of lake	DO value	Observation No. of lake	DO value
1	5.1	26	4.9
2	5.6	27	6.6
3	5.3	28	5.7
4	5.7	29	5.4
5	5.8	30	5.9
6	6.4	31	5.6
7	4.3	32	6.7
8	5.9	33	5.4
9	5.4	34	4.8
10	4.7	35	6.4
11	5.6	36	5.8
12	6.8	37	5.3
13	6.9	38	5.7
14	4.8	39	6.3
15	5.6	40	4.5
16	6.4	41	5.6
17	5.9	42	6.2
18	6.0	43	4.2
19	5.5	44	5.2
20	5.4	45	5.8
21	4.4	46	6.1
22	5.1	47	5.1
23	5.6	48	5.9
24	5.8	49	5.5
25	5.7	50	4.7

distribution of the variable, they do not allow us to make precise statements about a variable or to compare several different distributions, except in the simplest of ways. Since the data are expressed as numbers, it is not surprising that mathematical or numerical descriptions of these distributions are the most useful and concise way of analyzing the data. Sections 2.2 to 2.5 discuss various statistical measures which can be used to summarize the important attributes of the distribution of an interval-valued variable. The DO data for the 50 lakes listed in Table 2-1 illustrate all the descriptive measures discussed in this chapter.

2.1. Tabular and Graphical Techniques

There are several ways of displaying a collection of data. At this time, let us restrict ourselves to methods useful for the description of a single variable such as the DO levels of the 50 lakes. In the last three chapters of this book, methods for illustrating and analyzing the *joint* distribution of two variables are described. For example, if the data included the DO level and the number of vacation homes along the shoreline of each lake, we would have two separate variables to analyze. Although we might be interested in each variable in its own right, we would probably also wish to examine the *relationship* between the oxygen level of a lake and the number of vacation homes on its shoreline. Are DO levels lower in lakes with more vacation homes? We return to methods for analyzing the joint distribution of two variables in Chapter 12.

Frequency Tables

In *univariate*, or single-variable, situations, there are several ways of summarizing or displaying our observations. The first step is usually to construct a *tally sheet*, or *frequency table*. Whereas nominal variables have ready-made classes that can be used to organize such a table, interval-valued variables do not. An initial decision must be made to define the classes or categories themselves. If one class were to be defined for each 0.1 mg of DO from 4.2 to 6.9, the frequency table of Table 2-2 would be generated. Each different value is associated with a *frequency*, or tally, which is simply the number of observations that take on the given value. Because of the gaps in the data, there are 28 categories for the 50 observations! This does not help much. This type of classification scheme might be useful for discrete, or integer-valued, variables with a limited range, but not in this case. How can the representation of the DO data be improved? The answer lies in grouping the observations.

Grouping is effected by using a different set of categories for the same data. If the oxygen data are grouped into classes using an interval of 0.5 unit beginning at 4.0 and ending at 7.0, then six classes are created. Table 2-3 summarizes the frequency table. Frequencies are expressed in two forms. First, they are expressed as raw counts in the fourth column. It is apparent, for example, that 30 of the observations have values between 5 and 6 and smaller numbers of lakes have values below 5 or above 6. In the last column, these frequencies are expressed as relative frequencies, or proportions of the total number of observations. The sum of the column of frequency counts should equal the total number of observations, in this case $n = 50$. The sum

TABLE 2-2
An Ungrouped Frequency Table

DO value, mg/L	Frequency	DO value, mg/L	Frequency
4.2	1	5.6	6
4.3	1	5.7	4
4.4	1	5.8	4
4.5	1	5.9	4
4.6	0	6.0	1
4.7	2	6.1	1
4.8	2	6.2	1
4.9	1	6.3	1
5.0	0	6.4	3
5.1	3	6.5	0
5.2	1	6.6	1
5.3	2	6.7	1
5.4	4	6.8	1
5.5	2	6.9	1

TABLE 2-3
A Frequency Table for DO Data

Class	Interval	Midpoint	Frequency count	Relative frequency
1	4.0–4.49	4.25	3	.06
2	4.5–4.99	4.75	6	.12
3	5.0–5.49	5.25	10	.20
4	5.5–5.99	5.75	20	.40
5	6.0–6.49	6.25	7	.14
6	6.5–6.99	6.75	4	.08
			50	1.00

of the relative frequency column is always 1.00. Expressed in this form, the oxygen data of Table 2-1 are much easier to interpret.

How are the interval widths and the number of intervals in a frequency table selected? A few simple rules should be kept in mind.

RULE 1   Use intervals with simple bounds.

It is easier to use intervals with a common width of 0.5, 1.0, 10, 100, or 1000 rather than peculiarly selected widths such as 7.28, 97.63, or 729.58. Simplicity is the key. Similarly, the frequency table proves much easier to interpret if the lower bound or class limit of the first interval is a suitably chosen round number. If the interval width is conveniently chosen, this leads to regularly placed, easy-to-interpret class limits and midpoints. The values selected in Table 2-3 serve this purpose. A good first step in constructing a frequency table is to order the data from lowest to highest value. Then select a lower limit on the first class at a round number just below the lowest observation and an upper limit just above the highest-valued observation. In this case 4.0 is slightly below 4.2, and 7.0 is slightly greater than the maximum value of 6.9. Given these limits, we have a range of $7.0 - 4.0 = 3$ units. Potential interval widths of 0.25, 0.5, 0.75, and 1.0 all lead to easily understood class limits and midpoints.

RULE 2   The intervals should not overlap and must include all observations.

An observation can appear in one and only one class. Such classes are termed *mutually exclusive* since placement in one class precludes placement in any other interval. Do not specify overlapping interval widths such as 3–5 and 4–6. An observation of 4.5 would appear in both intervals. Although it is permissible to have categories such as <3 (less than 3) or 5+ (greater than 5, or >5), these should be avoided ordinarily. If they are selected because there is an extreme-valued observation well above or below all other observations, then they are surely misused. The difficulty of such open intervals is that it is impossible to tell what the highest or lowest values might be. As we see below, this important characteristic of the data should not be hidden.

To avoid ambiguity and to define mutually exclusive categories, select class limits carefully. Note that the upper bound of the first interval, 4.49, does not equal the lower bound of the second interval, 4.50. An observation of exactly 4.50 must be placed in the second class. If both limits are defined as 4.50, it is not clear exactly where a value of 4.50 is placed. Defining limits with one more decimal place than the most precise observation in the data set is a good general rule.

RULE 3 All intervals should be the same width.

It may seem obvious why this rule is important. It certainly simplifies matters. An even more important reason than simplicity comes to light when a graph of the numbers in a frequency table is constructed. As we see below, very misleading graphical representations of data can be made if this rule is not followed.

RULE 4 Select an appropriate number of classes.

Usually the most difficult decision to make is the number of classes. Some statisticians offer exact mathematical rules for determining the number of classes in any given case. One of these rules suggests that the approximate number of classes is equal to 5 times the logarithm to the base 10 of the number of observations. In this case, this leads to $5 \times \log 50 = 5(1.699) = 8.495$ and either 8 or 9 classes. Other statisticians suggest there should be between 6 and 16 or even 6 and 25 different classes depending on the sample size. Clearly, the more observations there are, the more classes can be constructed. But none of these rules can be considered inviolate.

There are two things to keep in mind. First, when two items are grouped, there is an indication that they are considered to be alike. As a general rule, then, never select a number of groupings or an interval width that forces two observations which cannot be considered alike into the same group. For example, it does not make sense to group a lake with a DO value about 5.0 with one having a value below 5.0, since this is the breakpoint between lakes meeting the public water supply standard and lakes not meeting this standard. Similarly, it may be unwise to group temperatures above and below the freezing point. It is usually good practice to respect important natural breakpoints. As a general rule, it is advisable to group only observations for which the differences in value can safely be ignored. Decisions of this type can usually be made only in the context of the specific data set involved. The precision and accuracy of the data play important roles in this decision.

Second, grouping, or aggregation, leads to a trade-off between the gains from simplicity and the loss of information. A frequency table with six classes implies we have replaced the total number of observations, in this case 50, with 12 new numbers—the 6 midpoints and 6 frequencies of the classes. There is thus a considerable gain in simplicity. However, some information is lost. Consider the first interval. If we know only the midpoint of 4.25 and the frequency of 3, then a good estimate of the sum of the 3 observations in this interval is $3 \times 4.25 = 12.75$. Part of the information loss is the exact values of the three observations in the interval. Yet, we are still able to approximate characteristics of the three observations. In fact, the three observations in the first interval sum to $4.2 + 4.3 + 4.4 = 12.90$. This is very close to the guess made by using the information in the frequency table. In this case, the gains in simplicity resulting from the reduction of the data to six classes seem to offset the slight loss of accuracy that also occurs. However, serious problems can occur if the observations are grouped into too few intervals. This becomes apparent in the analysis of the graphical counterparts to frequency tables—histograms.

Frequency Histograms and Frequency Polygons

Often it is easier to understand the information given in a frequency table when it is presented in graphical form. There appears to be a grain of truth in the adage "A picture is worth 1000 words." A special form of bar graph known as a *histogram* is used for this purpose. Figure 2-1 depicts the histogram for the DO data of Table 2-3. The X axis contains a scale of DO delimiting the six classes. Notice the small space that is left on each side of the axis below the lower bound of 4.0 and above the upper bound of 7.0. The vertical, or Y, axis can be labeled with absolute (or raw) frequencies, with relative frequencies, or with both. In Figure 2-1 the left side of the Y axis is labeled with values of absolute frequency, and the right-hand side shows relative frequencies. Relative frequency histograms are quite useful for comparing the distributions of different variables. The relative frequencies can be expressed as proportions, or percentages.

The height of a bar is proportional to the frequency in the class interval. It is important that the intervals be of equal width so that the histogram does not give a misleading impression of the data. Suppose, for example, that the last two categories are grouped to create a class of 6.0 to 6.99. The resultant histogram, shown in Figure

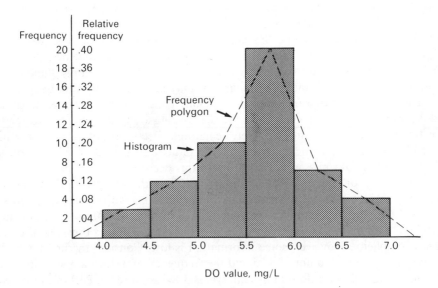

FIGURE 2-1. Histogram and frequency polygon.

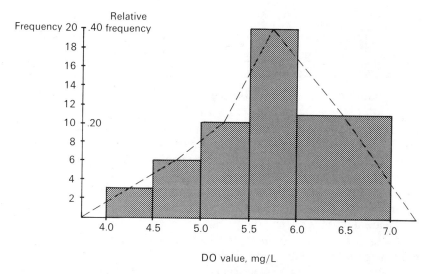

FIGURE 2-2. Histogram with unequal intervals.

2-2, gives the impression that most of the lakes have sufficient DO (greater than 5 mg) whereas it is clear from Table 2-3 and Figure 2-1 that this is not so. There is another advantage to selecting equal intervals. Assume that the width of classes is standardized to be 1 unit and frequencies are expressed as proportions. The total area under the histogram (area = height × width) is

$$.06(1) + .12(1) + .20(1) + .40(1) + .14(1) + .08(1) = 1.0 \qquad (2-1)$$

The misleading histogram of Figure 2-2 does not have this property.

There is an alternative way of presenting a frequency distribution. In many instances, bar graphs or histograms give the impression that the variable is not really continuous. For nominal variables this may not be a problem, and we often graphically summarize such data by using bar graphs with spaces between the bars. But for interval-valued data, we can often give a better representation of the distribution of a continuous variable if we construct a *frequency polygon*. A frequency polygon is obtained by connecting the midpoints at the top of each interval. By convention, we usually extend the endpoints of the frequency polygon beyond the upper and lower bounds to what would normally be the midpoint of the next interval. In Figure 2-2 the frequency polygon begins at 3.75 and ends at 7.25. By using this convention, it is easy to show that the area under the histogram and the area of the frequency polygon must be equal. For every triangular area not in the histogram but under the frequency polygon, there is a corresponding triangle above the frequency polygon but beneath the histogram. Thus, both representations have equal areas.

We can think of both these graphical representations as crude approximations to smooth curves. If we are graphing the distribution of a sample, we can think of this sample as an approximation to a more general distribution—the distribution of the variable for the population. As a consequence, it is usual to smooth the histogram and frequency distribution to a simpler representation, as in Figure 2-3. This curve

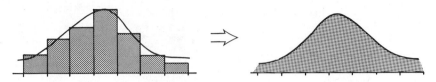

FIGURE 2-3. Histogram as a simplfication of a continuous distribution.

is only a graphical approximation to what we believe the actual population to be. It is an ideal shape. If there are a large number of observations in a sample and we construct a histogram with many intervals, we could probably generate a close approximation to this smooth curve.

What do histograms or smoothed frequency polygons tell us about the variable? Characteristic shapes of these histograms can tell us important facts about the variable we are analyzing. A few things can be easily checked.

1. *Is the distribution peaked or rectangular?* A peaked distribution such as in Figure 2-1 suggests that many of the observations are centered on some typical value of this variable. A typical DO value seems to be about 5.75 for the 50 lakes in this sample. If the histogram has no prominent peaks and the heights of all the bars are more or less equal, then the distribution is rectangular. This implies that there is no such typical value and any value in the range is almost equally likely.

2. *How many different peaks are there?* Obviously, if there is only one peak, then there is one typical value. Such a peak is termed a *mode*, and a distribution with one mode is said to be *unimodal*, as in Figure 2-4(*a*). In some cases there may be a

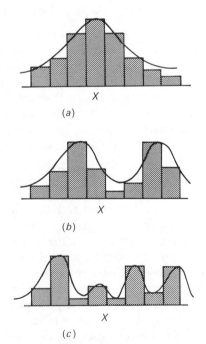

FIGURE 2-4. (*a*) Unimodal, (*b*) bimodal, and (*c*) multimodal distributions.

second peak. It may or may not be as noticeable as the first one. Histograms of this type, such as in Figure 2-4(*b*), are termed *bimodal*. Or the distribution may be *multimodal*, as in Figure 2-4(*c*). Bimodal distributions can occur if two separate factors contribute to the distribution of the variable. For example, the distribution of heights of all 18-year-olds might contain two modes, one for females and one for males. The distribution of heights for each sex taken individually would probably be unimodal. Many inferential statistical techniques are invalid when they are applied to variables with bimodal and multimodal distributions.

3. *Is the distribution symmetric about some central point?* A *symmetric distribution* is one in which each half of the distribution is a mirror image of the other half. When this is not the case, the distribution is said to be *skewed*. If the tail of the distribution is on the right, or toward the positive side of the horizontal axis, the distribution is said to be *positively skewed*. If the tail is on the left, it is *negatively skewed*. These three cases are contrasted in Figure 2-5. For many inferential statistical tests a symmetric distribution is required, but a few "tricks" can be used to reduce the extent of skewness of many distributions. The histogram of dissolved oxygen illustrated in Figure 2-3 appears to be slightly negatively skewed.

When the number and width of the intervals are chosen, it is important to give a representative impression of the distribution of the variable. The visual impression of a histogram is quite sensitive to the interval width chosen. Poorly selected interval widths can mask important characteristics of the distribution. Modality, skewness, and/or the existence of extreme values should be highlighted by the histogram and not hidden. Such features can be hidden by choosing too few, wide intervals.

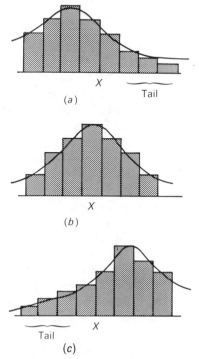

FIGURE 2-5. (*a*) Positive skew, (*b*) symmetric, and (*c*) negative skew distributions.

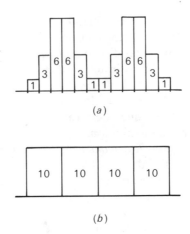

FIGURE 2-6. (a) Bimodal variable; (b) rectangular distribution.

Consider the two histograms of Figure 2-6. In Figure 2-6(a) the bimodal nature of the variable is clearly illustrated with the use of 12 intervals. In Figure 2-6(b) the same data are grouped into four classes, and the result is a perfectly rectangular distribution. Which is more realistic? Clearly, it is Figure 2-6(a). Similar cases can be constructed to show the reduction of extreme skewness or to hide the existence of an extreme-valued observation.

Cumulative Table and Ogive

For some purposes it is convenient to summarize and illustrate data in a slightly different way. Suppose, for example, that we are interested in determining the number or proportion of lakes with DO values less than or equal to some given level. How many lakes have DO values less than 5.0? What proportion of the lakes have DO levels less than 5.5? Quick answers to these questions can be derived by portraying the data in *cumulative* form, as in Table 2-4. The cumulative frequency column is calculated by summing the frequencies of all intervals below and including the interval under consideration. At the end of the first interval, for a value of $X = 4.49$, there are three observations. There are nine observations below a value of 4.99—three in the first interval and six in the second interval. The table is completed in this way. The last entry in the cumulative frequency column is always equal to the total number of observations, or the sample size, n. In Table 2-4 the last entry in the fourth column is 50.

Sometimes it is useful to express this cumulative frequency column in relative terms, as in the last column. In the first row we calculate the cumulative percentage as the cumulative frequency divided by the total number of observations, multiplied by 100. For the first interval, $\frac{3}{50} \times 100 = 6$ percent; for the second interval, $\frac{9}{50} \times 100 = 18$ percent; and so on. Expression of this information as relative frequency percentages is extremely helpful when we have a large number of observations and the simple cumulative frequencies are not as easily interpretable. Suppose, for example, there are 2200 lakes and the cumulative frequency at less than a DO level of 5.0

TABLE 2-4
A Cumulative Frequency Table

Class	Interval	Midpoint	Frequency		Cumulative frequency	Cumulative relative frequency
1	4.0–4.49	4.25	3 ⎫	$3 + 6 = 9$	3	.06
2	4.5–4.99	4.75	6 ⎬	⟶ 9	9	.18
3	5.0–5.49	5.25	10 ⎭	$3 + 6 + 10 = 19$ ⟶ 19	19	.38
4	5.5–5.99	5.75	20		39	.78
5	6.0–6.49	6.25	7		46	.92
6	6.5–6.99	6.75	4		50	1.00
			‾‾ 50			

is 550. This is much easier to interpret if we say 25 percent of the observations have DO levels below 5.0 mg/L.

As you may have already guessed, we can also use graphical devices to represent the information in a cumulative frequency table. Figure 2-7 displays the *cumulative frequency histogram* and the *cumulative frequency polygon*, or *ogive*. The cumulative histogram is drawn in exactly the same manner as a simple histogram except that the height of the bars is proportional to the cumulative frequency at the end of the interval. Normally, the ogive is drawn by connecting the lower bound of the first interval to the cumulative frequency at the end of the first interval, at the end of the second interval, and so on. The last point is always at the end of the last

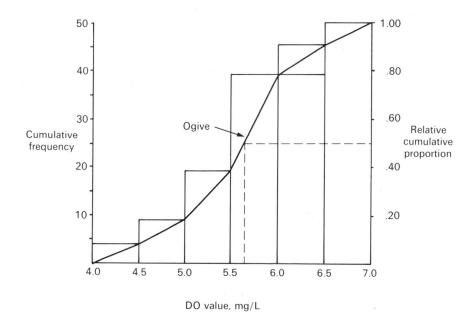

FIGURE 2-7. Cumulative frequency histogram and ogive.

interval with the cumulative frequency equal to the total number of observations. It is often convenient to label two separate vertical axes, one for the cumulative frequency and the other for the relative or percentage cumulative frequency. This is an alternative format to labeling a single axis with both measures, as in Figure 2-1.

We can use the ogive to derive an estimate of the number or percentage of observations less than some given value. Suppose we are interested in the DO level that is exceeded by 50 percent of the observations. Extend a horizontal line from the cumulative frequency percentage axis at 50 percent until it hits the ogive, and then drop the line vertically from this point to the DO axis. We estimate that 50 percent of the lakes have a DO value less than about 5.6 or 5.7. This procedure is illustrated in Figure 2-7.

Since the ogive and frequency polygon are both derived from the same data, it is not surprising that the shapes of these curves are closely related. For example, a peaked frequency polygon such as Figure 2-3 is always associated with a more or less S-shaped ogive such as Figure 2-7. Figure 2-8 illustrates generalized or smoothed

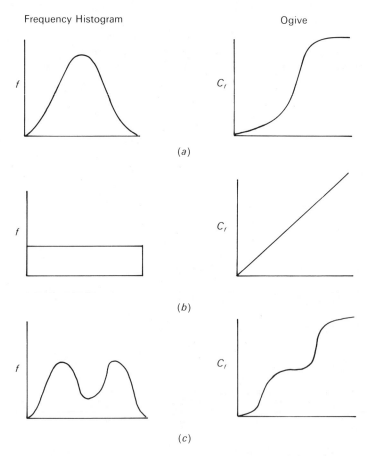

FIGURE 2-8. On left side are frequency histograms; on right, ogives.

frequency polygons and the corresponding ogives. Rectangular or near-rectangular frequency polygons have straight-line ogives, and peaked distributions have one S-shaped segment for each peak.

Other Graphical Devices

Computer packages for statistical analysis of data usually offer several graphical display alternatives. Some are appropriate for nominal variables, others are appropriate for ordinal and interval variables, and a few can be used for all types of variables. Figure 2-9 displays the alternatives available from the Statistical Analysis System (SAS), one of the more versatile packages available. Figure 2-9(*a*) and (*b*) depicts the usual forms of a histogram—vertical and horizontal bar graphs. Unlike in Figure 2-1, there is usually some space between the bars. The restriction of computer line printers to a limited number of printing positions sometimes results in a moderate loss of accuracy in these diagrams. For example, in a vertical bar graph one row of asterisks may correspond to five observations. Frequencies must be rounded to the nearest five for display purposes. This problem does not usually occur when plotting devices are used instead of the line printer.

Nominal variables can be summarized as bar graphs, as pie charts [as in Figure 2-9(*c*)], as star charts [as in Figure 2-9(*d*)], or as block diagrams [as in Figure 2-9(*e*)]. Appendix 2B contains an SAS program to generate these graphs as well as the descriptive statistics described in Sections 2.2 to 2.5.

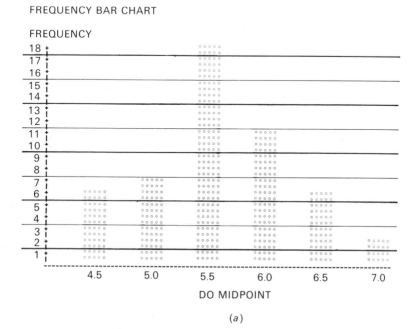

FREQUENCY BAR CHART

FREQUENCY

(a)

FIGURE 2-9. Graphical display alternatives produced by SAS.

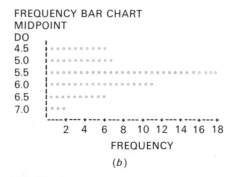

FREQUENCY BAR CHART
MIDPOINT
DO

(b)

FREQ PIE CHART OF DO

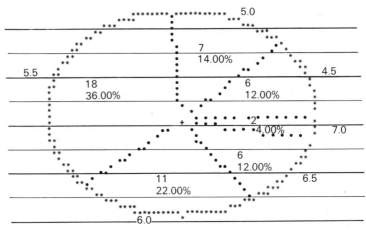

(c)

CENTER = 0   FREQ STAR CHART OF DO   OUTSIDE = 18

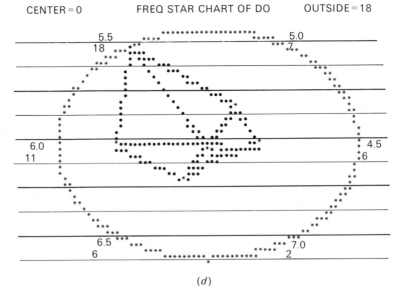

(d)

FIGURE 2-9 (continued).

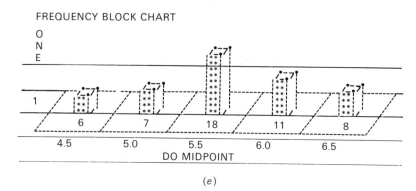

(e)

FIGURE 2-9 (continued).

2.2. Measures of Central Tendency

To make precise statements about the distribution of a variable, quantitative measures, or descriptive statistics, are used. Although the tabular and graphical devices described in Section 2.1 can tell us a great deal about a variable and even allow us to make simple comparisons about two or more variables, they are never precise. Numerical statistics are an even more compact way of summarizing interval-scale variables, enabling exact comparisons between pairs of variables to be made. One important characteristic of a distribution is the location of the typical value, or center, of the variable. In fact, there are several different measures of central tendency, each having certain advantages and disadvantages as well as specific mathematical properties. Each measure usually specifies a single number that can be considered the center of a set of observations. The fact that we can compute several different measures of central tendency implies we have to fully understand each one in order to know when each is an appropriate summary statistic. Although each measure is in some sense the "average" of a set of numbers, none is considered universally superior to all others. As we shall see, however, there is one which is more commonly used.

Midrange

The *range* of a variable is simply the difference between the highest- and lowest-valued observations. The *midrange* lies halfway between these two observations. In other words, the midrange is the arithmetic average of the extreme observations. Although it is easy to calculate, it has the disadvantage of giving inordinate emphasis to the two most extreme observations. Neither of these can be considered typical. All other observations are totally disregarded. Because of this drawback, this measure is seldom used. For the DO data, the maximum is 6.9, the minimum is 4.2, the range is 2.7, and the midrange is $(6.9 + 4.2)/2 = 5.55$.

Mode

Strictly speaking, the *mode* of a distribution is the value of the variable that appears most frequently. For many continuous interval-scale variables there are no exact repeated values in the data. We could say there is no mode, or we might even say that there are n different modes. This measure is not very useful in this case. If the data are grouped, the *modal category*, or *modal class*, is defined as the interval with the highest frequency; the midpoint of the interval is often termed the *crude mode*. Unfortunately, this approach does not define a *unique* mode or interval since the value depends on the number and location of the intervals selected in the grouping procedure. Loosely speaking, the value associated with a peak of a histogram or frequency polygon is termed a *mode*. In this sense there may be one or more modes.

When is the mode useful? Under certain conditions this measure is useful for nominal variables and discrete-valued interval variables. For nominal variables the mode is the category with the greatest frequency. Note that the mode is not really a measure of central tendency in this case, since a nominal variable is not ordered and hence can have no center. For ordinal and discrete interval variables, the notion of a mode is useful for cases in which there is one value with a dominating frequency. Suppose a set of 200 households yields the following tabulation of household size:

Number of household members	Frequency
1	15
2	30
3	100
4	25
5	30

The mode is 3 persons per household, and it seems to be a good summary measure for this distribution. To many, it is preferable to saying there are 3.125 members in the average household.

Median

The *median* of a set of observations is the value of the variable that divides the observations so that one-half are less than or equal to the median and one-half are greater than or equal to it. To compute the median, it is again useful to order the data so that the smallest value is first, the next smallest is second, and so on. Table 2-5 contains the ordered values of the 50 DO values. Since the median should divide the set into two equal parts, there should be 25 observations in each set. The first set contains observations 1 to 25, and the second set contains observations 26 to 50. The median is the $(n + 1)/2$ observation. Since $n = 50$, the median is the value of the 25.5th observation. Obviously when n is even, we must have some rule to follow. Normally, the median is defined as the average of the middle two observations. In this case, the median is the average of the 25th and 26th terms, or $(5.6 + 5.6)/2 = 5.6$. When there are an odd number of observations, $(n + 1)/2$ is always a whole number and defines the middle term. Suppose that we have the following five ordered observations: 2, 2, 3, 5, and 8. The median is the middle term, or $(5 + 1)/2 = 3$d term, and is equal to 3.

The median is often termed a *positional* measure since it locates the position of an observation relative to a set of other observations. Just as it is possible to define

TABLE 2-5
Table 2-1 Expressed in Order of Increasing Value

4.2
4.3
4.4
4.5
4.7
4.7
4.8
4.8
4.9
5.1
5.1
5.1
5.2 ← 1st quartile = 25th percentile = P_{25}
5.3
5.3
5.4
5.4
5.4
5.4
5.5
5.5
5.6
5.6
5.6
5.6
5.6 ⎤ ← Median = 2d quartile = 5th decile = 50th percentile = P_{50}
5.6
5.7
5.7
5.7
5.7
5.8
5.8
5.8
5.8
5.9
5.9
5.9 ← 3d quartile = 75th percentile = P_{75}
5.9
6.0
6.1
6.2
6.3
6.4
6.4
6.4
6.6
6.7
6.8
6.9

the median, or middle term, it is also possible to define *quartiles* as the values which divide the set of observations into quarters. The *first quartile* divides the lower half into two equal sets. The *median* is the *second quartile*, and the *third quartile* divides the upper half into two equal sets. Thus, 25 percent of the observations are below the first quartile, 50 percent are below the median, and 75 percent are below the third quartile. Table 2-5 indicates that the first quartile of the DO data is the value of the $(25 + 1)/2 = 13$th observation and is equal to 5.2. Similarly, the third quartile is found to be 5.9.

We can generalize this procedure to define *deciles*, which divide the observations into tenths, and *percentiles*, which divide the data into hundredths. The median is the second quartile, the fifth decile, and the 50th percentile. The nth percentile is usually specified as P_n, and the three quartiles are Q_1, Q_2, and Q_3. Therefore $P_{25} = Q_1 = 5.2$, $P_{50} = Q_2 = 5.6$, and $P_{75} = Q_3 = 5.9$. Positional terms are most commonly associated with educational testing procedures and are used to locate a student among a large set of student scores for an examination. Percentiles seem to have been infrequently applied in geographical research.

Mean

The *mean* of a set of observations is the most commonly used measure of central tendency. In most instances the term *mean* is the shortened version of *arithmetic mean*, and it is the sum of the values of all observations, divided by the number of observations. In common use this is simply the average of a set of observations. The mean DO value of the 50 lakes is $(4.2 + 4.3 + \cdots + 6.7 + 6.8 + 6.9)/50 = 279/50 = 5.58$. The three dots, or ellipses, are shorthand for noting that the summation omits some terms in the set. The middle 45 observations of DO are omitted. To simplify many of the formulas in descriptive and inferential statistics, it is convenient to utilize a system of notation. If the variable is denoted by the uppercase X and a subscript i is used to denote individual observations, then we can rewrite the mean of a sample as $(X_1 + X_2 + \cdots + X_n)/n$. An even more efficient way of writing this makes use of sigma notation.

The symbol $\Sigma_{i=1}^n$ (read "sigma for i equal to 1 to n") is a mathematical shorthand which has considerable use in statistics. It is known as the *summation operator* since it means that the items following the symbol are to be summed through all successive values of the *index of summation i*. Below (or following as subscript) the sigma, the index of summation specifies the start of the summation, in this case $i = 1$. This is the usual case, but it is possible to start the summation at any point within a series of observations. Above (or following as subscript) the sigma, the index defines the last observation in the summation; here $i = n$. Using this notation, we can write

$$\sum_{i=1}^n X_i = X_1 + X_2 + \cdots + X_n \tag{2-2}$$

The sigma operator greatly simplifies the specification of many different operations in statistics. Students unfamiliar with summation notation should consult Appendix 2A. Even those who are somewhat experienced with its use will find it a useful review.

With this notation it is possible to rewrite the definition of the mean as

$$\bar{X} = \frac{\sum_{i=1}^{n} X_i}{n} \tag{2-3}$$

where the symbol $\bar{X}$ (read "X bar") denotes the mean of a set of observations. At times, it is necessary to distinguish the mean of a *sample* from the mean of a *population*. For a sample, the observations are numbered consecutively from 1 to n, and the mean is denoted by $\bar{X}$. For a finite population consisting of N observations, the formula for the mean is

$$\mu = \frac{\sum_{i=1}^{N} X_i}{N} \tag{2-4}$$

where μ is the lowercase Greek letter mu. These two formulas are not interchangeable. There are many different sample means $\bar{X}$. Depending on which n observations are selected from a population or universe with N members, we get a different value for $\bar{X}$. However, if we compute the mean from all N observations, there is only one possible mean μ. Generally, a sample of size n is only a small portion of the members of the population, so that $n < N$.

Two important conventions have been introduced in these formulas which are used throughout the book. First, lowercase n always refers to a sample size, and uppercase N refers to the size of a finite population. Second, characteristics of populations are defined by Greek letters such as μ, but samples are not. These distinctions become particularly important when we try to make inferences about a population from a sample. For example, we might wish to use $\bar{X}$ to estimate a value for μ. In particular, we may wish to determine how close we might be to the mean of the population with samples of increasing size. When $n = N$ and all members of the population are included in the sample, $\bar{X} = \mu$. We return to this issue and see the exact relationship between the sample and population means in Chapter 7.

It is also possible to approximate the value of $\bar{X}$ for a set of observations by using only the information from a frequency table. Normally, this is called the *grouped approximation to the mean*. Denote the classes or intervals in a frequency table as $k = 1, 2, \ldots, K$, the frequency of the kth class as f_k, and the midpoint of the kth interval as X_k. The mean can then be approximated by

$$\bar{X} = \frac{\sum_{k=1}^{K} f_k X_k}{n} \tag{2-5}$$

This approximation relies on the assumption that all the observations in any one class are equal to the midpoint or else symmetrically spread about it. To the extent that the set of observations meets this assumption, Equation (2-5) will closely approximate Equation (2-3). By using the information in Table 2-3, Equation (2-5) estimates the mean as

$$\bar{X} = \frac{3(4.25) + 6(4.75) + 10(5.25) + 20(5.75) + 7(6.25) + 4(6.75)}{50} = 5.59 \qquad (2-6)$$

Note how close this value is to the ungrouped mean of 5.58 obtained from using Equation (2-3). Unless the data are available only in grouped form, there is little reason to use Equation (2-5). Before the advent of electronic calculators and high-speed computers, the grouped approximation to the mean was a useful shortcut to computing $\bar{X}$. Currently computers can perform exact calculations by using Equation (2-3) so rapidly that the savings from using Equation (2-5) are insignificant. In fact, the time spent by the computer in sorting the observations would probably outweigh any savings in computation time derived from using the grouped method.

Choosing an Appropriate Measure of Central Tendency

For the oxygen data of Table 2-1 the following measures of central tendency are obtained:

Mean $\bar{X} = 5.58$ Median $= 5.60$
Mode $= 5.60$ Midrange $= 5.55$

All are very similar, although this is not necessarily a usual outcome. Let us review the derivation of each measure from a histogram. The mode is associated with a peak of the distribution, and the median is the value that splits the distribution into two equal areas. The midrange divides the length of the distribution into two equal parts. What about the mean? Think of the histogram as a set of weights placed along a board in proportion to the height of the bars. If we were to place a wedge under the board, then it would balance only when placed at the location of the mean. The mean is the "center of gravity," or balancing point, of the distribution illustrated in Figure 2-10.

The mean is the most commonly reported measure of central tendency. It has one distinct advantage over all other measures—it is sensitive to a change in any of the observations. To illustrate, consider the set of five observations 2, 2, 3, 5, and 8. The mean is 4, the mode is 2, and the median is 3. Suppose the last observation is changed from 8 to 18. The median is still 3, the mode is still 2, but the mean changes from $\bar{X} = 4$ to $\bar{X} = 6$. Although we can show that a change in one observation *can* change the value of the other measures of central tendency, any modification of an

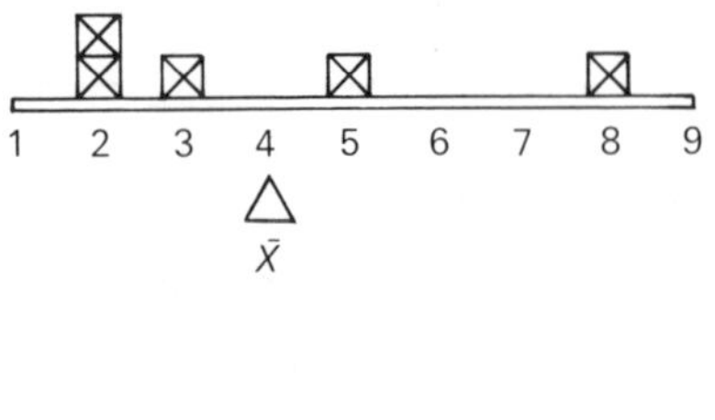

FIGURE 2-10. Mean as the center of gravity.

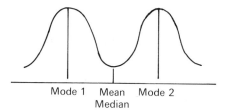

Mode 1   Mean   Mode 2
         Median

FIGURE 2-11. Measures of central tendency in a bimodal distribution.

observation *must* change the mean. However, in several instances the mean is not a good measure of central tendency:

    1. *For bimodal distributions.* Consider Figure 2-11. Note that the mean and median are coincident, but neither picks up the "typical value" of this variable. Neither would the midrange, which is at the same location. Since all these measures provide one value, they can never be useful in this case. The best summary statistic is clearly the mode.

    2. *For skewed distributions.* The mean is not very useful for variables that are either positively or negatively skewed. Figure 2-12 illustrates the relative location of the mean, median, and mode for skewed distributions. For a positively skewed

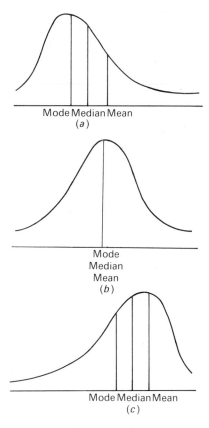

Mode Median Mean
(*a*)

Mode
Median
Mean
(*b*)

Mode Median Mean
(*c*)

FIGURE 2-12. Measures of central tendency for skewed distributions.

distribution, the mean is greater than the median which is greater than the mode. For a negatively skewed distribution the reverse is true: the mean is less than the median which is less than the mode. Although the median is pulled toward the tail of the distribution, it is not nearly as sensitive as the mean. Note that for symmetric distributions the three are equal. As a distribution deviates from this shape, the three become increasingly different. The median is the best measure in this case.

3. *For a distribution with an extreme value (or a small number of extreme values).* We can think of a distribution with an extreme value as highly skewed. The median is also the best measure in this case. Consider again the data 2, 2, 3, 5, and 8. If 8 is changed to 200, then the mode remains at 2, the median stays at 3, but the mean increases to 44. Quite clearly the mean is no longer even close to any observation and could hardly be termed typical. The median is central to four of the five observations.

4. *For nominal and ordinal data.* Truly nominal data have no central value since they are unordered. It is impossible to compute a mean for ordinal data since we do not know the distances between the ranks. We could not add a rank of 5 and a rank of 3 to get a mean of 4 without assuming that the data are interval, not ordinal. However, since the median is a positional measure, we can locate the median for an ordinal variable.

In sum, the mean is the most commonly used measure of central tendency. It is sensitive to all changes in the distribution and makes use of all information from a set of observations. Unless one of these four conditions applies, it is the best measure of central tendency.

Properties of the Mean and Median

There are also three important properties of the mean and median:

PROPERTY 1 The sum of the deviations of each observation from the mean is zero.

This property can be expressed even more compactly as

$$\sum_{i=1}^{n} (X_i - \bar{X}) = 0 \qquad (2\text{-}7)$$

and it indicates another sense in which the mean is the center of a set of observations. The property is easily proved by using the simple rules for summation discussed in Appendix 2A. Using Rule 1, we can express Equation (2-7) as

$$\sum_{i=1}^{n} X_i - \sum_{i=1}^{n} \bar{X} = 0 \qquad (2\text{-}8)$$

And since $\bar{X}$ is a constant (for any given set of observations), we can write $\sum_{i=1}^{n} \bar{X} = n\bar{X}$ from Rule 2. Finally we note from our definition of the mean in Equation (2-3) that

$$n\bar{X} = n\left(\frac{\sum_{i=1}^{n} X_i}{n}\right) = \sum_{i=1}^{n} X_i \qquad (2\text{-}9)$$

and Equation (2-8) can be rewritten as

$$\sum_{i=1}^{n} X_i - \sum_{i=1}^{n} X_i = 0$$

This property is easily verified by using any set of observations. To keep matters simple, consider again the set of observations 2, 2, 3, 5, and 8 for which $\bar{X} = 4$. Summing the deviations yields $(2-4) + (2-4) + (3-4) + (5-4) + (8-4) = -2 + (-2) + (-1) + 1 + 4 = 0$.

PROPERTY 2 The sum of squared deviations of each observation from the mean is a minimum, that is, less than the sum of squared deviations from any other number.

Formally, we say

$$\sum_{i=1}^{n} (X_i - M)^2 \qquad (2\text{-}10)$$

is minimized when $M = \bar{X}$. This is often called the *least squares property of the mean*. The quantity

$$\sum_{i=1}^{n} (X_i - \bar{X})^2$$

is termed the *total variation of variable X*. Although it is slightly more difficult to prove, this property is easy to illustrate. Using the same set of five observations, we calculate the sum of squared deviations as follows:

Observation	$\bar{X} = 4$	Median = 3
2	$(2-4)^2 = 4$	$(2-3)^2 = 1$
2	$(2-4)^2 = 4$	$(2-3)^2 = 1$
3	$(3-4)^2 = 1$	$(3-3)^2 = 0$
5	$(5-4)^2 = 1$	$(5-3)^2 = 4$
8	$(8-4)^2 = 16$	$(8-3)^2 = 25$
	Sum = 26	Sum = 31

Note that the sum of squared deviations is 26 for the mean and 31 for the median. Any other value would necessarily lead to a sum greater than 26. This property suggests yet another sense in which the mean is the center of a distribution.

A closer examination of Equation (2-10) reveals why the mean is a poor measure of central tendency when the set of observations is highly skewed or contains an extreme value. To minimize Equation (2-10), the values of M must be selected that minimizes squared deviations. A deviation of $X - M = 1$ is counted as $1^2 = 1$, but a deviation of $X - M = 5$ is counted as $5^2 = 25$ and given 25 times as much weight. Squaring the deviations penalizes larger deviations more than smaller ones. As a result, the mean is "drawn" toward extreme values, or values in the tail of the distribution. In this way the high penalties associated with large deviations are minimized.

PROPERTY 3 The sum of *absolute* deviations is minimized by the median.

To minimize the expression

$$\sum_{i=1}^{n} |X_i - M| \qquad (2\text{-}11)$$

the value of M should equal the median. It is in this sense that the median lies in the center of a distribution. This can also be illustrated by using the small sample of $n = 5$ observations:

Observation	Median = 3	Mean $\bar{X} = 4$
2	$|2 - 3| = 1$	$|2 - 4| = 2$
2	$|2 - 3| = 1$	$|2 - 4| = 2$
3	$|3 - 3| = 0$	$|3 - 4| = 1$
5	$|5 - 3| = 2$	$|5 - 4| = 1$
8	$|8 - 3| = 5$	$|8 - 4| = 4$
	Sum = 9	Sum = 10

The sum of absolute deviations from the median is 9 and from the mean is 10. Because it has been drawn toward the extreme value of 8 in order to minimize the squared deviations, the mean is in an inferior position to the median for absolute deviations. Although the median is 5 units away from this extreme observation, it is penalized far less by using this criterion than by using the sum of squared deviations.

Besides explaining why the mean and the median are superior measures of central tendency in particular cases, these properties have interesting interpretations when we consider spatial data in Chapter 3. They have important implications for locating the center of a spatial point pattern.

2.3. Measures of Dispersion

The various measures of central tendency discussed in Section 2.2 are very useful for calculating a summary statistic which represents an average, or typical, value of variable X. However, these statistics measure only one characteristic of a frequency distribution. Compare Figure 2-13(*a*) and (*b*). Both variables have identical means but much different *dispersions*, or *spread*, about the mean. Variable Y is much more compact than variable X. Let us now turn to several different measures to evaluate the spread, or dispersion, of a distribution.

Range

In defining the midrange of a variable, we also defined the range as the difference between the highest- and lowest-valued observations. It turns out that the range of the oxygen data is $6.9 - 4.2 = 2.7$. This approach can be criticized on the grounds

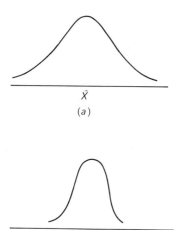

$\bar{X}$

(a)

$\bar{Y}$

(b)

FIGURE 2-13. Variables have same means but different dispersions.

that if takes into account only the two extreme observations and ignores the dispersion of all other observations. When is the range useful? Most computer programs for statistical analysis output the maximum and minimum values of the variable, and a few output the range. In some instances they provide important, easy-to-understand information. The minimum DO value gives a fairly good estimate of the lower bound on current lake oxygen levels. However, when one or both are extreme values and are unusual observations—and this often happens in geographical research—the range gives a very misleading measure of dispersion. Also, larger samples from a population invariably have greater ranges since they are more likely to contain the rare or unusual members of the population.

Quartile Deviation

The *quartile deviation* is one-half of the *interquartile range*:

$$\text{Quartile deviation} = \frac{P_{75} - P_{25}}{2} \qquad (2\text{-}12)$$

Here P_{75} is the 75th percentile, or third quartile; P_{25} is the 25th percentile, or first quartile; and the difference between the two is the interquartile range. The quartile deviation is $(5.9 - 5.2)/2 = 0.35$. The percentiles are given in Table 2-5. Because this measure ignores the extreme observations and utilizes the middle half of the observations, it is generally a more stable measure of dispersion. Different samples from the same population may have widely different ranges but are likely to have fairly similar quartile deviations. Unfortunately neither the range nor the quartile deviation takes full advantage of the entire set of observations and their values. Three other measures of dispersion take full advantage of the actual values of all observations. All are based on the deviations of each observation to the mean $\bar{X}$. The mean is utilized as the center of the distribution since it is the single most satisfactory measure of central tendency.

Mean Deviation

As formalized in Property 1, the sum of the deviations about the mean is zero. This is true because the positive deviations offset the negative deviations. To get around this, we take the absolute values of these deviations and define the *mean deviation* as follows:

$$\text{Mean deviation} = \frac{\Sigma_{i=1}^{n}|X_i - \bar{X}|}{n} \tag{2-13}$$

    The calculation of the mean deviation for the DO data is illustrated in Table 2-6. The second column contains the signed deviations and sums to 0.0. The absolute deviations sum to 24.36, leading to a mean (or average) absolute deviation of 24.36/50 = .4872. There is one disadvantage to the mean deviation. Many statistical results depend on easy algebraic manipulation of formulas such as Equation (2-13). Note how we were able to prove that the deviations summed to zero by manipulating Equation (2-7). The algebraic simplification of absolute value terms is quite cumbersome and generally does not lead to simple results. For this reason an alternative measure of dispersion, the *standard deviation*, is more commonly used.

TABLE 2-6
Work Table for Mean Deviation, Standard Deviation, and Variance

Observation No.	X_i	$X_i - \bar{X}$	$|X_i - \bar{X}|$	$(X_i - \bar{X})^2$	X_i^2
1	5.1	−0.48	0.48	0.2304	26.01
2	5.6	0.02	0.02	0.0004	31.36
3	5.3	−0.28	0.28	0.0784	28.09
4	5.7	0.12	0.12	0.0144	32.49
5	5.8	0.22	0.22	0.0484	33.64
6	6.4	0.82	0.82	0.6724	40.96
7	4.3	−1.28	1.28	1.6384	18.49
8	5.9	0.32	0.32	0.1024	34.81
9	5.4	−0.18	0.18	0.0324	29.16
10	4.7	−0.88	0.88	0.7744	22.09
11	5.6	0.02	0.02	0.0004	31.36
12	6.8	1.22	1.22	1.4884	46.24
13	6.9	1.32	1.32	1.7424	47.61
14	4.8	−0.78	0.78	0.6084	23.04
15	5.6	0.02	0.02	0.0004	31.36
16	6.4	0.82	0.82	0.6724	40.96
17	5.9	0.32	0.32	0.1024	34.81
18	6.0	0.42	0.42	0.1764	36.00
19	5.5	−0.08	0.08	0.0064	30.25
20	5.4	−0.18	0.18	0.0324	29.16
21	4.4	−1.18	1.18	1.3924	19.36
22	5.1	−0.48	0.48	0.2304	26.01
23	5.6	0.02	0.02	0.0004	31.36
24	5.8	0.22	0.22	0.0484	33.64

TABLE 2-6 (continued)

Observation No.	X_i	$X_i - \bar{X}$	$|X_i - \bar{X}|$	$(X_i - \bar{X})^2$	X_i^2
25	5.7	0.12	0.12	0.0144	32.49
26	4.9	−0.68	0.68	0.4624	24.01
27	6.6	1.02	1.02	1.0404	43.56
28	5.7	0.12	0.12	0.0144	32.49
29	5.4	−0.18	0.18	0.0324	29.16
30	5.9	0.32	0.32	0.1024	34.81
31	5.6	0.02	0.02	0.0004	31.36
32	6.7	1.12	1.12	1.2544	44.89
33	5.4	−0.18	0.18	0.0324	29.16
34	4.8	−0.78	0.78	0.6084	23.04
35	6.4	0.82	0.82	0.6724	40.96
36	5.8	0.22	0.22	0.0484	33.64
37	5.3	−0.28	0.28	0.0784	28.09
38	5.7	0.12	0.12	0.0144	32.49
39	6.3	0.72	0.72	0.5184	39.69
40	4.5	−1.08	1.08	1.1664	20.25
41	5.6	0.02	0.02	0.0004	31.36
42	6.2	0.62	0.62	0.3844	38.44
43	4.2	−1.38	1.38	1.9044	17.64
44	5.2	−0.38	0.38	0.1444	27.04
45	5.8	0.22	0.22	0.0484	33.64
46	6.1	0.52	0.52	0.2704	37.21
47	5.1	−0.48	0.48	0.2304	26.01
48	5.9	0.32	0.32	0.1024	34.81
49	5.5	−0.08	0.08	0.0064	30.25
50	4.7	−0.88	0.88	0.7744	22.09

$n = 50$
$n - 1 = 49$

$\sum_{i=1}^{n} X_i = 279$ $\sum_{i=1}^{n} (X_i - \bar{X}) = 0$ $\sum_{i=1}^{n} |X_i - \bar{X}| = 24.36$ $\sum_{i=1}^{n} (X_i - \bar{X})^2 = 20.02$ $\sum X_i^2 = 1576.84$

$$\text{Mean} = \frac{\sum_{i=1}^{n} X_i}{n} = 5.58$$

$$\text{Mean deviation} = \frac{\sum_{i=1}^{n} |X_i - \bar{X}|}{n} = \frac{24.36}{50} = .4872$$

$$\text{Variance} = s^2 = \frac{\sum_{i=1}^{n} (X_i - \bar{X})^2}{n - 1} = \frac{20.02}{49} = .4086$$

$$\text{Standard deviation} = s = \sqrt{\frac{\sum_{i=1}^{n} (X_i - \bar{X})^2}{n - 1}} = \sqrt{\frac{20.02}{49}} = .6392$$

Or from computational formulas:

$$s^2 = \frac{1}{n-1}\left(\sum X_i^2 - n\bar{X}^2\right) = \frac{1}{49}[1576.84 - 50(5.58)^2] = .4086$$

$$s = \sqrt{\frac{1}{n-1}\left(\sum X_i^2 - n\bar{X}^2\right)} = \sqrt{\frac{1}{49}[1576.84 - 50(5.58)^2]} = .6392$$

Standard Deviation

For a sample of observations the *standard deviation* is defined as the square root of the mean squared deviations about the mean

$$s = \sqrt{\frac{\sum_{i=1}^{n}(X_i - \bar{X})^2}{n-1}} \qquad (2\text{-}14)$$

where s is the commonly accepted notation for the standard deviation of a sample. For a population, the standard deviation is defined as

$$\sigma = \sqrt{\frac{\sum_{i=1}^{N}(X_i - \mu)^2}{N}} \qquad (2\text{-}15)$$

where σ, lowercase Greek sigma, is the usual symbolic representation. The notation is always $\bar{X}$ and s for samples and μ and σ for populations.

There are two significant differences between Equations (2-14) and (2-15). First, the sample standard deviation s has the squared deviations summed over the n different observations, while the squared deviations of the population standard deviation are summed over all N members. Second, the denominator for s is $n-1$, but it is N for σ. When n is very large (say, 100 or more), the difference between s and σ is quite small; and when $n = N$, the values of s and σ are equal. That is, when the sample contains all members of the population, there can be no difference between s and σ.

But why do samples have $n-1$ in the denominator? It turns out that if we divide by n, we always *underestimate* the true population standard deviation σ. By dividing by $n-1$, the standard deviation s becomes larger and the underestimate is corrected. A more detailed explanation is given in Chapter 7, where we deal with the *statistical estimation* of parameters of populations. However, this anomaly can also be explained by using two much simpler arguments.

First, let us recall the second property of the mean, which states that $\sum_{i=1}^{n}(X_i - M)^2$ is minimized for any set of observations when $M = \bar{X}$. That is, substituting any other number besides $\bar{X}$ for M will necessarily result in a larger sum. Even if we substitute the true population mean μ, at best we can do as well as with $M = \bar{X}$ and generally we will do worse. However, if we know μ, it makes sense to estimate s as

$$\sqrt{\frac{\sum_{i=1}^{n}(X_i - \mu)^2}{n}} \qquad (2\text{-}16)$$

When we do not know μ, we replace μ with $\bar{X}$ to obtain

$$\sqrt{\frac{\Sigma_{i=1}^{n}(X_i - \bar{X})^2}{n}} \tag{2-17}$$

But since the sum of the squared deviations around $\bar{X}$ must be less than that around μ, Equation (2-17) must be smaller than Equation (2-16). It turns out that by replacing n by $n-1$ we divide by a smaller denominator and get a larger value for s. So the correction of n to $n-1$ turns out to be just right.

There is also a second reason for the division by $n-1$. One other important property of the mean is that $\Sigma_{i=1}^{n}(X_i - \bar{X}) = 0$. Because of this identity, only $n-1$ of the n deviations are independent quantities. If we know $n-1$ of the deviations, we can always use Equation (2-7) to determine the last one. Consider once again the five observations 2, 2, 3, 5, and 8, with $\bar{X} = 4$. The first four deviations sum to $(2-4) + (2-4) + (3-4) + (5-4) = -2 + (-2) + (-1) + 1 = -4$. For Equation (2-7) to hold, the last deviation must be equal to 4. This is verified by calculating $8 - 4 = 4$. In general, then, among all the quantities $X_1 - \bar{X}, X_2 - \bar{X}, \ldots, X_n - \bar{X}$, there are only $n-1$ *independent* quantities. This is another reason why we divide by $n-1$ and not n. Thus the sample mean absolute deviation should also be divided by $n-1$ for *estimation* purposes.

Equation (2-15) does not represent the most computationally efficient formula for determining the standard deviation. Use of Equation (2-15) requires N different subtractions (one for each observation), N different squares, and one square root operation. By using simple rules for summations, we can show that Equation (2-15) is equivalent to

$$\sigma = \sqrt{\frac{1}{N}\sum_{i=1}^{N} X_i^2 - \bar{X}^2} \tag{2-18}$$

This is shown in Appendix 2C. Equation (2-18) requires $N+1$ squares (one for each observation and one for the mean), one subtraction, one multiplication, and one square root. This compares favorably with the computational requirements of Equation (2-15). The analogous formula for s is

$$s = \sqrt{\frac{1}{n-1}\left(\sum_{i=1}^{n} X_i^2 - n\bar{X}^2\right)} \tag{2-19}$$

The calculation of the standard deviation for the oxygen data is illustrated in Table 2-6. At the bottom of the table the numerical equivalence of Equations (2-14) and (2-19) is shown for the DO variable.

For grouped data, s can be approximated by using

$$s = \sqrt{\frac{1}{n-1}\left(\sum_{k=1}^{k} f_k X_k^2 - \frac{\sum_{k=1}^{k} f_k X_k}{n}\right)} \qquad (2\text{-}20)$$

The required calculations are illustrated for the grouped DO data of Table 2-3 in Table 2-7. In this instance, the value of s estimated by using Equation (2-20), or .6349, differs only in the third and fourth digits from the true value of $s = .6392$. Because it provides an approximation, the formula for s given by Equation (2-20) should be used if the data are available only in grouped form or if an approximate value for s is all that is needed. It should *not* be used when the goal is to make inferences about a population.

Variance

A final measure of dispersion is the *variance*, which is simply the mean, or average, squared deviation:

$$\text{Sample } s^2 = \frac{1}{n-1}\sum_{i=1}^{n}(X_i - \bar{X})^2 \qquad (2\text{-}21)$$

$$\text{Population } \sigma^2 = \frac{1}{N}\sum_{i=1}^{N}(X_i - \mu)^2 \qquad (2\text{-}22)$$

TABLE 2-7
Work Table for Grouped Standard Deviation

f_k	X_k	$f_k X_k$	$f_k X_k^2$
3	4.25	12.75	54.1875
6	4.75	28.50	135.3750
10	5.25	52.50	275.6250
20	5.75	115.00	661.2500
7	6.25	43.75	273.4375
4	6.75	27.0	182.2500
50		279.5	1582.1250

$$\left(\sum_{k=1}^{K} f_k X_k\right)^2 = (279.5)^2 = 78,120.25$$

$$s = \sqrt{\frac{1}{49}\left(1582.1250 - \frac{78,120.25}{50}\right)} = .6344$$

$$s^2 = .4024$$

The sample variance has a divisor of $n - 1$, and the population variance has a divisor of N. Variants of these formulas for computational purposes are easily derived as the squares of Equations (2-18) and (2-19). The variance is seldom used as a descriptive summary statistic since the squaring of the deviations often leads to a value of s^2 well out of the range of the observations. It proves very difficult to interpret. By taking the square root of the variance to obtain the standard deviation, we more or less compensate for the initial squaring of the deviations. The standard deviation is usually much easier to interpret. For inferential purposes, however, the variance is an extremely important measure of a variable.

Interpretation and Use of the Standard Deviation

Both the standard deviation and the variance give an indication how typical of a whole distribution the mean actually is. The larger these measures, the greater the spread of the observations and the less typical the mean. When all observations are equal to the mean, all deviations are zero and both s and s^2 are zero. Just like the arithmetic mean, the standard deviation is very sensitive to extremes. Because it involves the squares of deviations, the standard deviation gives more relative weight to large deviations. A highly skewed distribution or one with a few extreme observations is not effectively summarized by the mean and standard deviation. The median and quartile deviation are more representative summary statistics in this case.

Together, the mean and standard deviation can be used to locate individual observations within the distribution of a variable as a whole. Consider Figure 2-14. On the X axis we can mark the values of the variable which are 1 and 2 standard deviations above and below the mean. As an alternative to the percentile technique, we can now locate individual observations or values of X, using this convention. For example, we might speak of an observation or value of X which is between 1 and 2 standard deviations above the mean, or between the mean and 1 standard deviation below the mean, and so on. Also, we can think of observations that lie more than a certain number of standard deviations above or below the mean as representing atypical observations.

An important theorem by the Russian mathematician Chebyshev provides a means of using the standard deviation to interpret the dispersion in a data set.

THE CHEBYSHEV THEOREM

Given a set of n observations $X_1, X_2, \ldots, X_n$, then for all $k > 1$ at least $1 - 1/k^2$ of the observations are within k standard deviations of their mean $\bar{X}$.

This remarkable theorem holds true for any data set. Therefore, when $k = 2$, we can say that at least $1 - \frac{1}{4} = \frac{3}{4}$ of the observations in the sample lie between $\bar{X} - 2s$ and $\bar{X} + 2s$. For the DO data, $\bar{X} - 2s = 5.58 - 2(.6392) = 4.3016$ and $\bar{X} + 2s = 5.58 + 2(.6392) = 6.8584$. We expect a minimum of $\frac{3}{4}(50) = 37.5$ observations in this range. An examination of Table 2-5 reveals that $\frac{47}{50}$, or 92 percent, of the observations lie in this range.

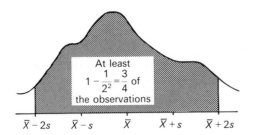

FIGURE 2-14. Illustration of the Chebyshev theorem.

Coefficient of Variation

To compare the dispersion of two frequency distributions having different means, we usually cannot simply compare the two standard deviations. Of course, if the means of the two variables are equal, then the variable with the smaller standard deviation is less dispersed. If, however, the means of the two variables are unequal, then it is misleading to rely on the standard deviations alone. A variable with a mean value in the thousands is likely to have a larger standard deviation than a variable with a mean value less than 100. The *relative variability* of a frequency distribution is defined as the ratio of the standard deviation to the mean. This measure is termed the *coefficient of variation* (CV) and is defined

$$CV = \frac{s}{\bar{X}} \text{ or } \frac{\sigma}{\mu} \tag{2-23}$$

Since both $\bar{X}$ and s are measured in the same units (that is, units of variable X), their quotient must be a dimensionless measure. So we can use CV to compare variables measured in different units or at different scales.

To illustrate the advantages of the coefficient of variation, examine the following rainfall data collected from two climatological stations over a 60-yr period:

	Annual rainfall	
Parameter	Station *A*	Station *B*
$\bar{X}$	92.6	38.8
s	16.6	9.1
CV	.179	.235

Note that even though the standard deviation of station *A* is larger than that of station *B*, a comparison of the coefficients of variation suggests that station *A* has the least variable rainfall. Relatively less dispersed variables will have lower coefficients of variation. The lower limit to this measure is zero.

2.4. Other Descriptive Measures

It is also possible to develop numerical measures for other characteristics of the shape of frequency distributions. Two of the more commonly measured characteristics are skewness and kurtosis. *Skewness* measures the degree of asymmetry in a distribution, and *kurtosis* measures the degree of peakedness. To develop numerical measures of these properties, let us first define the moments of the distribution about the mean. The *first moment* about the mean is defined as $\Sigma_{i=1}^{n} (X_i - \bar{X})^1$, the *second moment* as $\Sigma_{i=1}^{n} (X_i - \bar{X})^2$, etc. The first moment of a frequency distribution is always zero. This follows directly from Property 1, discussed in Section 2.2. The second moment about the mean is the variance, defined in Section 2.3.

The *third moment* about the mean $\Sigma_{i=1}^{n} (X_i - \bar{X})^3$ measures the skewness of a distribution. Note that all deviations for data values that are less than the mean will have negative cubes, and those above the mean will have positive cubes. If the distribution is symmetric, then the positive and negative values will offset one another and the third moment will be zero. However, if the distribution has a tail on the right, then a few large positive deviations will dominate the sum (when they are cubed) and the net result is always positive. This is why such distributions are termed *positively skewed*. However, a distribution with a tail on the left has a negative third moment since a few large negative deviations dominate the sum. Such distributions are *negatively skewed*.

Unfortunately, we cannot use the third moment to compare the skewness of different variables since it is not standardized. Two different relative measures based on the moments of a distribution are useful. If we label the moments π_1, π_2, etc., then we can define one measure of skewness as

$$\text{Relative skewness } 1 = \frac{\pi_3^2}{\pi_2^3} \tag{2-24}$$

where the division of the third moment squared by the second moment cubed standardizes the index. As a consequence of squaring the third moment, the sign is lost. It is easier to interpret this measure when it is calculated as

$$\text{Relative skewness } 2 = \frac{\pi_3}{\sqrt{\pi_2^3}} \tag{2-25}$$

This retains the sign of the third moment. A positive-valued index indicates positive skewness, and a negatively skewed index indicates negative skewness. Neither measure is stable, particularly for small samples of observations. Instead, a simpler measure based on the difference between the mean and the median is used:

$$\text{Pearson's } \gamma = \frac{3(\bar{X} - \text{median})}{s} \tag{2-26}$$

This is much simpler to calculate. When the distribution is symmetric, the mean and the median coincide and the index yields zero. When $\bar{X} >$ median, the distribution is positively skewed; when $\bar{X} <$ median, it is negatively skewed. Recall that the coincidence of the mean and the median is used as an indicator of skewness in Section 2.2. For the DO variable, this third skewness measure yields $3(5.58 - 5.60)/.6392 = -.09$. As the histogram suggests, these data are only slightly negatively skewed.

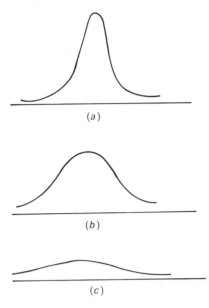

(a)

(b)

(c)   **FIGURE 2-15.** Kurtotic distributions.

Similarly, we can define a measure of kurtosis by using the fourth moment about the mean π_4. When it is standardized as

$$\text{Relative kurtosis} = \frac{\pi_4}{\pi_2^2} \qquad (2\text{-}27)$$

this measure can be used to distinguish among *leptokurtic distributions* [relative kurtosis less than zero, Figure 2-15(a)], *mesokurtic distributions* [relative kurtosis is zero, Figure 2-15(b)], and the *platykurtic distributions* [relative kurtosis greater than zero, Figure 2-15(c)]. Again this measure is very sensitive to the existence of extreme observations and is relatively unstable for small samples.

2.5. Exploratory Data Analysis

The graphical devices and summary statistics presented in the first four sections of this chapter represent what might be called the conventional, or traditional, core of elementary descriptive statistics. Virtually every textbook for introductory courses in statistics includes all this material. In addition, most of the techniques are programmed into the standard statistical computer packages used to analyze data sets. Over the past 10 years, a wealth of novel, and some say ingenious, methods of analyzing data have been developed. This rather recent thrust in statistics is now termed *exploratory data analysis* (EDA), after John Tukey's influential 1977 book of the same name. The motivation for developing these techniques is to improve our ability to uncover features of a set of numbers we wish to understand—features which may go unnoticed if traditional techniques are employed. Conventional textbooks in elementary statistics often give the false impression that there is a *single, correct* way to analyze data. In EDA there is an explicit recognition that there may be *several different* ways to confront data, each with appropriate strengths and weaknesses.

EDA is a practical philosophy of data analysis. Rather than having the techniques guide students in the analysis of data, the data themselves guide students in their choice of appropriate techniques. The goal is understanding; but the data lead, not follow, the statistician. Because they are so recent, EDA techniques are not well known or widely available in computer packages. Some say that many of the advantages of EDA are lost when it is forced into the mold of computer programming algorithms. EDA techniques are exploratory and, as such, are often not capable of being cast into simple rules of thumb. The analysis is part art and part science. As experience with the techniques is gained, perhaps more "rules" may emerge.

Consider a collection of data values of some variable, or what exploratory data analysts prefer to call a *batch*. Normally, the data are not termed a *sample* since EDA does not concern itself with inferential techniques. The first step in conventional approaches is to organize and reduce a large number of data values into a convenient framework, usually by constructing a histogram and frequency tableau. The first step in EDA has the same purpose but uses different techniques. These graphical techniques of EDA are known as *displays*. The first display to be constructed is usually the stem-and-leaf display. The purpose of this display is to illustrate

1. The range of the values within the data
2. Where concentrations of values occur
3. How many concentrations of data values there are within the batch
4. Whether there are gaps in the values
5. Whether there is symmetry in the batch
6. Whether there are extreme values that differ markedly from the remaining values in the batch
7. Any other data peculiarities

This is very similar to the rationale for constructing a frequency table and histogram. The unique feature of the stem-and-leaf display is that the actual digits which make up the data values in the batch are used to construct the display. It is thus possible to reconstruct the original data from the stem-and-leaf display. The digits are used to guide the sorting and display process.

Stems and Leaves

The first step in the construction of a stem-and-leaf display is to divide the digits of the data values in the batch into three classes: sorting digits, display digits, and digits that can be ignored. Suppose that the data consist of a set of elevations between 12,000 and 20,000 ft above sea level. Let us examine several ways of splitting a typical observation such as 15,328. One way would be

Sorting digits:	Display digits:	Digits to be ignored:
15	3	28

In this case there are two sorting digits, the hundreds' digit is the display digit, and the last two digits are ignored. There is some choice in the number of digits to be assigned and the number to be ignored. It is possible to use all the digits and ignore none. For example, a perfectly legitimate split would be

Sorting digits:	Display digits:	Digits to be ignored:
1532	8	none

Just as we must make a choice for the length of the interval in constructing a histogram to best illustrate properties of the variable, so, too, must the correct choice of display digits be made in stem-and-leaf constructions. When the digits are divided 15–3–28, we are saying that the elevation in hundreds of feet is important, but tens of feet and feet are unimportant. If we divide the digits 1532–8–0, we are indicating that the last digit is important in the display. This might be desirable if all the observations are between 15,320 and 15,330. The effects of digit classification on the appearance of stem-and-leaf displays are illustrated in the examples of this section.

The leading, or sorting, digits are called the *stem* of the observation, and the display digit is referred to as the *leaf*. Only four steps are required to specify an algorithm for the construction of a stem-and-leaf display:

1. Sort the data from lowest to highest value.

2. Examine the data and select a pair of digits to divide the data values into a stem, leaf, and ignored portion.

3. In a column, list *all possible* sets of leading digits, from lowest to highest. It is necessary to include stems that may not be represented in the particular batch of data being analyzed. The first stem is always the lowest in the batch, and the last stem is always the highest stem of the values in the batch. It is unnecessary to go beyond these bounds.

4. For *each* observation or data value, write the display digit on the stem line corresponding to its sorting digits. When all data values in the batch are put into the display, there should be one leaf for each observation.

Although this may seem complicated, a quick examination of the overall appearance of stem-and-leaf displays from several examples will clarify the procedure.

EXAMPLE 2-1. As a first example, let us portray the DO batch of Table 2-1 as a stem-and-leaf display. Begin by sorting the data values from lowest to highest, as in Table 2-8. The data are divided at the decimal point, leaving only three stems—4, 5, and 6. The tenths' digit is the display digit in the leaves of the table, and no digits are ignored. The stem-and-leaf display from this format is illustrated in Table 2-9. There are nine leaves for stem 4, corresponding to the values 4.2, 4.3, 4.4, 4.5, 4.7, 4.7, 4.8, 4.8, and 4.9. There are 30 leaves for stem 5 and 11 leaves for stem 6. This is not a good example of a stem-and-leaf display. There are too few stems for the 50 leaves. It is difficult to adequately interpret the values in the batch. Nevertheless, we can still see that this batch of values is symmetric, ranges from 4.2 to 6.9, and is concentrated at the values of 5.0 to 5.9. Moreover, it is possible to retrieve the actual data values in the batch simply by matching each of the 50 leaves in the display with the stem value of the appropriate row. This operation is not possible with an ordinary histogram. A number of ways of improving the appearance of this display are illustrated in subsequent variations of the basic stem-and-leaf display.

TABLE 2-8
Dissolved Oxygen Values of Table 2-1, Sorted
by Value, mg/L

Rank	Value	Rank	Value
1	4.2	26	5.6
2	4.3	27	5.6
3	4.4	28	5.7
4	4.5	29	5.7
5	4.7	30	5.7
6	4.7	31	5.7
7	4.8	32	5.8
8	4.8	33	5.8
9	4.9	34	5.8
10	5.1	35	5.8
11	5.1	36	5.9
12	5.1	37	5.9
13	5.2	38	5.9
14	5.3	39	5.9
15	5.3	40	6.0
16	5.4	41	6.1
17	5.4	42	6.2
18	5.4	43	6.3
19	5.4	44	6.4
20	5.5	45	6.4
21	5.5	46	6.4
22	5.6	47	6.6
23	5.6	48	6.7
24	5.6	49	6.8
25	5.6	50	6.9

TABLE 2-9
Stem-and-Leaf Display for the Data of Table 2-1

Stems	Leaves
4	234577889
5	111233444455666666777788889999
6	01234446789

EXAMPLE 2-2. Table 2-10 contains a list of 40 houses, their addresses, and their dates of construction. These houses are all contained within a three-square-block area of a city and are the subjects of an intensive study of housing maintenance. The stem-and-leaf display for this batch of construction dates is shown in Table 2-11. In this case, there are three digits in the stem, and the last digit is the leaf. What does this stem-and-leaf display reveal? First, we note two data extremes—one a house

TABLE 2-10
Addresses and Construction Dates of a Batch of Houses

Observation No.	Address	Date constructed
1	67 Pine Street	1924
2	73	1924
3	77	1924
4	82	1924
5	85	1924
6	91	1924
7	92	1936
8	96	1946
9	1100 Fifth Street	1945
10	1102	1934
11	1104	1934
12	1105	1936
13	1106	1868
14	1113	1980
15	1128	1911
16	1131	1955
17	1143	1933
18	1156	1956
19	1178	1958
20	1191	1960
21	1198	1961
22	221 Magnolia Boulevard	1914
23	228	1962
24	264	1962
25	279	1935
26	282	1932
27	291	1951
28	297	1952
29	300	1953
30	1433 Reading Road	1908
31	1434	1947
32	1440	1947
33	1443	1947
34	1451	1932
35	1453	1946
36	1457	1946
37	1479	1910
38	1482	1938
39	1486	1937
40	1498	1931

TABLE 2-11
Stem-and-Leaf Display for Data of Table 2-10

Stems	Leaves
186	8
187	
188	
189	
190	8
191	014
192	444444
193	12234456678
194	5666777
195	123568
196	0122
197	
198	0

Note. UNIT = 1. And 186 8 represents 1868.

constructed four decades earlier than the second oldest house in the batch, the other built almost two decades after the second newest house in the neighborhood. Second, examine the dates of construction of houses in the 1920s. All were built in 1924. This peculiarity would not be picked up in a conventional histogram. But, by using the actual digits of the data values in the display, this potentially important characteristic of the batch becomes visible. We might also note that all construction in the 1940s was in the postwar period.

Another useful addition to a stem-and-leaf display is to indicate the units of the data values, particularly where they may not be obvious. Table 2-11 is labeled UNIT = 1, and an example relating the stem-and-leaf split to the actual data value given as 186 8 represents 1868. When the data include a decimal point, this UNIT specification becomes mandatory. A heading for the display of Example 2-1 illustrated in Table 2-12 would be UNIT = 0.1, where 4 7 represents 4.7.

TABLE 2-12
A Two-Line Stem-and-Leaf Display of the Lake DO Batch

Stems	Leaves
4 *	234
4 ·	577889
5 *	1112334444
5 ·	55666666777788889999
6 *	0123444
6 ·	6789

Note. UNIT = 0.1. And 4 2 represents 4.2.

TABLE 2-13
A Five-Line Stem-and-Leaf Display of DO

Stems	Leaves
4 T	23
4 F	45
4 S	77
4 ·	889
5 *	111
5 T	233
5 F	444455
5 S	6666667777
5 ·	88889999
6 *	01
6 T	23
6 F	444
6 S	67
6 ·	89

Note. UNIT = 0.1. And 4 2 represents 4.2.

Multiple Lines per Stem

Sometimes the stem-and-leaf display for a batch can be ineffective if it is too stretched out or too compressed. The stem-and-leaf display of the DO batch represents a poor display. Note that all the leaves for each stem appear on a single line. There are 20 leaves on stem 5. There are two other ways of constructing a display to overcome this problem. First, we can use two lines per stem, doubling the number of lines in the display. Often this helps alleviate the appearance of overcrowding in the display. On the first stem we place the leaves 0, 1, 2, 3, and 4. This line is indicated by an asterisk in the last stem digit. The second line is indicated by a dot and contains the leaves 5, 6, 7, 8, and 9. Table 2-12 illustrates a revised version of the stem-and-leaf display of the DO batch. The symmetry of the batch is now more apparent than in Table 2-9. An alternative of five lines per stem is illustrated in Table 2-13. In this case the five lines for each stem are denoted by asterisks for leaves 0 and 1, by a T for leaves 2 and 3, by an F for leaves 4 and 5, by an S for leaves 6 and 7, and by a dot for leaves 8 and 9. The symmetry of the batch is again prominent, as is the concentration of DO values between 5.4 and 5.9. In this instance there appears to be a definite advantage of moving from a display of one to two lines per stem, but limited gains are achieved by extending the display to five lines per stem. The two-line display illustrates the symmetry of the batch, the concentration of values between 5.5 and 5.9, and the number of repeated values in this range.

Listing Stray Values

Data values that are far removed from the remainder of the batch are known as *strays*. That a batch contains such values seems a sufficiently important feature to

TABLE 2-14
A Reexpression of Table 2-11, Using
Convention for Strays

Stems	Leaves
LO	1868
190	8
191	014
192	444444
193	12234456678
194	5666777
195	123568
196	0122
HI	1980

Note. UNIT = 1. And 190 8 represents 1908.

warrant a special designation in a stem-and-leaf display. The stem-and-leaf display of the house construction dates summarized in Table 2-11 contains two such data values, 1868 and 1980. To include them in the stem-and-leaf display requires four additional stems for 187, 188, 189, and 197, but there are no data values, and therefore leaves, for these stems. Once it is determined that there are strays in a batch, it is useful to designate them with the labels HI and LO depending on the end of the display to which they belong. To emphasize the gap between these values and the rest of the batch, one blank line is left between the list and the rest of the display. Table 2-14 is a reexpression of Table 2-11, using this convention.

Positive and Negative Values

If the data batch contains both positive and negative values, then the stems near zero require special designation. Numbers slightly less than zero are attached to stem -0, and numbers slightly greater than zero are given stem $+0$. For example, the display

UNIT = 0.1	
-1 2 Represents -1.2	
-1	2
-0	7 8
$+0$	4 5 6
1	3

contains the values -1.2, -0.7, -0.8, 0.4, 0.5, 0.6, and 1.3. A value of exactly 0.0 can be put on either 0 stem. If there are several different 0.0 data values, then it is best to divide them as equally as possible between the two zero stems. This tends to preserve the appearance of the display.

2.6. Summary

In this chapter we have been concerned with describing a set of observations, using both graphical methods and numerical statistics. Both methods are extremely useful for describing sets of data which are statistical populations or samples from these populations. When the data under consideration represent a sample, numerical measures are preferred since they lend themselves to inferential modes of analysis better than graphical techniques. Many of these descriptive statistics are used extensively in the chapters of this text dealing with inferential statistics.

Among the graphical techniques, the frequency distribution histogram is the most commonly used method for summarizing a set of data. Other graphical techniques can also be useful as long as they provide both an easily understood and an accurate portrait of the data. A quick glance at these figures provides a great deal of information about the shape of a distribution. Cumulative displays such as ogives can also be used effectively in certain situations.

The average or typical value of a variable can usually be best summarized by the arithmetic mean, although the mode and the median are preferred in particular situations. For making inferences, the mean is preferred. For purposes of measuring the variability in a set of data, the standard deviation is the preferred descriptive measure, although its square, the variance, is most useful in inferential statistics. Under certain conditions, the quartile deviation proves a better descriptive measure than the standard deviation.

The standard deviation and the mean are the two best summary measures of a variable. They can also be used to locate an observation within a frequency distribution. The dispersion of observations in a frequency distribution can be effectively estimated by using the standard deviation in conjunction with Chebyshev's theorem.

Measures of skewness and kurtosis are not often reported in the descriptive summary of a variable. Most measures are extremely laborious to calculate and are particularly sensitive to the existence of extreme-valued observations. Graphical techniques give a quick visual impression of the extent of skewness and kurtosis. Often the existence of skewness is detected either by comparing the values of the mean, median, and mode or by using a simple measure such as Pearson's γ.

Appendix 2A. Review of Sigma Notation

Just as we use plus to denote addition or minus to denote subtraction, we can use the symbol Σ (pronounced sigma) to denote the repeated operation of addition. The symbol Σ is known as the *summation operator* since it literally means "take the sum of." In conjunction with the summation operator, we normally utilize an index of summation to specify the elements to be summed. For example,

$$\sum_{i=1}^{4} X_i$$

is interpreted as the sum of X_1 to X_4, or $X_1 + X_2 + X_3 + X_4$. If $X_1 = 1$, $X_2 = 7$, $X_3 = 9$, and $X_4 = 16$, then $\Sigma_{i=1}^{4} X_i = 1 + 7 + 9 + 16 = 33$. Any lowercase letter can be used as an index of summation. Usually, the variable i is used in a single summation, although j and k are sometimes used.

The sigma operator can also be used when an undetermined number of elements are to be summed. Thus, $X_1 + X_2 + \cdots + X_n$ can be expressed as $\Sigma_{i=1}^n X_i$. The length of the summation depends on the value of n. When there is no possibility of confusion, it is frequently convenient to suppress the index of summation and the limits of the summation. We may thus write ΣX when we mean the sum is to be taken over "all" values of X. The specific interpretation of *all* should be evident from the context of the summation. In this text, we find it convenient to suppress these indices when the meaning is $\Sigma_{i=1}^n X_i$.

Several rules for simplifying summations periodically are used in this text. All the following rules can be easily verified by using elementary algebra.

RULE 1 If X and Y are two variables, then

$$\sum_{i=1}^n (X_i + Y_i) = \sum_{i=1}^n X_i + \sum_{i=1}^n Y_i$$

To see this, expand the left-hand side:

$$\sum_{i=1}^n (X_i + Y_i) = (X_1 + Y_1) + (X_2 + Y_2) + \cdots + (X_n + Y_n)$$

$$= (X_1 + X_2 + \cdots + X_n) + (Y_1 + Y_2 + \cdots + Y_n)$$

$$= \sum_{i=1}^n X_i + \sum_{i=1}^n Y_i$$

RULE 2 If k is a constant and X is a variable, then

$$\sum_{i=1}^n kX_i = k \sum_{i=1}^n X_i$$

Again, expand the left-hand-side:

$$\sum_{i=1}^n kX_i = kX_1 + kX_2 + \cdots + kX_n$$

$$= k(X_1 + X_2 + \cdots + X_n)$$

$$= k \sum_{i=1}^n X_i$$

RULE 3 For any two variables X and Y,

$$\sum_{i=1}^n X_i Y_i = X_1 Y_1 + X_2 Y_2 + \cdots + X_n Y_n$$

Note that this does not equal $(\Sigma_{i=1}^{n} X_i)(\Sigma_{i=1}^{n} Y_i)$. Consider the following simple example:

i	X	Y
1	2	4
2	6	8
3	10	12

First, we calculate $\Sigma_{i=1}^{3} X_i Y_i = 2 \times 4 + 6 \times 8 + 10 \times 12 = 176$. Then we calculate $(\Sigma_{i=1}^{3} X_i)(\Sigma_{i=1}^{3} Y_i) = (2 + 6 + 10) \times (4 + 8 + 12) = 18 \times 24 = 432$.

RULE 4 If k is a constant, then

$$\sum_{i=1}^{n} k = nk$$

Note that

$$\sum_{i=1}^{n} k = \underbrace{k + k + \cdots + k}_{n \text{ times}} = nk$$

RULE 5 If m is an exponent and X a variable, then

$$\sum_{i=1}^{n} X_i^m = X_1^m + X_2^m + \cdots + X_n^m$$

RULE 6 If m is an exponent and X a variable, then

$$\left(\sum_{i=1}^{n} X \right)^m = (X_1 + X_2 + \cdots + X_n)^m$$

Suppose that $n = 3$, $m = 2$, $X_1 = 2$, $X_2 = 6$, and $X_3 = 10$. Then

$$\sum_{i=1}^{3} X_i^2 = 2^2 + 6^2 + 10^2 = 140$$

$$\left(\sum_{i=1}^{3} X_i \right)^2 = (2 + 6 + 10)^2 = 324$$

In general, we see from Rule 5 and Rule 6 that

$$\sum_{i=1}^{n} X_i^2 \neq \left(\sum_{i=1}^{n} X_i \right)^2$$

We can also use double summations with two separate subscripts. When we encounter the expression $\sum_{i=1}^{n} \sum_{j=1}^{n} X_{ij}$, we work from the inside out. First, we use the j subscript and then the i subscript, so that

$$\sum_{i=1}^{n} \sum_{j=1}^{n} X_{ij} = \sum_{i=1}^{n} (X_{i1} + X_{i2} + \cdots + X_{in})$$

$$= (X_{11} + X_{12} + \cdots + X_{1n}) + (X_{21} + X_{22} + \cdots + X_{2n})$$

$$+ \cdots + (X_{n1} + X_{n2} + \cdots + X_{nn})$$

If X is an array or matrix of the form

$i = 1$	X_{11}	X_{12}		$\cdots$	X_{1n}
$i = 2$	X_{21}	X_{22}		$\cdots$	X_{2n}
	$\cdot \ \cdot \ \cdot$	$\cdot \ \cdot \ \cdot$	$\cdot \ \cdot \ \cdot \ \cdot$	$\cdot \ \cdot \ \cdot$	$\cdot \ \cdot \ \cdot$
$i = n$	X_{n1}	X_{n2}		$\cdots$	X_{nn}

then the expression $\sum_{i=1}^{n} \sum_{j=1}^{n} X_{ij}$ leads to the sum of all the entries in the matrix. Note that we can also use summation notation to compute the sum of certain rows or columns of this matrix. For example,

$$\sum_{j=1}^{n} X_{2j} = X_{21} + X_{22} + \cdots + X_{2n}$$

specifies the sum of the second row. Similarly, the sum of column 1 is

$$\sum_{i=1}^{n} X_{i1} = X_{11} + X_{22} + \cdots + X_{n1}$$

Appendix 2B. SAS Program for Bar Chart, Pie Chart, Star Chart, Block Diagram, and Descriptive Statistics

Program title	DATA POLLUTE;
Variable name	INPUT DO;
Input from terminal	CARDS;
	5.1
	5.6
Data	, . . . ,
	5.5
	4.7
	;
Prints input data	PROC PRINT DATA = POLLUTE;
Calls for chart procedure	PROC CHART DATA = POLLUTE;
Vertical and horizontal	VBAR DO;
bar charts	HBAR DO;

Pie chart	PIE DO;
Star diagram	STAR DO;
Block diagram	BLOCK DO/LEVELS = 5;
Descriptive statistics	PROC MEANS DATA = POLLUTE;
Skewness and kurtosis statistics	PROC MEANS DATA = POLLUTE;
	SKEWNESS KURTOSIS;

Appendix 2C. Derivation of Computational Formulas for σ and σ^2

To show the equivalence of Equations (2-15) and (2-18), let us expand the numerator of the expression inside the radical:

$$\sum_{i=1}^{N} (X_i - \bar{X})^2 = (X_1 - \bar{X})^2 + (X_2 - \bar{X})^2 + \cdots + (X_N - \bar{X})^2$$

$$= (X_1^2 - 2X_1\bar{X} + \bar{X}^2) + (X_2^2 - 2X_2\bar{X} + \bar{X}^2) + \cdots + (X_N^2 - 2X_N\bar{X} + \bar{X}^2)$$

$$= (X_1^2 + X_2^2 + \cdots + X_N^2) - 2\bar{X}(X_1 + X_2 + \cdots + X_N)$$

$$+ (\bar{X}^2 + \bar{X}^2 + \cdots + \bar{X}^2)$$

$$= \sum_{i=1}^{N} X_i^2 - 2\bar{X}\left(\sum_{i=1}^{N} X_i\right) + N\bar{X}^2$$

The middle term is now simplified by using the identity

$$N\bar{X} = \frac{N \sum_{i=1}^{N} X_i}{N} = \sum_{i=1}^{N} X_i$$

so that

$$\sum (X_i - \bar{X})^2 = \sum_{i=1}^{N} X_i^2 - 2N\bar{X}^2 + N\bar{X}^2$$

$$= \sum_{i=1}^{N} X_i^2 - N\bar{X}^2$$

This is now divided by the denominator of Equation (2-15), N, to yield

$$\frac{1}{N} \sum_{i=1}^{n} X_i^2 - \bar{X}^2$$

A similar simplification can be undertaken for s and s^2.

REFERENCE

John W. Tukey, *Exploratory Data Analysis* (Reading, Ma.: Addison-Wesley, 1977).

FURTHER READING

The topics covered in this chapter are normally included in virtually all statistics textbooks. Students sometimes find it useful to consult other textbooks to see a simpler, more difficult or detailed, or just a different presentation of some topic. A selection of the textbooks intended for geography students is listed below. Several of these books use a different notation from that used in this text. Students should recognize that equivalent formulas may appear different for this reason only. Examples of the use of simple descriptive statistics in published research in geography are included in virtually all journals of geography, particularly in British and North American journals. However, in many cases, the use of descriptive statistics may be only a minor part of a discussion of research results that employs other quite sophisticated mathematical or statistical techniques. Some of the problems at the end of this chapter suggest the range of application of these techniques in geography. The range of application is as wide as the discipline itself.

D. Ebdon, *Statistics in Geography* (Oxford, England: Oxford University Press, 1977).
G. B. Norcliffe, *Inferential Statistics for Geographers* (London: Hutchinson, 1977).
J. Silk, *Statistical Concepts in Geography* (London: G. Allen, 1979).
P. Taylor, *Quantitative Methods in Geography* (Boston: Houghton Mifflin, 1977).
R. B. G. Williams, *Introduction to Statistics for Geographers and Earth Scientists* (London: Macmillan, 1984).

Students interested in the techniques of exploratory data analysis should consult the Tukey (1977) reference cited above or Velleman and Hoaglin:

P. F. Velleman and D. C. Hoaglin, *Applications, Basics and Computing of Exploratory Data Analysis* (Boston: Duxbury, 1981).

PROBLEMS

1. A large city is divided into 60 police precincts. The number of burglaries in the last 12 months in each precinct is as follows:

200	251	182	191	219	195	224	171	171	204	205	186	221	193	206
225	170	242	200	231	196	188	219	180	224	208	205	184	236	182
207	209	193	225	209	194	219	176	236	234	160	186	203	201	190
201	173	213	258	200	221	180	209	259	172	161	211	241	211	181

a. Construct a frequency table of these data. Justify your choice of categories.
b. Draw a frequency polygon and histogram.
c. Calculate the following descriptive statistics for these data: mean, variance, median, mode, and Pearson's coefficient of skewness.
d. Calculate the mean and variance of these data, using the formulas for grouped data. How do they compare to the actual values computed by using all 60 observations?
e. Summarize your findings about the distribution of the number of burglaries apparent from your graphs and calculations in parts (a) to (c).

2. Corn yields in bushels per acre in 60 counties are as follows:

74	76	67	90	72	74	56	26	96	91	86	96	70	78	65	58	47	35	99	82
81	88	72	82	60	72	67	45	34	98	76	70	85	71	80	83	70	60	42	93
96	75	80	75	92	90	95	92	59	97	81	63	82	79	78	90	88	61	78	79

   a. Calculate the mean, median, mode, and standard deviation for these data.
   b. Construct a frequency distribution, using six classes.
   c. Use the frequency distribution to approximate the mean, median, and standard deviation. Compare your answers to those obtained in part (a).

3. A magazine publisher determines the market penetration in each of 40 major markets, expressed as the proportion of total households in the market subscribing to one of the magazines produced by the company. The market penetration data are as follows:

.31	.50	.20	.18	.30	.25	.20	.23	.19	.25
.44	.62	.22	.29	.50	.28	.30	.36	.14	.72
.50	.75	.45	.31	.45	.50	.60	.21	.28	.61
.29	.42	.26	.48	.60	.48	.60	.13	.28	.60

Complete parts (a) to (c) of Problem 2.

4. Explain the meaning of the following terms and concepts:
   a. Frequency table
   b. Distribution of a variable
   c. Unimodal, bimodal, and
      multimodal distributions
   d. Univariate descriptive statistics
   e. Median
   f. Percentile
   g. Quartile
   h. Range
   i. Interquartile range
   j. Coefficient of variation
   k. Midrange

5. Explain the relationship between a frequency polygon and an ogive.

6. When is the mean not a good measure of central tendency? Give an example of a variable that might be best summarized by some other measure of central tendency.

7. Consider the following five observations: -7, 3, 2, 13, and -1.
   a. Calculate the mean.
   b. Determine the median.
   c. Show that Properties 1, 2, and 3 hold for this set of numbers.

8. For the data in Problem 1, find the proportion of values in the intervals $\mu \pm 2\sigma$, $\mu \pm 2.25\sigma$, and $\mu \pm 2.5\sigma$. Apply the Chebyshev theorem to determine the minimum proportion of observations in these ranges. Is the theorem verified for these data?

9. Repeat Problem 8 for the data of Problem 2.

10. Repeat Problem 8 for the data of Problem 3.

11. Within how many standard deviations about the mean will there be at least (a) 60 percent of all observations, (b) 80 percent of all observations, and (c) 90 percent of all observations in a set of data?

12. Write an SAS program to draw a histogram, and calculate simple descriptive statistics for the data of Problem 1, 2, or 3.

13. Construct histograms for the data in Problem 1, using two new interval widths. Which representation is better? Why?

14. Use a skewness measure to determine whether the data in Problem 1, 2, or 3 are skewed to the right or left.

15. The following data describe the distribution of annual household income in three regions of a country:

Region	$\bar{X}$	s
A	$36,000	$16,000
B	$28,000	$13,000
C	$23,000	$11,000

In which region is income the most evenly spread? The least evenly distributed?

16. The presence of arsenic, chromium lead, or silver in concentrations greater than 0.05 mg/L is usually sufficient to cause rejection of a water supply for use as drinking water. Consider the following sets of measurements, in milligrams per liter, taken in the last year at some location.

Month	Arsenic	Chromium	Lead	Silver
Jan.	0.8	0.7	0.2	0.4
Feb.	0.7	0.7	0.1	0.1
Mar.	0.9	0.8	0.2	0.1
Apr.	0.5	0.5	0.5	0.5
May	0.7	0.6	0.6	0.5
June	0.2	0.1	1.4	1.8
July	0.2	0.3	1.3	1.3
Aug.	0.2	0.2	1.3	1.4
Sept.	0.1	0.2	1.0	1.1
Oct.	0.2	0.1	0.2	0.1
Nov.	0.1	0.2	0.1	0.1
Dec.	0.6	0.6	0.2	0.2

a. Examine the data to see whether there are any patterns in these concentrations by substance and time period.
b. Design a stem-and-leaf display that emphasizes the patterns you have uncovered in part (a). Can you think of a way of portraying all the data in one display?
c. Examine alternative formats for the display.

17. The monthly rainfall, in millimeters, during March at a particular climatic station for the last 40 years is as follows:

0.0	1.8	0.0	19.6	0.0	6.1	5.1	7.9	30.5	9.7
11.7	24.1	37.8	6.3	3.0	22.9	14.2	28.7	0.5	12.2
12.3	12.4	128.2	68.9	17.0	14.1	3.0	5.1	0.0	7.8
5.0	6.2	42.4	12.7	24.1	2.3	5.6	14.1	28.6	13.2

a. Construct a complete stem-and-leaf display, using the tenths' digit as the display digit. Evaluate the display.

b. Construct a complete stem-and-leaf display in which the tenths' digit is ignored. Evaluate the display.

c. Improve both displays, using the convention for strays.

d. What does the stem-and-leaf display indicate about rainfall at this location?

3

Descriptive Statistics for
Spatial Distributions

The graphical devices and summary measures used in conventional descriptive statistics, as well as the newer techniques developed in exploratory data analysis, can be used to effectively summarize many of the data utilized by geographers. In Chapter 2, for example, a number of techniques of ordinary descriptive statistics are applied to a set of 50 observations of lake-dissolved oxygen values. Although such methods can tell us a lot about the frequency distribution of a variable, in some cases this is only part of the story. There are many instances in geographical research where the *spatial distribution* of a variable is of primary concern. How can we describe *the map* of some variable in an efficient, precise, and simplified way? *Where* are the high values? *Where* are the low values? Is the variable *evenly distributed* across the map or *concentrated* in a single or small number of locations? Is there any *pattern*, or is the map in some sense *random*? To answer these questions, geographers often use several techniques derived from descriptive statistics, but having particular properties or interpretations when applied to spatial data.

Let us begin by identifying some of the different representations of spatial data commonly used by geographers. Four different types are recognized in Figure 3-1: areal data, point data, network data, and directional data. *Areal data* include numerical observations based on areal subdivisions of some region. As a map, areal data can be presented in many ways. The simplest two-dimensional representations are the *choropleth* map, Figure 3-1(*a*), and the *contour* map, Figure 3-1(*b*). These two forms can also be given three-dimensional representations. The *stepped statistical surface* illustrated in Figure 3-1(*c*) is the three-dimensional equivalent of the choropleth map. The value or magnitude of each areal unit is mapped in the third dimension. This often improves the interpretability of the spatial pattern of the map. Where the phenomenon depicted on the map varies continuously, such as rainfall, it is accepted practice that the three-dimensional representation be an *isarithmic map*, such as the continuous smoothed surface in Figure 3-1(*d*).

A second major class of spatial data is *point data*. The dot map illustrated in Figure 3-1(*e*) is the most commonly used cartographic display for point data. The analysis of these point distributions, including the search for both pattern and underlying process, has been a topic of continuing interest to geographers. The dots may represent some plant species, farmhouses, towns, villages and cities, or even earth-

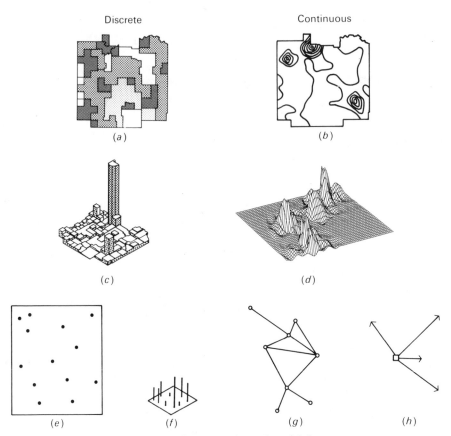

FIGURE 3-1. Representations of spatial data.

quake epicenters. When it is possible to associate a magnitude or frequency with each dot, a three-dimensional portrayal of the information is also possible, as in Figure 3-1(f). Two other types of data sometimes generated in geographical research are *network data*, which arise particularly in the analysis of drainage systems and transport networks, and *directional data*, which occur in studies examining the orientation of geographic phenomena. These two types are illustrated in Figure 3-1(g) and (h), respectively.

The distinction between continuous and discrete areal data and point data is sometimes blurred. Cartographers sometimes argue that a continuous areal representation is appropriate if the phenomenon exists everywhere on the map, both at and between observation points. This argument appears to be valid for many physical phenomena such as rainfall and temperature, but is less compelling for a variable such as population density. To map this variable in a continuous form on an isarithmic map, the assumption must be made that population exists everywhere and not at discrete points. Remember that it is possible to convert data expressed in one way to most other formats, if we are willing to make a few assumptions.

First, let us consider one way in which point data can be converted to areal data. The area around each point is associated to it in such a way that mutually exclusive and collectively exhaustive areal subdivisions are created. A common method is the construction of *Thiessen polygons*. These areas or polygons are drawn in order to satisfy the condition that *any location on the map associated with a point is closer to that point than to any other point on the map*. These polygons were defined by climatologist A. H. Thiessen to create regions around rainfall stations in such a way that station totals could be weighted by their surrounding areas to compute an "average" rainfall for the entire area. His solution is based on the simple logic that observed rainfall at any location is likely to be most similar to the *nearest* location for which a rainfall total is known.

Thiessen polygons can be constructed by using a very simple algorithm:

1. Join each point to all neighboring points.
2. Bisect these lines.
3. Draw the regions.

It is characteristic of the construction that the bisectors created in step 2 will either meet in 3s at a single point or else end at the border of the map. The construction is illustrated for a five-point problem in Figure 3-2. The bisectors meet at three different

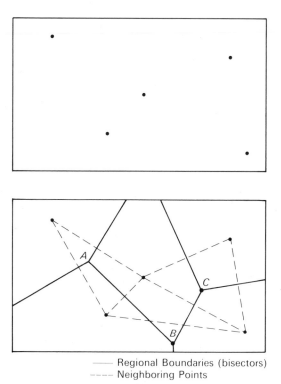

——— Regional Boundaries (bisectors)
---- Neighboring Points

FIGURE 3-2. Conversion of point to areal data by using Thiessen polygons.

intersections surrounding the central point, *A*, *B*, and *C*. The other ends of the bisectors terminate at the map boundary. Besides the original application in climatology, Thiessen polygons also find application in human geography. Suppose the points represent competing retail outlets in a city. Thiessen polygons define market areas that minimize distance traveled by consumers. Each goes to the closest retail outlet.

Other data conversions are also possible. For instance, areal data can be converted to point data by selecting a representative point in each area. The *center of gravity* is an obvious choice. The implicit assumption is that all the activity or phenomenon is concentrated at this point and not spread throughout the area. Point data can also be converted to a set of isolines by using the process of *interpolation*. The process is illustrated in Figure 3-3. In this figure, the rainfall totals for four points, *A*, *B*, *C*, and *P*, are given as well as their locations on a map. The position of the isoline with a value of 50 between point *P* and the three points *A*, *B*, and *C* can be determined by simple geometry. Since point *A* has a value of 55, the difference between the rainfall at points *A* and *P* is 10. Therefore, on the line *AP*, the isoline of 50 should lie midway between *A* and *P*. Between *B* and *P*, the split is in the ratio 5 : 7, and between *C* and *P*, the split along *CP* is in the ratio 5 : 1. The isoline of 50 is then approximated as a smooth curve running through these points. This is a linear interpolation, since it assumes the value of the phenomenon changes linearly between the observed points. Other *nonlinear* interpolations are also possible.

Although the types of data generated from maps are capable of being translated to other representations, the statistical methods used to analyze such maps developed from different traditions and were used for different purposes. These methods will be presented on the basis of the classification made explicit in Figure 3-3, with the understanding that many of the methods can be applied to data expressed in more than one way. As we shall see, many of these methods can be applied to aspatial data as well.

3.1. Areal Data

Data are frequently published for discrete areal units such as states or provinces, counties, census tracts within cities, and many other administrative areas. This

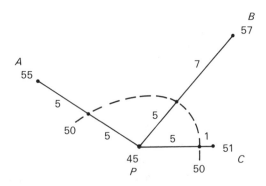

FIGURE 3-3. Linear interpolation of an isoline from point data.

section describes both graphical and statistical summaries of such data. These methods are therefore applicable particularly to choropleth and stepped statistical surface maps.

Location Quotients

The location quotient is most frequently used in economic geography and locational analysis, but it has much wider applicability. The location quotient (LQ) is an index for comparing an area's share of a particular activity with the area's share of some basic or aggregate phenomenon. Suppose, for example, that the following data are taken from a larger data set which details the employment structure of a region divided into four areas:

Area	Manufacturing		Service		Total	
	Employment	Percentage	Employment	Percentage	Employment	Percentage
A	5	5	40	10	150	15
B	70	70	220	55	500	50
C	15	15	60	15	200	20
D	10	10	80	20	150	15
	100	100	400	100	1000	100

Is employment in manufacturing or services concentrated in some area(s), or is it evenly distributed across the map? Location quotients compare the distribution of an activity to some *base*, or *standard*, in this case to total employment. The question can be reexpressed as follows: Is manufacturing or service employment more or less concentrated than total employment?

DEFINITION: LOCATION QUOTIENT
The location quotient for a given activity for area i is the ratio of the percentage of the total regional activity in area i to the percentage of the total base in area i. If A_i is equal to the level of the activity in area i and B_i is the level of the base, then

$$LQ_i = \frac{A_i/\Sigma\, A_i}{B_i/\Sigma\, B_i} \qquad (3\text{-}1)$$

The numerator of Equation (3-1) is the percentage of the activity in area i, and the denominator is the percentage of the base. A location quotient is thus the ratio of two percentages and is therefore dimensionless.

The location quotient for manufacturing in area A is $\frac{5}{100}/\frac{150}{1000} = \frac{5}{15} = 0.333$. That is, region A has 5 percent of the manufacturing employment of the region but 15 percent of the total employment in all sectors. On the basis of its share of total employment, we would expect region A to have 15 percent of the manufacturing employment. It has only 5 percent, or 0.333, as much as would be expected. Location quotients can be interpreted by using the following conventions:

1. If LQ > 1, this indicates a relative concentration of the activity in area i, compared to the region as a whole.
2. If LQ = 1, the area has a share of the activity in accordance with its share of the base.
3. If LQ < 1, the area has less of a share of the activity than is more generally, or regionally, found.

The complete table of location quotients for these two employment sectors indicates which areas of the map fall into these categories:

	Location quotient	
Area	Manufacturing	Services
A	0.333	0.667
B	1.400	1.100
C	0.750	0.750
D	0.677	1.333

For manufacturing, the location quotients reveal a concentration in area B and less than expected shares in each of the three other areas. For services, however, the distribution is much less concentrated, with relative concentrations in areas B and D and less than expected shares in A and C. Spatial patterns can be revealed by mapping these location quotients.

The selection of the base or standard distribution used in the denominator of location quotients is subject to choice. Usually, if the activities are part of some aggregate, then the aggregate is used as the base. In this example, it makes sense to compare the concentration of various industrial sectors to the total employment in all sectors. However, it is also possible to use area populations as the standard of comparison. In this instance, we would be comparing the areal distribution of sectoral employment to the areal distribution of population. Or, the base activity could also be defined as the actual land area in each of the areas of the region.

Coefficient of Localization

One of the drawbacks to the use of the location quotient is that one value is calculated for each area in the region being analyzed. For a city with perhaps 300 or more census tracts, this would be a very inefficient form of summary, even if the location quotients themselves were mapped. As an alternative, the coefficient of localization (CL) can be calculated. It describes the relative concentration of an activity by a single number.

DEFINITION: COEFFICIENT OF LOCALIZATION

The coefficient of localization (CL) is a measure of the relative concentration of an activity in relation to some base. To calculate CL:

1. Calculate the percentage share of the regional activity in each area, $A_i/\Sigma_{i=1}^{n} A$.
2. Calculate the percentage share of the regional base in each area, $B_i/\Sigma_{i=1}^{n} B$.
3. Subtract the value in step 2 from that in step 1, and add *either* all the positive differences or all the negative differences.
4. Divide by 100.

Using the sectoral employment in manufacturing or services, we can calculate the coefficient of localization as follows:

Area	Percentage of manufacturing employment	Percentage of total employment	Difference +	Difference −
A	5	15		10
B	70	50	20	
C	15	20		5
D	10	15		5
	100	100	20	20

Therefore, the CL for manufacturing is equal to $\frac{20}{100} = .20$. Similarly, the coefficient of localization for serivces is .10. Note that the CL calculation requires that the areal percentages of both the activity and the base to add to 100.

The CL ranges from 0 to 1. This differs from the location quotient which has a lower limit of zero but an upper limit approaching positive infinity. If CL = 0, the percentage distribution of the activity is evenly spread over the region in exact accordance with the base. The only numerical way this can occur is if the areal percentages for the activity are exactly equal to the areal percentages for the base. As CL approaches 1, the activity becomes increasingly concentrated in one region. The results indicate that both services and manufacturing are relatively evenly spread over the areas, but, as we might expect, service employment is more evenly spread.

Lorenz Curve

The Lorenz curve is another way to index the distribution of a variable among spatial units, although it is most commonly used to measure the extent of inequality of a variable distributed over aspatial categories. For example, if all residents of a country have the same income, then income should be distributed perfectly equally. To the extent that there exist some rich and some poor individuals, there is some inequality in income distribution. There is an obvious analogy to areal data. An activity may be concentrated in one or a few areas, or it may be spread throughout the region. The Lorenz curve is a graphical display of the degree of inequality.

As with location quotients and the coefficient of localization, the Lorenz curve compares the areal distribution of some activity to some base distribution. The Lorenz curve is constructed by using the following rules:

1. Calculate the location quotients for the various areas in a region. Reorder the areas in decreasing order of their location quotients.
2. Cumulate the percentage distributions of both the activity and the base in the order determined in step 1.
3. Graph the cumulated percentages for the activity and the base, and join the points to produce a Lorenz curve.

The necessary calculations for the manufacturing employment data are summarized in Table 3-1, and the Lorenz curves for both manufacturing and service employment are illustrated in Figure 3-4.

There are several important properties of a Lorenz curve. If the activity is evenly distributed across the areas of the region, that is, in proportion to the base, then the Lorenz curve is a straight line following the 45° diagonal shown in Figure 3-4. The more concentrated the activity, the further the Lorenz curve is from the diagonal. In the limiting case, the curve follows the X axis to the point (100, 0) and then proceeds vertically to the point (100, 100). Except for the case where the activity is perfectly evenly distributed, the slope of the Lorenz curve is always increasing. This follows directly from the ordering of the areas by LQ. The Lorenz curves in Figure 3-4 suggest that manufacturing employment is more concentrated than service employment. The Lorenz curve is favored by many who prefer graphical summaries to statistical measures.

The most common summary measure of inequality used in conjunction with the Lorenz curve is the *Gini coefficient*, or, as it is sometimes called, the *index of dissimilarity*.

TABLE 3-1
Work Table for Lorenz Curve Calculations for Manufacturing Data

Step 1. The location quotients for the four areas are: A, 0.333; B, 1.400; C, 0.750; D, 0.667. So the order of the areas in the cumulative table should be B, C, D, and A.

Step 2.

Area	LQ	Percentage of manufacturing employment	Percentage of total employment	Cumulative percentage Manufacturing	Total
B	1.400	70	50	70	50
C	0.750	15	20	85	70
D	0.667	10	15	95	85
A	0.333	5	15	100	100
		100	100		

Step 3.   See Figure 3-4.

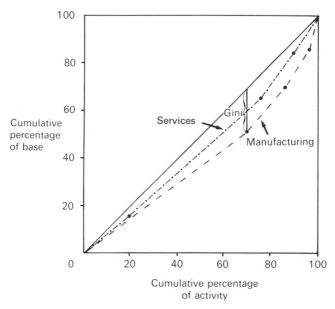

FIGURE 3-4. Lorenz curves for industrial data.

DEFINITION: GINI COEFFICIENT, OR INDEX OF DISSIMILARITY
The Gini coefficient, or index of dissimilarity, is defined graphically as the maximum vertical deviation between the Lorenz curve and the diagonal.

The range of the Gini coefficient is from 0 to 100 percent. There are two other ways of calculating the value of this index. First, it can be determined by identifying the largest difference between the cumulated percentages of the activity and the base. From Table 3-1, the maximum difference between the two cumulative columns for manufacturing and total employment is $70 - 50 = 20$. This is exactly equal to the vertical deviation between the Lorenz curve and the diagonal at this point. The point (70, 50) differs from the diagonal (70, 70) by 20 percent.

　　Second, we might note that the CL for manufacturing was previously calculated as .20. The Gini coefficient can be easily obtained by multiplying the coefficient of localization by 100. For manufacturing we see that $(.20)(100) = 20$, and for service employment $(.10)(100) = 10$. Another equivalent method for calculating the Gini coefficient is to take one-half of the sum of the *absolute value* of the differences between the uncumulated percentage distributions of the activity and the base. Using the data for the manufacturing sector given in Table 3-1, we calculate the Gini coefficient as $\frac{1}{2}(|70 - 50| + |15 - 20| + |10 - 15| + |5 - 15|) = \frac{1}{2}(20 + 5 + 5 + 10) = 20$. Smaller values of both the Gini coefficient and the CL indicate similarity between the areal distribution of the base and the activity. As they increase, the similarity between the base and the activity decreases. This is why the Gini coefficient is sometimes called the index of dissimilarity.

　　The Lorenz curve and Gini coefficient can also be used to measure the degree of similarity of the percentage distributions of any two activities, neither of which is necessarily the base. For example, it is possible to compute the Gini coefficient between manufacturing and service employment distributions. Using the original data introduced for calculating location quotients, we see the Gini coefficient is $\frac{1}{2}(|5 - 10| + |70 - 55| + |15 - 15| + |10 - 20|) = \frac{1}{2}(30) = 15$. This application of the Gini coefficient is utilized in urban social geography to compare the areal distributions of ethnic groups in cities. The similarity of the areal distribution of ethnic groups is a useful indicator of their degree of integration or assimilation into the host society. Over time, as assimilation occurs, the residential segregation of many ethnic groups becomes less pronounced. Gini coefficients can be used to test this hypothesis.

All the procedures described for areal data have one common problem which limits their utility in assessing the similarity of two maps: They are inextricably tied to the exact areal subdivisions used in their calculations and can be interpreted only in this light. As we see in Section 3.4, this leads to several different specific problems. The values of the coefficients are very sensitive to the size and areal definitions used as the basis for their calculations. These issues reappear in the topics presented in Chapters 11 and 12.

3.2. Point Data

All the techniques used to describe areal distributions presented in Section 3.1 are applicable to spatial as well as aspatial data. Unfortunately, each of the measures fails to explicitly incorporate the spatial dimension through a variable related to one of the fundamental spatial concepts—distance, direction, or relative location. This is not true of the statistical methods designed for the analysis of point data. Distance is either explicitly or implicitly included within these measures. This branch of statistics is termed, appropriately enough, *geostatistics*.

The first step in analyzing a set of point data is to overlay the map with a Cartesian grid and determine the coordinates of each point on the map. Figure 3-5 illustrates this procedure for a point distribution with nine observations. Each of the nine points is given an X and Y coordinate on the basis of a 100×100 grid placed over the map. The origin of the grid is usually placed at the southwest corner of the map, so that all the coordinates have positive values. Observation 1, for example, has coordinates (20, 40). The set of nine observations can thus be represented by the following point coordinate data:

Observation No.	X Coordinate	Y Coordinate
1	20	40
2	30	60
3	34	52
4	40	40
5	44	42
6	48	62
7	50	10
8	60	50
9	90	90

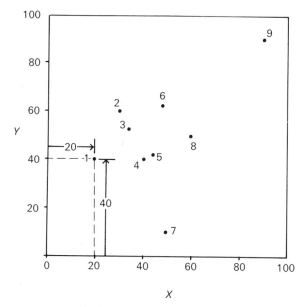

FIGURE 3-5. Obtaining grid coordinates from a dot map.

Depending on the type of point distribution involved, we might add a third variable to these data. For example, if the dots represent towns, villages, and cities, we might associate a "weight," such as the population, to each of the points. In any case, at least two, and possibly three, variables characterize the dot map. The first step in geostatistics is to summarize this map in an efficient way. The standard statistical concepts of central tendency and variability are commonly used for this purpose. Each concept has a distinctly spatial interpretation.

Measures of Central Tendency

The first question to answer is, What is the center, or middle, or average, of this point distribution? Translated into geostatistics, the question is more appropriately phrased, Where on the map is the center of this point distribution? Five different measures can be used to identify the center of the map:

1. Mean center
2. Weighted mean center
3. Manhattan median
4. Euclidean median
5. Weighted medians

Since the properties of these five measures are so often incorrectly stated and insufficiently identified in the geographical literature, they are discussed in detail in this section.

MEAN CENTER

The mean center can be thought of as the "center of gravity" of a point distribution and is a simple generalization of the familiar arithmetic mean. It is remarkably easy to calculate.

DEFINITION: MEAN CENTER

Let (X_i, Y_i), $i = 1, 2, \ldots, n$, be the coordinates of a given set of n points on a map. The mean center of this point distribution is defined as $(\bar{X}, \bar{Y})$ and is given by

$$\bar{X} = \sum_{i=1}^{n} \frac{X_i}{n} \quad \text{and} \quad \bar{Y} = \sum_{i=1}^{n} \frac{Y_i}{n} \tag{3-2}$$

Note that the mean center defines a point or location on the map with coordinates $(\bar{X}, \bar{Y})$. Using the coordinate data for the set of nine points of Figure 3-5, we find the mean center has coordinates

$$\bar{X} = \frac{20 + 30 + 34 + 40 + 44 + 48 + 50 + 60 + 90}{9} = 46.22$$

$$\bar{Y} = \frac{40 + 60 + 52 + 40 + 42 + 62 + 10 + 50 + 90}{9} = 49.56$$

This location is identified by the circular symbol in Figure 3-6.

One simple application of the mean center is to trace the center of gravity of a

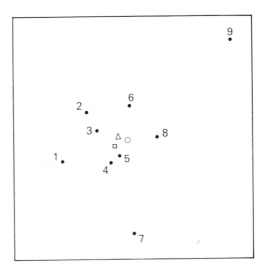

○ Mean center
△ Manhattan median
☐ Euclidean median

FIGURE 3-6. Measures of central tendency for point pattern of Figure 3-5.

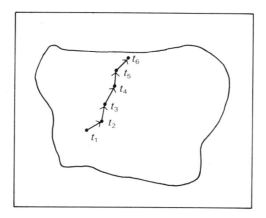

FIGURE 3-7. Tracing the movement of the mean center over six periods.

population distribution over time. To do this, the mean center of the population dot map is calculated for a series of maps of the same region at regular intervals of time. The movement of the mean center over the study period summarizes the change in the distribution of population in the region. For the hypothetical region illustrated in Figure 3-7, for example, the movement of the mean center indicates a systematic northerly trend.

WEIGHTED MEAN CENTER

The mean center can also be generalized to include the case for which each of the points on the map has an associated frequency, magnitude, or "weight." For example, suppose each of the points in Figure 3-5 represents a city with a given population. The location of the mean center should be drawn toward those points with the largest populations and away from those points with the smallest populations. If the populations of the cities represented by the points in Figure 3-5 are 10, 20, 10, 20, 10, 80, 10, 90, and 100, respectively, we expect the mean center to be drawn toward the locations of observations 6, 8, and 9. Since these three observations all lie in the northwest corner of the map, the weighted mean center will be drawn in this direction. The values of both the X and Y coordinates of the center should be greater than the coordinates of the mean center, (46.22, 49.56).

DEFINITION: WEIGHTED MEAN CENTER
Let (X_i, Y_i), $i = 1, 2, \ldots, n$, be the coordinates of a set of n points, and let w_i be the weight attached to the ith point. The weighted mean center has coordinates $(\bar{X}_w, \bar{Y}_w)$ given by

$$\bar{X}_w = \frac{\sum_{i=1}^{n} w_i X_i}{\sum_{i=1}^{n} w_i} \quad \text{and} \quad \bar{Y}_w = \frac{\sum_{i=1}^{n} w_i Y_i}{\sum_{i=1}^{n} w_i} \tag{3-3}$$

The weighted mean center for the nine points of Figure 3-5 is determined in the following way:

$$\bar{X}_w = \frac{10\cdot20+20\cdot30+10\cdot34+20\cdot40+10\cdot44+80\cdot48+10\cdot50+90\cdot60+100\cdot90}{10+20+10+20+10+80+10+90+100}$$

$$= \frac{21,120}{350} = 60.34$$

and

$$\bar{Y}_w = \frac{10\cdot40+20\cdot60+10\cdot52+20\cdot40+10\cdot42+80\cdot62+10\cdot10+90\cdot50+100\cdot90}{10+20+10+20+10+80+10+90+100}$$

$$= \frac{21,900}{350} = 62.57$$

As expected, the weighted mean center with coordinates (60.34, 62.57) is located to the northwest of the mean center (46.22, 49.56).

The formulas for calculating the mean center, (3-2), and weighted mean center, (3-3), are equivalent to the formulas used to calculate the arithmetic mean, (2-3), and grouped mean, (2-5). There is also one important property of the arithmetic mean which is shared with the mean center and has interesting implications for a spatial distribution. Recall that the arithmetic mean of a set of numbers minimizes $\sum_{i=1}^{n}(X_i - \bar{X})^2$. This property, the so-called least squares property of the mean, has been explained in Chapter 2. The equivalent property for the mean center is that $(\bar{X}, \bar{Y})$ minimizes

$$\sum_{i=1}^{n} (X_i - \bar{X})^2 + (Y_i - \bar{Y})^2 \tag{3-4}$$

Note that, by the Pythagorean theorem, the distance between points (X_i, Y_i) and $(\bar{X}, \bar{Y})$ is

$$d = \sqrt{(X_1 - \bar{X})^2 + (Y_i - \bar{Y})^2}$$

Therefore, the mean center has the property that it minimizes the sum of *squared distances* from itself to the n other points. Also, like the mean, the mean center is very sensitive to the existence of extreme observations. Extreme observations in a point distribution are distant points set apart from the others. To see this sensitivity, let us augment the nine points of Figure 3-5 with a tenth point having coordinates (1000, 1000). The new mean center is (141.6, 144.6). This point lies between the existing observations and the new extreme point.

MANHATTAN MEDIAN

The concept of the median can also be applied to point distributions. For a set of n observations, the median is defined as the "middle," or $[(n + 1)/2]$th observation in an ordered array of values of X. How can we find the middle, or median, of a point distribution? The spatial median in this sense is the point of intersection of two perpendicular lines, one which divides the distribution of points in a north-south direction into two equal parts and the other which divides it into two equal parts in

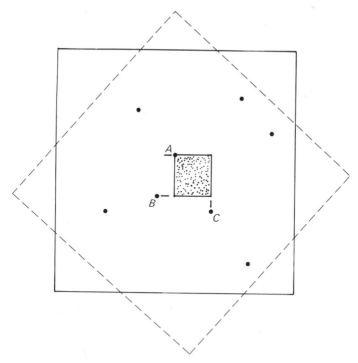

FIGURE 3-8. Manhattan median for an even number of points.

an east-west direction. Consider again the nine points of Figure 3-5. In the north-south direction, the line must pass through observation 8 since it has four points below it (observations 1, 4, 5, and 7) and four observations above it (observations 2, 3, 6, and 9). A horizontal line through observation 8 can thus be drawn. In an east-west direction, observation 5 is the middle point. A vertical line is drawn through this observation. The intersection of these two vertical lines is shown as the triangular symbol in Figure 3-6. It is close to, but not coincident with, the mean center.

There are two distinct problems with the Manhattan median defined above. First, let us see what happens when we try to determine the median for a set with only eight observations, such as in Figure 3-8. Using this definition, we would have to find a line in the north-south direction having four observations on each side. As in Figure 3-8, *any vertical line* between points *A* and *C* will have four points on either side. Similarly, *any horizontal line* between points *A* and *B* divides the set into two equal parts. The shaded area in Figure 3-8 encloses all points having the property of the Manhattan median. It is obviously not a unique point. What went wrong? It turns out that the Manhattan median is *never* unique when there are an even number of points and is *always* unique when there are an odd number of points.

The second problem with this measure of central tendency can also be illustrated with Figure 3-8. Suppose the coordinate system of the map is shifted to the position shown by the dashed lines. What happens to the shaded area defining the median? It must be rotated in the same way. This will define another rectangular area—for the same points! The location of the Manhattan median is therefore not unique under axis rotation. This is somewhat undesirable from a statistical point of view.

The Manhattan median shares one important property with the median of a frequency distribution. From (2-11), we know that the median minimizes the sum of the absolute deviations between itself and the other n points in the frequency distribution. The equivalent property for the Manhattan median is formalized as follows:

DEFINITION: MANHATTAN MEDIAN

Let (X_i, Y_i), $i = 1, 2, \ldots, n$, be the coordinates of a set of n points. The Manhattan median (X_m, Y_m) minimizes

$$\sum_{i=1}^{n} |X_i - X_m| + |Y_i - Y_m| \tag{3-5}$$

That is, (X_m, Y_m) minimizes the sum of the absolute Manhattan deviations from itself to the other n points of the distribution.

Why is it called the Manhattan median? Imagine we are trying to locate some facility in a city where travel is limited to the north-south and east-west directions. It is impossible to travel "as the crow flies." As shown in Figure 3-9, the Manhattan "distance" between the two points (X_i, Y_i) and (X_m, Y_m) is $|X_i - X_m| + |Y_i - Y_m|$. The absolute value signs are used to ensure that the calculated distance is nonnegative. This measure, or *metric*, of spatial separation bears a close resemblance to the movement possibilities in a dense rectangular grid street network of a large city, in particular Manhattan. Hence the name *Manhattan metric* and the term *Manhattan median*. Travel is possible only along the north-south and east-west directions. Note that the Manhattan distance between two points is always greater than the distance calculated

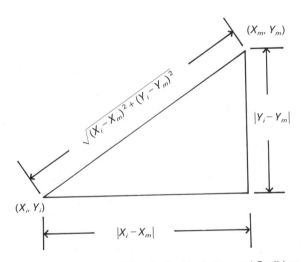

FIGURE 3-9. Distances between points in the Manhattan and Euclidean metrics.

by using the Pythagorean theorem, except when the two points lie on a north-south or east-west line. We term the distance metric calculated by using the Pythagorean theorem the *Euclidean metric*.

EUCLIDEAN MEDIAN

Knowing that the Manhattan median minimizes the sum of Manhattan distances from itself to the n points of a spatial distribution, we might ask, Which point minimizes the Euclidean distances from itself to these same n points?

DEFINITION: EUCLIDEAN MEDIAN

Let (X_i, Y_i), $i = 1, 2, \ldots, n$, be the coordinates of a set of n points. The Euclidean median has coordinates (X_e, Y_e) and is defined as the location that minimizes

$$\sum_{i=1}^{n} \sqrt{(X_i - X_e)^2 + (Y_i - Y_e)^2} \tag{3-6}$$

Unfortunately, there is no direct method of determining the coordinates (X_e, Y_e). Appendix 3A describes one iterative numerical algorithm developed by Kuhn and Kuenne (1962) which can be used to solve this problem. Only small problems can be solved without the aid of a minicomputer or programmable calculator. For the nine points of Figure 3-5, the location of the Euclidean median is found to be (43.98, 42.05). As illustrated in Figure 3-6, the location of the Euclidean median (the square symbol) is very close to, but distinct from, the mean center and Manhattan median.

WEIGHTED MEDIANS

Just as the mean center can be generalized to the weighted mean center, so, too, can the Euclidean and Manhattan medians. The weighted Manhattan median is defined to have equal weights above and below, to the left and to the right. Unlike its unweighted counterpart, problems of nonuniqueness are not nearly so common. The weighted Euclidean median has drawn much more attention from geographers.

DEFINITION: WEIGHTED EUCLIDEAN MEDIAN

Let (X_i, Y_i), $i = 1, 2, \ldots, n$, be the coordinates of a set of n points distributed in the plane. To each of these points there is an attached weight w_i. The weighted Euclidean median (X_{we}, Y_{we}) minimizes

$$\sum_{i=1}^{n} w_i \sqrt{(X_i - X_{we})^2 + (Y_i - Y_{we})^2} \tag{3-7}$$

In other words, the distances between the median and each point are weighted by the value w_i. These weights are defined in the context of the problem at hand. This problem can also be solved by using the iterative algorithm proposed by Kuhn and Kuenne. The locations of the weighted mean center, Manhattan median, and

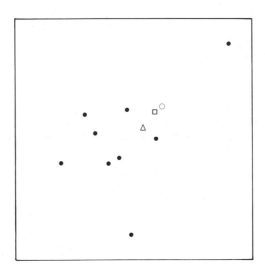

○   Weighted mean center
△   Weighted Manhattan median
□   Weighted Euclidean median

FIGURE 3-10. Weighted medians for point distribution of Figure 3-5.

Euclidean median for the points in Figure 3-5 are illustrated in Figure 3-10. The weights used in each case are the populations set for the weighted mean center problem. Again, all three locations are reasonably close, although we would expect divergences when there are extreme points (or weights) in the spatial distribution.

Interpreted in one way, the weighted Euclidean median can be shown to be the solution to the classical location problem of Alfred Weber. *Weber's problem* is to find the best, or optimal, location for a factory—one which minimizes the sum of transport costs between the factory and two sources of raw materials and between the factory and the market. The situation is illustrated in Figure 3-11. If we make the additional assumptions that (1) transportation costs are a linear function of distance traveled, (2) the weights attached to the points represent the weights of raw material required to produce 1 ton of the finished product, and (3) the weight at the market is 1, then the optimal location for the factory coincides with the weighted Euclidean median. The location of the factory represents the resolution of the three forces "pulling" the factory toward each of the three locations.

The problem can also be viewed as a public facility location problem. Given the locations of groups of facility users, for example, school children, where is the best location for a public facility such as a school? The weights might be defined as the number of school-age children living on a block. The objective of minimizing (3-7) can be interpreted as minimizing the total distance traveled by school children. Note that this also minimizes the average distance traveled by the children. The weighted mean center minimizes the square of the distances traveled by the school children.

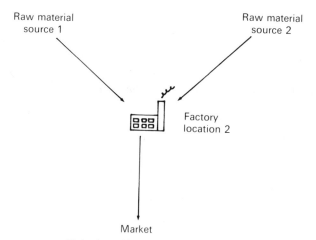

FIGURE 3-11. Weber's problem as the weighted Euclidean median.

Since we are interested in the length of their walk to school, not the *square* of the length of their trip, the Euclidean median is the best solution.

The problem of locating a spatial median is a cornerstone of many private and public facility location problems. One generalization that has drawn a great deal of attention from geographers is the complex problem of locating an entire system of facilities, say five schools, within a spatial distribution of potential users. This is one variant of what are now termed *location-allocation problems*. Once a set of facilities is *located*, consumers or partrons are *allocated* to the appropriate, usually the closest, facility. The optimal locations for a system of five facilities are illustrated in Figure 3-12. This simple extension of the Weber problem is known as the *multiple-source Weber problem*. There is now a rich body of literature on both the theoretical aspects of location-allocation problems and their applications to school, hospital, and other facility systems.

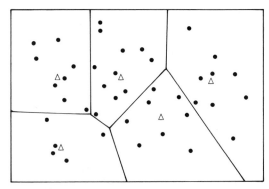

FIGURE 3-12. Optimal locations for a system of five facilities.

Measures of Dispersion

The second important characteristic of a spatial distribution is its dispersion. Measures of dispersion in descriptive statistics are usually based on the notion of a deviation, the differences in value of an observation from a central value such as the mean or median. For spatial distributions the notion of a deviation is the actual distance between an observation and the central point.

STANDARD DISTANCE

The *standard distance* (SD) is the spatial equivalent to the standard deviation. Distances between each observation and the mean center are squared and summed, and this sum is divided by the number of observations. The easiest way to calculate the standard distance of a point distribution is to use the coordinates of each point directly in the following formula:

$$SD = \sqrt{\frac{\Sigma_{i=1}^{n} (X_i - \bar{X})^2}{n} + \frac{\Sigma_{i=1}^{n} (Y_i - \bar{Y})^2}{n}} \tag{3-8}$$

This formula can be rewritten as

$$SD = \sqrt{\sigma_X^2 + \sigma_Y^2}$$

More dispersed point patterns will have large standard distances. For example, consider the two point distributions in Figure 3-13. The distribution in Figure 3-13(b) is

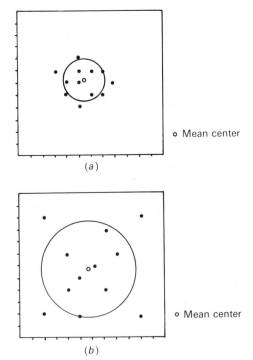

(a)

o Mean center

(b)

o Mean center

FIGURE 3-13. Dispersion of point patterns using standard distance.

clearly more dispersed than the point distribution in (a). A useful graphical device is to draw circles with a radius equal to the standard distance around the mean center of the distribution. It is also possible to use weighted observations by computing a standard distance around the weighted mean center. Together with the mean center, the standard distance can be used to compare and contrast point distributions. Both measures are very sensitive to extreme observations.

QUARTILIDES

Dispersion about the Manhattan median center can be graphically displayed by using the spatial equivalent of the interquartile range. A *quartilide* divides a point distribution into quarters. Surrounding the median are eastern, western, northern, and southern quartilides. Together these define a rectangle, whose size clearly depends on the dispersion of the set of points. An example is shown in Figure 3-14 for a distribution with $n = 16$ points. In this case, problems of nonuniqueness arise because the median of a set of points with an even numer of observations is an area. The median should have eight points to the east and eight to the west. Many lines satisfy this criterion; the one selected here lies *midway* between the two central observations. Similarly, the line locating the median in a north-south direction lies midway between the central points in this direction. The same convention is used to define the quartilides. The eastern quartilide lies midway between the fourth and fifth easternmost points, the western quartilide lies midway between the fourth and fifth westernmost points, and so on. The larger the rectangle defined by these quartilides, the more dispersed the set of points. The construction seems to have been used seldom in the geographical literature.

AN EMPIRICAL APPROACH

A useful empirical approach to describe the dispersion of a point pattern is to graph the cumulative frequency of points around some central location in bands of distance. For example, the point data of Figure 3-15(a) are summarized in this manner in Figure 3-15(b). It is easy to see that 80 percent of the points lie within 6 mi of the mean center, 20 percent within 2 mi, and so on.

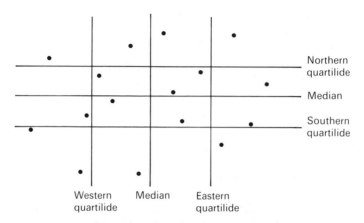

FIGURE 3-14. Dispersion of a point distribution using quartilides.

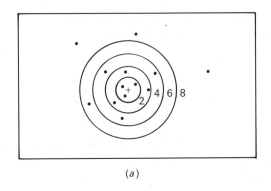

(a)

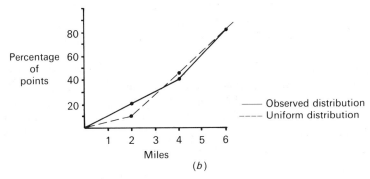

(b)

FIGURE 3-15. Summarizing the dispersion of a point distribution. (a) Concentric circles with distances; (b) graphical comparison to uniform distribution.

One useful benchmark is the uniform distribution—the percentages that would be expected if the points were *evenly distributed* around the mean center. The number of points within a given distance from a point should be proportional to the area within that given distance. Since the area of a circle is proportional to the radius squared, the uniform distribution is not represented as a straight line on the graph in Figure 3-15(b). What would a uniform distribution having 80 percent of its observations within 6 mi be like? Since the area within a distance of 4 mi is $\pi(4^2) = 16\pi$ and the area with 6 mi is $\pi(6^2) = 36\pi$, we would expect $16\pi/(36\pi) = \frac{4}{9}$ as many points within this radius. Similarly, a uniform distribution would have one-ninth as many points within 2 mi as it does within 6 mi. By comparison, then, the observed point distribution is more concentrated than a uniform distribution within the first 3 mi and less concentrated within the next 3 mi.

Although the measures of central tendency and dispersion can help us to make simple comparisons between different point distributions, they are incapable of assessing point *patterns*. Some point patterns are uniform, consisting of a more or less even distribution of points on the map. Other patterns are very clustered, with many points located in one area of the map and large areas without any points. A random

pattern contains elements of both clustered and uniform patterns. More sophisticated methods can be used to assess the pattern of a point distribution along this
continuum.

3.3. Directional Statistics

Direction is another fundamental measurement that can be extracted from maps. Just
as there exist some unusual characteristics in statistical measures based on distance,
so are there peculiar properties of directional measurements.

Measurement of Direction

Directions are measured as angles. In Figure 3-16, for example, the direction A from
the origin is based on the 30° angle it makes with the north bearing. Or, equivalently,

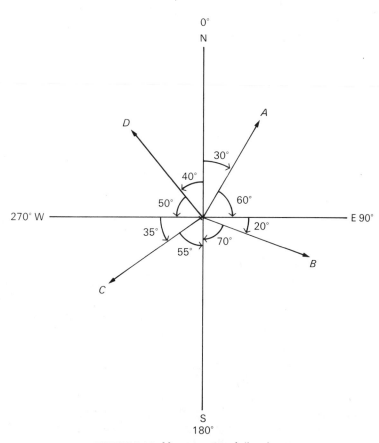

FIGURE 3-16. Measurement of direction.

the direction of A could just as easily be referenced by the 60° angle with the east bearing. Direction is thus a *relative* measure dependent on the frame of reference used to take the bearing. The usual frame of reference is to take true north as 0° and record all directions as bearings from true north in a clockwise direction. The direction of A is therefore 30°, B is 110°, C is 235°, and D is 320°. Direction is based on a circular frame of reference with limits [0, 360]. Because of the circular nature of this measurement system, two directions which are remarkably similar can have widely different bearings. Note that the directions represented by 31° and 32° are similar, but so are the two directions 0° and 359°. There is a 1° difference in both pairs of directions. It is thus impossible to translate these bearings into the simple number systems capable of being summarized by conventional descriptive statistics.

Graphical Summaries of Directional Data

Directional data can be presented graphically in five different ways. For raw or ungrouped data, the directions may be illustrated as points on a unit circle, or as vectors drawn from a common origin. Suppose the data consist of the following 10 bearings taken from a single location:

Observation No.	Bearing, 0° = N
1	60°
2	45°
3	230°
4	50°
5	220°
6	55°
7	225°
8	30°
9	40°
10	320°

These observations are summarized graphically in Figure 3-17. In (*a*) each direction is illustrated as a vector of unit length. Clearly the bearings tend to be predominantly northeasterly or southwesterly. This representation is more striking than the alternative representation in (*b*). Whenever there are a large number of observations, these summaries tend to be difficult both to draw and to interpret. In such instances it is common to group the observations and present the data in a *conventional histogram*, a *circular histogram*, or a *rose diagram*.

    The difficulty with the usual linear representation of the histogram is that the advantage of a circular display for what is inherently circular data is lost. This feature is illustrated with the wind direction data in Chapter 1. Two circular formats tend to be superior devices for portraying directional data. To illustrate their construction and interpretation, consider the following set of grouped data:

Observation No.	Compass direction	Number of migrants	Percentage
1	N	0	0
2	NNE	50	5.0
3	NE	100	10.0
4	ENE	200	20.0
5	E	125	12.5
6	ESE	75	7.5
7	SE	50	5.0
8	SSE	25	2.5
9	S	0	0
10	SSW	15	1.5
11	SW	35	3.5
12	WSW	275	27.5
13	W	25	2.5
14	WNW	25	2.5
15	NW	0	0
16	NNW	0	0
		1000	100.0

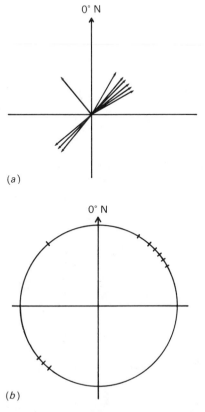

(a)

(b)

FIGURE 3-17. Summarizing ungrouped directional data.

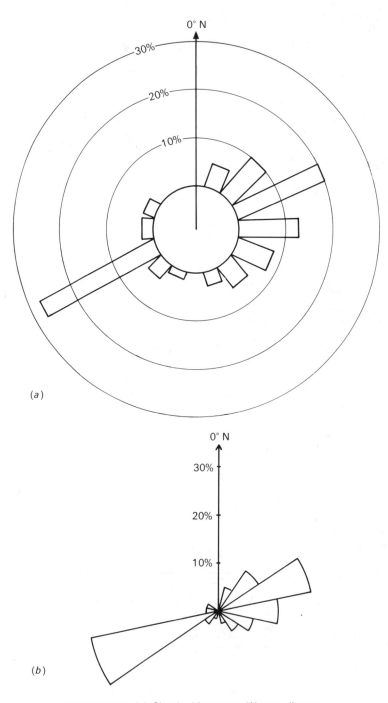

FIGURE 3-18. (a) Circular histogram; (b) rose diagram.

The directions are aggregated to the usual 16 compass directions (N, NNE, NE, etc.), $22\frac{1}{2}°$ subdivisions of the 360° in the compass. The frequencies represent the number of migrants from places within the compass heading to some particular city. Is there a directional bias to this migration stream?

To construct a circular histogram, first draw a reference unit circle, as in Figure 3-18(*a*). Intervals are placed on the perimeter of this reference circle. Just as in the usual linear histogram, there is some choice of interval width. In this case, the 16 compass directions in the tabulated data are used without further aggregation. Blocks are extended from the perimeter of the circle for each interval. The height of each block is scaled to the observed frequency (or percentage) in the category. The width of each block is equal to the length of the chord connecting the two points on the reference circle which delimit the interval. For purposes of interpretation, a series of concentric circles are drawn outside the reference circle and labeled with the appropriate frequency or percentage. In this example, concentric circles are drawn for 10, 20, and 30 percent of all migrants. The raw frequencies could also have been used in this histogram.

For the rose diagram of Figure 3-18(*b*), the class limits are radii extended from the origin. The *length* of each radius is proportional to the class frequency or percentage. The limit of the sector for each class is an arc centered on the origin. Or, if it is desired to have the *area* of the sector proportional to the class frequency or percentage, the square root of the frequency or percentage should be used to define radii lengths. A quick examination of Figure 3-18(*a*) or (*b*) reveals the strong directional trends of these migration flows. Migrants to this city come from places along a northeasterly to southwesterly corridor.

Directional Mean and Circular Variance

We have already noted that the usual linear number sequence cannot be used in making many calculations for directional data. Consider again the two pairs of angles (31°, 32°) and (0°, 359°). If we treat these as real numbers, we can compute the means of these pairs as 31.5° and 179.5°, respectively. This average seems intuitively reasonable for the first pair, but not for the second pair. The mean of 179.5° is almost due south, but the two directions are almost due north! Fortunately, we can solve this problem by using the concept of a vector and some elementary trigonometric relations. Appendix 3B briefly reviews sufficient elementary trigonometry for this problem.

A *vector* is completely defined by a magnitude (or length) and a direction. Let us assume that all our directional measurements have the identical length of 1 unit, that is, they are *unit vectors*. The *mean* of a set of vectors is defined as the *resultant* of vector addition. To locate the resultant, we simply place the vectors end to end, preserving their directions. The vector connecting the origin to the end of the last vector in the sequence is the resultant vector. The angle that this resultant makes with the 0°N bearing is the average, or mean, direction. The graphical solution to determine the directional mean is illustrated in Figure 3-19 for the four directional measurements of Figure 3-16. From the end of vector **A**, a vector of unit length is placed, with the bearing of vector **B**. Since the direction of vector **B** is 110°, **B** is drawn 20°

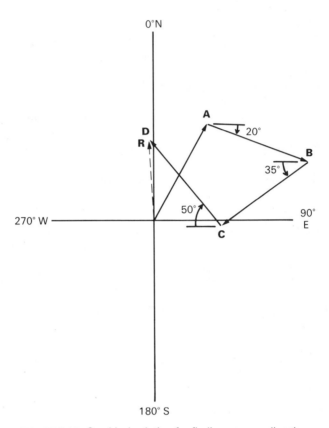

FIGURE 3-19. Graphical solution for finding average direction.

below the horizontal (or east bearing of 90°). Similarly, **C** is added to **B**, and finally **D** is added to **C**. The resultant **R** is found by connecting the origin to the end of this vector. It is shown as the dashed line in Figure 3-19. The directional mean, if read from a protractor on an accurate diagram, is almost true north, in fact 358.2°.

The difficulty with this method is obvious. It is extremely inefficient where there are a large number of directions to be added, and problems of precision usually arise. A direct method of calculating the directional mean uses some simple arguments from trigonometry.

Consider Figure 3-20. Two vectors **A** and **B** and a resultant **R** are drawn in the conventional XY plane. Vector **A** can be defined by the coordinates (X_A, Y_A) and vector **B** by the coordinates (X_B, Y_B). Clearly the coordinates of the resultant are $(X_A + X_B, Y_A + Y_B)$. Since both **A** and **B** are unit vectors, $\mathbf{OA} = \mathbf{OB} = 1$. Simple trigonometric relations can be used to determine θ_R, the direction of the resultant. Note that tangent of the resultant is given by the ratio of the opposite to the adjacent sides of the right-angle triangle based on θ_R:

$$\tan \theta_R = \frac{Y_A + Y_B}{X_A + X_B} \tag{3-9}$$

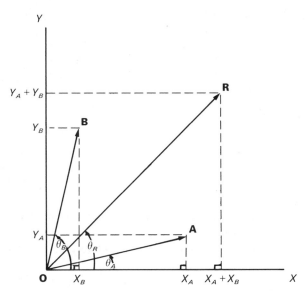

FIGURE 3-20. Trigonometric relations of a vector resultant.

Since $\mathbf{OA} = \mathbf{OB} = 1$, we know $X_A = \cos \theta_A$, $X_B = \cos \theta_B$, $Y_A = \sin \theta_A$, and $Y_B = \sin \theta_B$. Thus, (3-9) can be rewritten as

$$\tan \theta_R = \frac{\sin \theta_A + \sin \theta_B}{\cos \theta_A + \cos \theta_B} \qquad (3\text{-}10)$$

and it follows that

$$\theta_R = \arctan \frac{\sin \theta_A + \sin \theta_B}{\cos \theta_A + \cos \theta_B} \qquad (3\text{-}11)$$

Note that this arctangent is the ratio of the distance of both vectors in a *northerly direction* to the distance in an *eastward direction*.

For a set of n directions, the definition of directional mean can be formalized by a simple generalization of (3-11):

DEFINITION: DIRECTIONAL MEAN
Let $\theta_1, \theta_2, \ldots, \theta_n$ be a set of bearings or directions taken from a single origin. The directional mean, or bearing, of the resultant is

$$\theta_R = \arctan \frac{\sum_{i=1}^{n} \sin \theta_i}{\sum_{i=1}^{n} \cos \theta_i} \qquad (3\text{-}12)$$

Interestingly, it can be shown that the directional mean has the property

$$\sum_{i=1}^{n} \sin (\theta_i - \theta_R) = 0 \qquad (3\text{-}13)$$

That is, the sum of the sines of the angular deviations from each observation to the resultant is zero. This property is equivalent to a well-known property of the arithmetic mean given in Chapter 2: The sum of the deviations about the mean is zero.

In determining the arctangent given by (3-12), it is also necessary to fix the quadrant of the resultant bearing by observing the signs of $\Sigma_{i=1}^{n} \sin \theta_i$ and $\Sigma_{i=1}^{n} \cos \theta_i$. The easiest method of determining the resultant bearing is to compute θ_R by using the absolute value of the ratio and then to use the following corrections to specify the quadrant in which it lies.

1. If $\Sigma_{i=1}^{n} \sin \theta_i > 0$ ($+$) and $\Sigma_{i=1}^{n} \cos \theta_i > 0$ ($+$), then the ratio is $+/+$ and the value of θ_R can be used directly for the directional mean since the resultant must be in the first quadrant (i.e., between $0°$ and $90°$ or to the northeast).
2. If the ratio is $+/-$, then the bearing of the resultant is $180° - \theta_R$, or to the southeast, in the second quadrant.
3. If the ratio is $-/-$, then the bearing of the resultant is $180° + \theta_R$, or to the southwest, in the third quadrant.
4. If the ratio is $-/+$, then the bearing is $360° - \theta_R$ or to the northwest, in the fourth quadrant.

The calculations required to determine the directional mean of the four vectors in Figure 3-16 are given in Table 3-2. Note that the result confirms the answer obtained by using the graphical vector addition technique. Computing the directional mean for grouped data requires only a slight modification of (3-12):

$$\theta_R = \arctan \frac{\Sigma_{i=1}^{n} f_i \sin \theta_i}{\Sigma_{i=1}^{n} f_i \cos \theta_i} \qquad (3-14)$$

In this formula f_i is the frequency of the ith interval. This formula bears a close resemblance to the formula for the grouped mean of a set of observations of a variable presented in Chapter 2.

TABLE 3-2
Work Table for Calculation of Mean Direction

	Bearing $0° = N$	Cosine	Sine
A	30°	.86603	.50000
B	110°	−.34202	.93969
C	235°	−.57358	−.81915
D	320°	.76604	−.64279
		.71647	−.02225

$$\theta = \arctan \left(\frac{-.02225}{.71647} \right)$$
$$= \arctan .03106 = 1.8°$$
$$\text{Mean direction} = 360° - 1.8° = 358.2°$$

The directional mean shares many properties with the arithmetic mean. It is a poor measure of central tendency for a bimodal distribution. For a set of bimodal directional data, the directional mean usually falls between the two peaks. This can be a serious problem for directional data having one orientation, that is, two peaks differing by 180°. The migration data illustrated in Figure 3-18 have this characteristic, with strong peaks in the ENE and WSW directions. The directional mean lies between these two peaks of the distribution.

The variability of a sample of directional measurements is indicated by the length of the resultant vector **R**. In Figure 3-20, for example, the tip of the resultant vector lies far from the origin. This reflects the similarity of the directions **A** and **B**. If **A** and **B** were identical, the length of **R** would be 2, the sum of the lengths of the two unit vectors **A** and **B**. If **A** and **B** were in opposite directions, the resultant would be at the origin and the length of **R** would be 0. The length of the resultant vector **OR** can be determined in Figure 3-20 by using the Pythagorean theorem. It is the hypotenuse of the largest right-angle triangle and is equal to

$$\sqrt{(X_A + X_B)^2 + (Y_A + Y_B)^2} = \sqrt{\sin A^2 + \sin B^2 + \cos A^2 + \cos B^2}$$

In general, for n vectors

$$\mathbf{OR} = \sqrt{\left(\sum_{i=1}^{n} \sin \theta_i\right)^2 + \left(\sum_{i=1}^{n} \cos \theta_i\right)^2} \tag{3-15}$$

There is one problem with using this measure in this unstandardized form. Larger sample sizes can have longer resultant lengths than smaller samples without having less variability. To develop a standardized measure of variability it is necessary to account for differing sample sizes.

DEFINITION: CIRCULAR VARIANCE

Let θ_i, $i = 1, 2, \ldots, n$, be a set of directional measurements. The circular variance is defined as

$$S_0 = 1 - \frac{\mathbf{OR}}{n} \tag{3-16}$$

where **OR** is the resultant length and n is the sample size.

What are the bounds on S_0? When all the directional measurements are coincident, there is zero variability, $\mathbf{OR} = n$ and therefore $S_0 = 0$. When the resultant length $\mathbf{OR} = 0$, there is maximum variability and $S_0 = 1$. The bounds on S_0 are $[0, 1]$. Using the information in Table 3-2 for the directions in Figure 3-16, we see that $\mathbf{OR} = \sqrt{(.71647)^2 + (-.02223)^2} = .7168$ and $S_0 = 1 - .7168/4 = .8208$. This indicates a high degree of variability and confirms the visual impression of Figure 3-16. The most common use of S_0 is to compare the directional variability of different samples.

In some geographical studies, orientation, and not direction, is important. We say that a road has a north-south orientation, but traffic on the road has either a north or a south direction. If all angles less than 180° are doubled, then the directional data can be converted to orientation data. Then the directional mean and circular variance can be applied to these modified data. This is common practice in applications of directional statistics in physical geography such as drumlin or pebble orientations.

3.4. Descriptive Statistics and Spatial Data: Four Problems

There are four recurring problems which arise when virtually any descriptive statistic, including those specifically designed for the purpose, is applied to spatial data. So it is necessary to be extremely careful in interpreting the results of a statistical analysis based on locational observations. In fact, many of the recent advances in spatial statistics have been developed in response to the need to overcome the limitations imposed by these four problems. These four problems are often referred to as the boundary problem, the scale problem, the problem of modifiable units, and the problem of pattern.

Boundary Problem

The location of the boundary of the study area, as well as the placement of the internal boundaries in an areal design, is often a crucial question in geographical research. Poorly chosen designs can lead to several problems. Let us illustrate the potential difficulties with two examples. First consider the two point patterns of Figure 3-21. Both are identical and would yield identical values for any measure of central tendency or dispersion. However, these patterns are clearly different. The distribution in (*a*) could be described as a reasonably regular, dispersed pattern. In

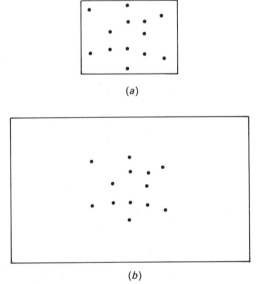

(*a*)

(*b*)

FIGURE 3-21. Boundary problem for point patterns.

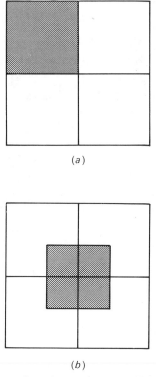

(a)

(b)

FIGURE 3-22. Boundary problem for areal patterns.

(b), the pattern could be described as clustered. This means that the standard distance or any other measure of dispersion cannot be interpreted independent of the study area.

As a second example, consider the problem of locating areal boundaries within some study area. Suppose the location of some phenomenon, for example a particular ethnic group in a city, is depicted as the shaded area in the two maps of Figure 3-22. Despite the fact that the "map" is the same in each case, the values of location quotients, coefficients of localization, and Gini coefficients would be markedly different for these two cases. All measures for Figure 3-22(a) would indicate significant areal concentration of the ethnic group, since the boundary of one zone completely encloses the neighborhood occupied by the ethnic group. For the boundaries in (b), these same measures would tend to indicate a rather even distribution of the ethnic group in this study area. In short, the map in (a) suggests ethnic *segregation*, and the map in (b) suggests *integration*. Although the results may not always be as dramatic as in this case, boundary locations can mask certain map patterns. The moral is clear—summary statistics can be interpreted only for the particular areal divisions upon which their calculation is based. If this framework is poorly chosen, the results may be at the minimum misleading and possibly even false. Where there is some control over the placement of areal boundaries, these problems can sometimes be minimized by the use of a fine areal breakdown within an appropriately delimited study area.

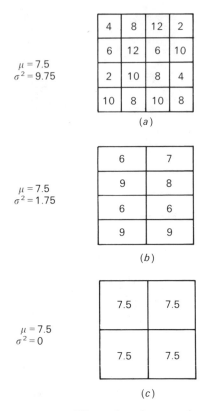

FIGURE 3-23. Effects of areal aggregation.

Scale Problem

A second problem is termed the *scale problem*, or the *areal aggregation problem*. Quite simply, the values for many descriptive statistics can vary systematically when increasingly aggregated areal data are used. An example of the effects of areal aggregation is in Figure 3-23. In (*a*), the areal distribution of some variable X is shown in an area with 16 cells. The mean value of the map is $\mu = 7.5$, and the variance is $\sigma^2 = 9.75$. Aggregate these data by joining neighboring cells to create eight zones, as in Figure 3-23(*b*). In each two-cell aggregation, the value of X for each zone is the average value of the two smaller cells from which it has been created. For example, the zone in the northwest corner of (*b*) has a value of $(4 + 8)/2 = 6$. What does change with this aggregation? The mean remains constant at 7.5, but the variance declines significantly, to $\sigma^2 = 2$. Much of the variation in X is now lost. Since geographers are often interested in spatial variation, this is particularly unfortunate. All observations on the eight-zone map are within a range $6 < X < 9$, but the values on the original map have a wider range of $2 < X < 12$. At an even higher level of aggregation, note that the four-district representation in Figure 3-23(*c*) has a variance of zero! This is an obvious distortion of the real situation.

    In many instances, then, spatial aggregation tends to reduce the variation depicted on a map. Comparisons of maps of a variable at different levels of aggregation must take into account the ramifications of this variance reduction. The problem becomes even more acute when we try to examine the *relationship* between two maps of different variables. The difficulties of aggregation in this problem of *correlation* are discussed in Chapter 11.

Problem of Modifiable Units

Even at the same scale of analysis, different areal definitions can also have a substantial impact on the values of most descriptive statistics. To see this, suppose the aggregation of Figure 3-23(*a*) into eight zones from 16 cells is accomplished by joining contiguous north-south rather than east-west neighbors, as in Figure 3-24(*a*), or by using a mixed pattern of north-south and east-west aggregations, as in Figure 3-24(*b*). Again, both systems lead to the same map mean of 7.5. However, although each of these aggregations has the same variance, it is higher than that of the eight-zone map of Figure 3-23(*b*). The effects of using modified areal units are not nearly so predictable as the effects of aggregation. Although all aggregations decrease the variance (it is possible to maintain the variance if the zones joined are exactly alike), some systems will result in significantly lower variances, other not. In this sense, the aggregations in Figure 3-24 are superior to the one in Figure 3-23(*b*). They both are more representative of the more detailed map in Figure 3-23(*a*). But what if the map in Figure 3-23(*a*) were not known? Can we have faith that the eight-zone map used in our analysis is truly representative of the actual variation of the variable across the map? We cannot be sure.

One conclusion we can draw from this analysis is that it is always better to join similar zones when we must aggregate the data. This will preserve the variation in the original map as much as possible. Since one of the "laws" of geography is that closer places are more alike than distant places, contiguous areal aggregations are likely to be less disruptive than aggregations of areas which are not close together.

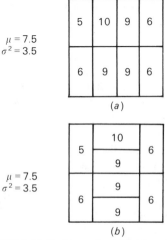

FIGURE 3-24. Problem of modifiable units.

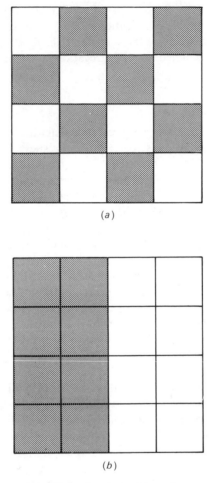

FIGURE 3-25. Illustrating the problem of map pattern.

Problem of Pattern

All the methods described in this chapter share one shortcoming: They are generally incapable of assessing the type of pattern which exists on a map. Consider the two contrasting patterns of Figure 3-25(a) and (b). Again, suppose the shaded areas represent residential concentrations of some particular ethnic group in this city. The coefficient of localization, Gini coefficient, or Lorenz curve would indicate a significant level of areal concentration of this ethnic group in both maps. There are eight zones completely populated by the group and eight zones where they are completely absent. Moreover, and even more important, because these measures are insensitive to pattern, both maps would have identically valued coefficients of areal concentration. But are these maps similar? Do both maps have about the same degree of concentration of this ethnic group? One could argue that there is much more segregation in the city mapped in Figure 3-25(*b*). Not only does the host population exclude them from their neighborhoods, but also the host population excludes them from nearby neighborhoods as well. Techniques capable of distinguishing between areal patterns are discussed in Chapter 11. Pattern in point distributions is discussed in Chapter 5.

3.5. Summary

Geographers share a common interest in the analysis of spatial distributions. Whether the data they collect are for areas, points, directions, or networks, specialized techniques can be used to summarize the maps of spatially varying phenomena. In fact, some of the techniques geographers often use to analyze spatial patterns are used to analyze aspatial variables. For areal data, the relative concentration of a variable across a map can be measured by the location quotient, the Gini coefficient, the coefficient of localization, or the Lorenz curve. For point data, measures of central tendency and dispersion such as the mean center and standard distance can be used to summarize these distributions. These measures have similar properties to conventional statistical measures such as the mean and standard deviation. Directional data, because of their circular framework, require special handling.

Virtually all these methods have several important shortcomings. First, they are extremely sensitive to the choice of boundaries. Second, they are not independent of the scale or degree of areal aggregation. Third, the values of many statistics can vary significantly depending on the choice of areal units depicted on the map. In fact, one could almost take some maps and determine the areal subdivisions that result in any desired value for some statistics! Finally, most methods are not capable of assessing the nature of the pattern exhibited on the map. More specialized methods presented in later chapters address several of these issues.

Appendix 3A. An Iterative Algorithm for Determining the Weighted or Unweighted Euclidean Median

One of the most efficient algorithms for solving the weighted or unweighted Euclidean median problem was developed by Kuhn and Kuenne (1962). The following algorithm is a simple variant of their method.

Let (X_i, Y_i), $i = 1, 2, \ldots, n$, be the set of n given points with weights w_i, $i = 1, 2, \ldots, n$. The location of the current estimate of the median in the tth iteration is (X^t, Y^t). As the algorithm proceeds, (X^t, Y^t) gradually converges to (X_e, Y_e). The algorithm stops whenever the estimates of the coordinates in successive iterations are less than some predetermined tolerance level TOL. For example, if we wish the coordinates of (X_e, Y_e) found by the algorithm to be within .01, we set TOL = .01. Since it is an approximating algorithm, the method is more efficient if a "good" starting point is selected. A good point is one close to the answer. The bivariate mean $(\bar{X}, \bar{Y})$ and weighted mean $(\bar{X}_w, \bar{Y}_w)$ are good and obvious choices for a starting point.

So $(X^1, Y^1) = (\bar{X}, \bar{Y})$. The algorithm can be described as the following sequence of steps:

1. Calculate the distance from each point (X_i, Y_i) to the current estimate of the median location

$$d_i^t = \sqrt{(X_i - X^t)^2 + (Y_i - Y^t)^2}$$

   where d_i^t is the distance from point i to the median during the tth iteration.
2. Determine the values K_i^t from $K_i^t = w_i/d_i^t$. Note that in the unweighted case all values of w_i are set equal to 1.
3. Calculate a new estimate of the median from

$$X^{t+1} = \frac{\sum_{i=1}^n K_i^t X_i}{\sum_{i=1}^n K_i^t} \qquad Y^{t+1} = \frac{\sum_{i=1}^n K_i^t Y_i}{\sum_{i=1}^n K_i^t}$$

4. Check to see whether the location has changed between iterations. If $|X^{t+1} - X^t|$ and $|Y^{t+1} - Y^t| \le \text{TOL}$, stop. Otherwise, set $X^t = X^{t+1}$ and $Y^t = Y^{t+1}$ and go to step 1.

The algorithm usually converges in a small number of iterations. For the nine points of Figure 3-5, for the unweighted case, the following steps summarize the results of the algorithm initialized with $\text{TOL} = .10$.

Iteration	X^t	Y^t
1	44.867	45.517
2	43.865	44.060
3	43.672	43.204
4	43.710	42.742

The location of the unweighted median is thus estimated as $(43.71, 42.74)$.

Appendix 3B. Trigonometric Review for Directional Statistics

Consider the angle θ defined in relation to the coordinate system of Figure 3-26. Here OP is a segment of length r, identifying a point P with coordinates (x, y) and making

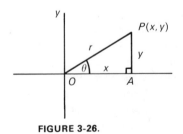

FIGURE 3-26.

sin θ

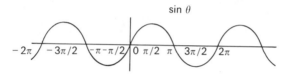

FIGURE 3-27.

an angle of θ with the x axis. Since OAP is a right-angle triangle with $OA = x$ and $AP = y$, by the Pythagorean theorem $r = \sqrt{x^2 + y^2}$. Trigonometric functions of an angle θ are defined in relation to the three quantities x, y, and r. These are known as the adjacent, opposite and hypotenuse, respectively, to the angle θ. The three principal trigonometric functions are defined as the following ratios:

$$\sin \theta = \frac{y}{x} \quad (\text{sine of } \theta)$$

$$\cos \theta = \frac{x}{r} \quad (\text{cosine of } \theta)$$

$$\tan \theta = \frac{y}{x} \quad (\text{tangent of } \theta)$$

As the angle of θ increases and P moves around the origin from $0°$ to $360°$, or equivalently from 0 to 2π radians, these three functions take on a range of values shown in Figures 3-27, 3-28, and 3-29, respectively. Notice that when $\theta = 0°$, $y = 0$, and therefore $\sin \theta = 0$. At $\theta = 90°$, y equals r and therefore $\sin \theta = 1$.

cos θ

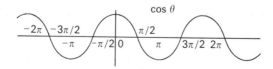

FIGURE 3-28.

tan θ

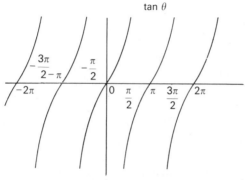

FIGURE 3-29.

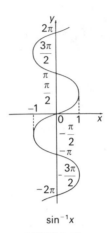

$$\sin^{-1}x$$

FIGURE 3-30.

The reciprocals of these three functions define three other important trigono-metric functions:

$$\csc \theta = \frac{1}{\sin \theta}$$

$$\sec \theta = \frac{1}{\cos \theta}$$

$$\cot \theta = \frac{1}{\tan \theta}$$

where $\csc \theta$ is the cosecant of θ, $\sec \theta$ is the secant of θ, and $\cot \theta$ is the cotangent of θ.

Inverse trigonometric functions are specified as $\sin^{-1} x$ (or arcsin x), $\cos^{-1} x$ (or arccos x), and so on. For $\theta = \sin^{-1} x$, for example, we are directed to find the angle θ with a sine of x. An example of an inverse trigonometric function is shown in Figure 3-30. Since the trigonometric functions are cyclical over a period of 360°, inverse trigonometric functions are many-valued functions and often quite difficult to work with.

REFERENCES

C. D. Harris, "The Market as a Factor in the Localization of Industry in the United States," *Annals, Association of American Geographers* 34 (1954) 315–348.

H. W. Kuhn and R. E. Kuenne, "An Efficient Algorithm for the Numerical Solution of the Generalized Weber Problem in Spatial Economics," *Journal of Regional Science* 4 (1962) 21–33.

FURTHER READING

Virtually all elementary statistical textbooks for geographers include a discussion of at least part of the material presented in this chapter. A useful first reference for geostatistics is a monograph by Neft (1966). Many of the early applications of geostatistics and debates on the interpretation of specific measures are reviewed in considerable detail. An introduction to location-allocation problems can be found in Abler, Adams, and Gould (1971) and in a recent review by Hodgart (1978). Advances in the statistical analysis of directional data are documented in the influential textbook of Mardia (1972), but this is an extremely advanced treatment for the beginning student. Geographical applications of graph theory and network analysis are reviewed in Tinkler (1979) and Haggett and Chorley (1969).

R. Abler, J. S. Adams, and P. R. Gould, *Spatial Organization: The Geographer's View of the World* (Englewood Cliffs, N.J.: Prentice-Hall, 1971).
P. Haggett and R. J. Chorley, *Network Analysis in Geography* (London: Edward Arnold, 1969).
R. L. Hodgart, "Optimizing Access to Public Services: A Review of Problems, Models, and Methods of Locating Central Facilities," *Progress in Human Geography* 2 (1978) 17–48.
D. V. Mardia, *Statistics of Directional Data* (London: Academic, 1972).
D. Neft, *Statistical Analysis for Areal Distributions* (Philadelphia: Regional Science Research Institute, 1966).
K. J. Tinkler, "Graph Theory," *Progress in Human Geography* 3 (1979) 85–116.

PROBLEMS

1. Test your knowledge of this chapter by explaining the meaning of the following terms:

   a. Spatial distribution of a variable
   b. Thiessen polygon
   c. Location quotient
   d. Coefficient of localization
   e. Lorenz curve
   f. Gini coefficient
   g. Mean center
   h. Standard distance
   i. Circular histogram
   j. Rose diagram
   k. Directional mean
   l. Circular variance

2. Differentiate between the concept of a Manhattan median and Euclidean median.

3. Consider the following 12 coordinate pairs and weights:

X Coordinate	Y Coordinate	Weight
50	80	7
30	70	4
50	70	5
60	70	6
40	60	4
70	70	6
50	60	2
80	60	2
40	50	1
60	50	1
0	50	1
50	40	1

a. Locate each point, using the given coordinates on a regular piece of graph paper.
b. Calculate (i) the mean center (assume all weights equal to 1), (ii) the weighted mean center, and (iii) the Manhattan median. Locate each on the graph paper.
c. Calculate the standard distance. Draw a circle with radius equal to the standard distance centered on the mean center.
d. (*Optional*) Use the Kuhn and Kuenne algorithm (Appendix 3A) to determine the Euclidean median.

4. Consider a geographic area divided into four regions, north, south, east, and west. The following table contains the distribution of employment in these four regions for three sectors of the economy:

	Economic sector			
	Primary	Manufacturing	Services	Total
North	2000	100	500	2600
South	100	300	200	600
East	300	800	600	1700
West	100	400	300	800
Total	2500	1600	1600	4700

a. Calculate location quotients for each region and each sector of the economy.
b. Calculate the coefficient of localization for each sector.
c. Draw Lorenz curves for each of the three sectors, superimposing them on a single graph.
d. Write a brief paragraph describing the spatial distribution of these activities, using the results of parts (a), (b), and (c).

5. Solve parts (a) to (d) of Problem 3, using the following coordinates and weights:

X Coordinate	Y Coordinate	Weight
20	100	1
70	90	1
100	100	1
40	70	2
80	70	1
60	60	2
40	40	4
50	50	3
70	40	6
20	20	5
50	20	2
100	20	1

6. The bearings $(0° = N)$ from a house to the retail outlets chosen by members of that household on their last 10 shopping trips are given in the following table:

Observation No.	Bearing, deg
1	150
2	320
3	180
4	300
5	120
6	340
7	170
8	360
9	90
10	30

a. Illustrate these bearings as vectors and as points on a unit circle in two separate graphs.
b. Calculate the directional mean.
c. Find the circular variance.

7. From the census of the United States or of Canada (or an equivalent source), generate the value of some variable for a region containing at least 30 subareas.
a. Calculate the mean and variance for the variable for these base units.
b. Group contiguous subareas in pairs so that there are 15 or more zones; calculate the mean and variance in the regions, using these larger zones.
c. Group contiguous zones identified in part (b) to get eight or more areas; find the mean and variance for this distribution.
d. Continue this procedure until there are only two subregions. At each step calculate the mean and variance.
e. The mean of the variable at each step should be equal. Why?
f. Construct a graph of the variance of the variable (Y axis) versus the number of subareas (X axis). What does it reveal?

8. For the same data used in Problem 7(b), generate five different groupings in which the subareas are joined in pairs. Calculate the variance for each grouping. Do your results confirm the existence of the problem of modifiable units?

9. What is the smallest possible value for a location quotient? When can it occur? What is the largest possible value for a location quotient?

10. Generate an example illustrating the sensitivity of both location quotients and coefficients of localization to the areal units on which their calculations are based. (*Hint*: Try a grouping experiment such as in Problem 8.)

11. What happens to a coefficient of localization if two areas with identical location quotients are joined into a single larger zone? What happens if two zones with different location quotients are grouped?

II

UNIVARIATE
INFERENTIAL
STATISTICS

4

Elementary Probability Theory

Everyone has some idea of what the word *probability* means. Everyday conversation contains numerous references to it: The *chances* of rain this holiday weekend are 50 percent. The *odds* are small that retail sales will be up this Christmas season. Or an unmarried male driver between the ages of 18 and 25 will *probably* have more accidents than an unmarried female in the same age bracket. All these statements invoke a notion of probability, although a very vague one. In this chapter we discuss some basic notions of probability theory in a more precise way. We show how to rigorously define the term *probability* as it is used in conventional statistical inference. Then we explain in detail the methods used to calculate the probability of an event occurring. Finally, we show how to add and multiply probabilities when two or more events are involved.

Probability theory is of interest in its own right, since it is the basis for a great number of decision-making activities. When several courses of action are open to a decision maker but the outcome of making a specific decision is not known with complete certainty, we can use probability theory to define rational choices or courses of action. This is known as *statistical decision theory*. In addition, and more central to our concerns, the theory of probability provides the logical foundation for *statistical inference*. Because these inferences are made from only a sample (or small portion of the members of some population), all statistical inferences are necessarily probabilistic. Therefore, a study of probability theory is essential if we are to make inferences about statistical populations.

4.1. Statistical Experiments, Sample Spaces, and Events

The basic concepts of probability theory are based on the notion of a statistical experiment, or random trial:

DEFINITION: STATISTICAL·EXPERIMENT
A statistical experiment, or random trial, is a process or activity in which one outcome from a set of possible outcomes occurs. Which outcome occurs is not known with complete certainty beforehand.

Consider, for example, the process of sampling from a population. Each time we select a single member of a population and record the value of some variable, we are performing a statistical experiment. The values of the variable for different members of the population represent the possible outcomes from the experiment. Using some type of randomization device, we select one member of the population and record the value of the variable. Which member is selected, and therefore what value is recorded, is unknown until the drawing actually occurs. Repeated trials of this experiment would yield a sample from a population.

DEFINITIONS: ELEMENTARY OUTCOMES AND SAMPLE SPACE
Each different outcome of an experiment is known as an elementary outcome.
The set of all elementary outcomes constitute the sample space.

Consider the statistical experiment of selecting a card from a shuffled deck of cards. The 52 elementary outcomes are the 52 individual cards in the deck. Together, these 52 outcomes represent the sample space of this experiment. Sometimes, it is not so easy to enumerate all possible outcomes of a statistical experiment. In such cases, a convenient way of generating the sample space is to use an *outcome tree*. Suppose we define an experiment as the toss of two coins. The elementary outcomes are shown in Figure 4-1. The first set of branches in the tree consists of the possible outcomes for the first coin, heads (H) or tails (T). The second set of branches contains the possible outcomes for the second coin. In total, there are four possible outcomes to this experiment. The outcome tree, or tree diagram, is a useful graphical device for portraying the sample spaces of small experiments. For larger experiments, such as

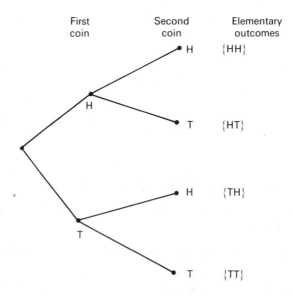

FIGURE 4-1. Outcome tree.

drawing two cards from a shuffled deck, this technique is neither practical nor helpful. In these instances, some of the counting rules discussed in Section 4.3 are especially useful.

There is some choice in how we define the elementary outcomes of an experiment. In the coin toss experiment, we could just as easily define the outcomes as (1) no heads, (2) one head, and (3) two heads. As we shall see in Section 4.2, it is easier to work with elementary outcomes which have an equal chance of occurring. Defined as {HH}, {HT}, {TH}, and {TT}, the four elementary outcomes are equally likely. When expressed as the number of heads, the probabilities of all three outcomes are not equal. There is a 1-in-4, or 25 percent, chance of getting either zero heads or two heads, but a 2-in-4, or 50 percent, chance of getting only one head.

After the elementary outcomes of an experiment have been defined, it is possible to examine collections of elementary outcomes, or *events*.

DEFINITION: EVENT
An event is a subset of the sample space of an experiment, a collection of elementary outcomes.

The particular subset used to define an experiment can include one or more elementary outcomes. Returning to the card-drawing experiment, we might define event A to be "the card drawn is a spade," event B to be "the card drawn is an ace," and event C to be "the card drawn is the ace of spades." Event A contains 13 elementary outcomes—the 13 spades in the deck; event B contains the four aces; and event C consists of only one outcome, the ace of spades. Because event C contains only one elementary outcome, the ace of spades is both an elementary outcome and an event. A *null event*, denoted $\emptyset$, is an event that contains no elementary outcomes. It cannot occur.

Set Notation and Terminology; Venn Diagrams

Another useful graphical device is the *Venn diagram*. Together with the terminology and notation of set theory, a Venn diagram can be used to portray the elementary outcomes of an experiment, define events, and illustrate the relationship between events. Let us denote the sample space of an experiment as S. The set S contains all the elementary outcomes of the experiment and can be represented as $S = \{E_1, E_2, \ldots, E_n\}$, where $E_1, E_2, \ldots, E_n$ are the n elementary outcomes of the experiment. In a Venn diagram, the sample space encloses all the elementary outcomes. As an example, consider the experiment in which a single draw is made from a deck of shuffled cards. The Venn diagram of Figure 4-2 illustrates the sample space for this problem. It contains all 52 elementary outcomes.

DEFINITION: EVENT SPACE
The event space of an experiment contains all elementary outcomes which constitute the event.

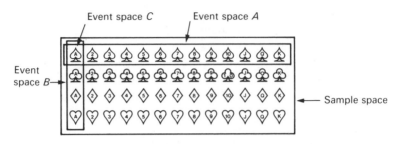

FIGURE 4-2. Venn diagram.

The three event spaces for events A, B, and C of the card selection example are illustrated in Figure 4-2.

Sometimes, we are interested in the elementary outcomes that are *not* part of some event A. In the card selection example, we might be interested in the set of elementary outcomes which are not aces.

DEFINITION: COMPLEMENTARY EVENT

The set of elementary outcomes not in an event space A constitute the complementary event to A. This complementary event is denoted $\bar{A}$.

As we shall see, the problem of determining the probability of some event A can occasionally be solved more simply by first determining the probability that event A does not occur. To do this, we would employ the complementary event $\bar{A}$.

Relationships between Events

Some questions of probability require us to examine the probability of two or more events, at least one of several events, or neither of two or more events occurring. To answer such questions, it is necessary to understand the possible ways in which two or more events can be related. Events can be compounded in two ways.

DEFINITION: EVENT INTERSECTION

The set of elementary outcomes that belong to both events A and B of a sample space is called the *intersection* of A and B and is denoted by $A \cap B$.

Figure 4-3 illustrates a Venn diagram in which two events G and H are defined over a sample space consisting of the elementary outcomes $\{E_1, E_2, \ldots, E_8\}$. Event G contains the elementary outcomes $\{E_1, E_2, E_3\}$, and event H consists of the elementary outcomes $\{E_3, E_4, E_5\}$. The intersection $G \cap H$ consists of the single elementary outcome E_3. The shaded area enclosing E_3 in Figure 4-3 is a graphical representation of $G \cap H$. As a second example, consider the two events A and B defined in the Venn diagram of Figure 4-2. The intersection $A \cap B$ contains all the elementary outcomes which are both aces and spades. The only elementary outcome meeting both criteria is the ace of spades. We can thus say that $C = A \cap B$ since event C is defined as the selection of the ace of spades.

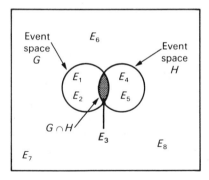

FIGURE 4-3. Venn diagram for the intersection of two events.

If the intersection of two events contains no elementary outcomes, that is, it defines a null event, then the events are said to be *mutually exclusive*.

DEFINITION: MUTUALLY EXCLUSIVE EVENTS
If events A and B are defined over a sample space and have no elementary outcomes in common, A and B are said to be mutually exclusive. In this case, $A \cap B = \emptyset$, where $\emptyset$ is the null event.

Venn diagrams for mutually exclusive and non-mutually-exclusive events are illustrated in Figure 4-4.

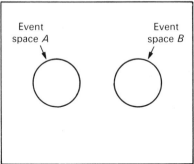

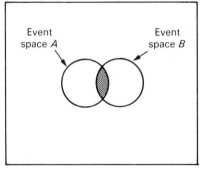

FIGURE 4-4. Venn diagrams for exclusivity.

Events A and B of the card selection experiment defined in Figure 4-2 are not mutually exclusive events since they share one outcome, the ace of spades. We have already noted that $A \cap B \neq \emptyset$. Let us now define event D as "the card drawn is a diamond." It is easily seen that $A \cap D = \emptyset$, so that these two events are mutually exclusive events. It is impossible to draw a single card and get both a spade and a diamond. Note also that $B \cap D \neq \emptyset$ and these two events are not mutually exclusive. The ace of diamonds is common to both events B and D.

What is the intersection of an event and its complement? Since an event and its complement are by definition mutually exclusive, it follows that $A \cap \bar{A} = \emptyset$.

There is a second way in which two events can be related.

DEFINITION: UNION

Let A and B be two events defined over a sample space S. The union of A and B, denoted $A \cup B$, is the set of all elementary outcomes that belong to *at least one* of events A and B.

Venn diagrams for the union of both mutually exclusive and non-mutually-exclusive events are shown in Figure 4-5. Consider once again the card selection experiment. The union $A \cup B$ contains 16 elementary outcomes—the 13 spades and the aces of diamonds, clubs, and hearts. The union $A \cup D$ contains 26 elementary outcomes—13 spades and 13 diamonds.

Non-mutually-exclusive events

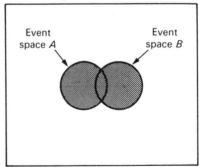

Mutually exclusive events

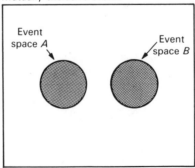

FIGURE 4-5. Venn diagram of the union of two events.

4.2. Computing Probabilities in Statistical Experiments

The outcome of a statistical experiment will be one, and only one, of the elementary outcomes of the sample space. To determine the probability that an event occurs in a single trial of an experiment, it is necessary to attach a value to each elementary outcome E_i of the experiment. This value is a *probability*— a measure of how likely it is that outcome E_i is the realized outcome of the experiment. Leaving aside for the moment the thorny question of how these values are obtained, let us first define three postulates that formalize certain properties of probabilities.

Probability Postulates

Denote the n elementary outcomes of the experiment as the set $S = \{E_1, E_2, \ldots, E_n\}$ and their assigned probability values as $P(E_1), P(E_2), \ldots, P(E_n)$.

POSTULATE 1

$$0 \le P(E_i) \le 1 \qquad \text{for } i = 1, 2, \ldots, n \tag{4-1}$$

For every elementary outcome in the sample space, the assigned probability must be a nonnegative number between 0 and 1 inclusive. If the occurrence of any outcome j is impossible, then $P(E_j) = 0$; and if outcome j is certain, then $P(E_j) = 1$.

POSTULATE 2

$$P(A) = \sum_{i \in A} P(E_i) \tag{4-2}$$

The second postulate is that the probability of any *event A* occurring is the sum of the probabilities assigned to the individual elementary outcomes which comprise event A.

POSTULATE 3

$$P(S) = 1$$
$$P(\emptyset) = 0 \tag{4-3}$$

The third postulate says that the probability value associated with the sample space must equal 1, and the probability value of a null event must be zero. Although these postulates are almost intuitive, it is possible to deduce a number of important results from them.

First, it is easily shown that the sum of the probabilities of all the elementary outcomes must equal 1:

$$\sum_{i=1}^{n} P(E_i) = 1 \tag{4-4}$$

This follows directly from Postulates 2 and 3. Since the sample space contains all the elementary outcomes, that is, $S = \{E_1, E_2, \ldots, E_n\}$, and from Postulate 2 we know that $P(S) = \sum_{i=1}^{n} P(E_i)$, it follows from Postulate 3 that $\sum_{i=1}^{n} P(E_i) = 1$.

Second, since any event A must contain a subset of the elementary outcomes of S, and since from Postulate 1 the probability of any elementary outcome occurring is between 0 and 1, it follows that

$$0 \leq P(A) \leq 1 \tag{4-5}$$

for any event A. In other words, every event has a probability value between 0 and 1. Also, if events A and B are mutually exclusive, then from Postulate 3

$$P(A \cap B) = 0 \tag{4-6}$$

This follows from the fact that the intersection of two mutually exclusive events must be the null set: $A \cap B = \emptyset$.

Definitions of Probability

The three probability postulates tell us that probability is a measure which, for any elementary outcome or event, is a number between 0 and 1. But how do we interpret this probability value? How do we determine these values in practice? Let us examine two different interpretations of probability. First, probability can be interpreted or defined by using a *subjective* method. In using a subjective method, the term *probability* refers to an individual's "degree of belief" that some event might occur. In this way we can assign probability values to events which occurred in the past, events which may occur in the future, and even unique events. For example, a doctor might suggest that a certain patient has a .5 probability of surviving another year. The problem with subjective probabilities such as this is that the value assigned to the event varies with the individual making the judgment. Another doctor might suggest that the probability of the patient surviving another year is .9. Subjective probabilities depend on the information available to the individual and the manner in which she or he evaluates that information. Since each person can arrive at his or her own answer for the probability value for an event, many parts of probability theory cannot utilize subjective probabilities. In some instances, however, we may be forced to rely on probability values estimated in this way.

In *objective* interpretations of probability, the probability value of an event is determined by examining the relative frequency of occurrence of the event in a repeated statistical experiment. Suppose, for example, we wish to determine the probability of a head occurring in the toss of a "fair" coin. An experiment is designed in which a coin is tossed a large number of times and the result recorded. If the coin is fair, we would expect the proportion of heads to be about $\frac{1}{2}$ or .5. The larger the number of trials, the closer this observed proportion should be to .5. This objective interpretation of the term *probability* can be formalized in the following definition of relative frequency:

DEFINITION: RELATIVE FREQUENCY INTERPRETATION OF PROBABILITY
Let event A be defined in some experiment. If the experiment is repeated N times and event A occurs in n of these trials, then $P(A) = n/N$. The relative frequency definition $P(A) = n/N$ defines the true probability of event A occurring in the limit, that is, when the number of trials approaches infinity.

Since experiments cannot be conducted indefinitely, the observed relative frequency in an extended set of trials can be used as an estimate of the probability of an event occurring.

Another objective method of determining probabilities uses deductive logic to assess the probabilities, usually based on arguments of symmetry or geometry. For example, the probability of getting a head in the toss of a fair coin might be deduced to be $\frac{1}{2}$ on the basis of the supposed symmetry of the coin. Similarly, the probability of drawing any single card in one draw from a shuffled deck of cards must logically be $\frac{1}{52}$. Whether based on the relative frequency or on deductive logic, the probabilities arrived at are objective. Because each person utilizing the approach would come to the same answer, these objective methods are fundamentally different from subjective interpretations of probability.

Computing Probabilities from Sample Spaces

Once the probability values are determined for the elementary outcomes of an experiment by an objective method, it is possible to compute the probability of occurrence of any event defined in the sample space. Very simply, the probability of any event occurring is the sum of the probabilities of the elementary outcomes that comprise the event. For example, once we have assigned the probability of $\frac{1}{52}$ to each elementary outcome of the card selection experiment, we can compute the probability of getting an ace as the sum of the probabilities of the four elementary outcomes: $P(\text{ace of spades}) + P(\text{ace of diamonds}) + P(\text{ace of clubs}) + P(\text{ace of hearts})$. This is equal to $\frac{1}{52} + \frac{1}{52} + \frac{1}{52} + \frac{1}{52} = \frac{4}{52} = \frac{1}{13}$.

Whenever all the elementary outcomes of an experiment are equally likely, that is, have the same probability value, the probability of an event occurring can be more easily determined. Suppose there are n equally likely elementary outcomes in the experiment and event A occurs if any one of m of these elementary outcomes occurs. Then

$$P(A) = \frac{m}{n} \qquad (4\text{-}7)$$

For example, the sample space S for the card selection experiment consists of the 52 elementary outcomes illustrated in Figure 4-2. The event space for event A, the selection of a spade, contains $m = 13$ elementary outcomes. Since $m = 13$ and $n = 52$, the probability of drawing a spade is $\frac{13}{52}$, or $\frac{1}{4}$. Similarly, the probability of drawing an ace is $\frac{4}{52} = \frac{1}{13}$ since $m = 4$ and $n = 52$.

If Equation (4-7) cannot be used to calculate event probabilities, in many situations the determination of the event probability can be particularly time-consuming. First, it is necessary to identify all elementary outcomes in the sample space. Then, the subset of the sample space which comprises the event space must be specified. Finally, the addition of the probabilities of the elementary outcomes in the event space yields the event probability. The simplification of this operation by using Equation (4-7) is obvious, particularly when there may be thousands or even millions of elementary outcomes in the sample and/or event spaces. In fact, it may even be difficult to generate a list of the elementary outcomes in a sample space. In such cases, *counting rules* can be used to compute both m and n.

4.3. Counting Rules for Computing Probabilities

To utilize Equation (4-7) to compute event probabilities, it sometimes proves useful to employ one or more counting rules. These counting rules are used to quickly enumerate the number of elementary outcomes in a sample or event space, without specifically listing each one. Let us consider four of the most commonly used counting rules.

Product Rule

RULE 1: PRODUCT RULE

Suppose there are r groups of objects. In group 1 there are n_1 objects, in group 2 there are n_2 objects, and so on. If we define the experiment to be the selection of one object from each of the r groups, then there are $n_1 n_2 \cdots n_r$ elementary outcomes in the experiment.

To illustrate the use of this counting rule, consider the following simple example.

EXAMPLE 4-1. A planner is interested in surveying retail patronage at three types of shoe stores: chain stores, discount stores, and specialty stores. He decides to sample one store of each type from the outlets in the city. If there are 10 chain stores, 7 discount stores, and 3 specialty stores in the city, from how many different possible combinations of stores can the planner choose?

There are $r = 3$ groups of objects with $n_1 = 10$, $n_2 = 7$, and $n_3 = 3$. If one store must be selected from each category, then there are $(10)(7)(3) = 210$ possible combinations of stores from which the planner can choose.

Combinations Rule

The second and third counting rules are concerned with a slightly different form of experiment. The experiment consists of selecting a subset of r objects from a set containing n objects, where $r \leq n$.

EXAMPLE 4-2. Suppose a city is considering adding two new libraries to its existing system. To minimize expenses, the libraries will be constructed on two of the five vacant properties currently owned by the city. How many different locational plans for the system of libraries are there?

Label the five sites available for construction as A, B, C, D, and E. The possible outcomes of the experiment are

$$
\begin{array}{cccc}
AB & BC & CD & DE \\
AC & BD & CE & \\
AD & BE & & \\
AE & & & \\
\end{array}
$$

Note that each of the possible plans contains two different sites. These are termed *combinations*.

DEFINITION: COMBINATION

A combination is a set of r distinguishable objects, irrespective of order.

The counting rule for combinations is concerned with the number of different combinations of size r that can be drawn from a master set of n objects.

RULE 2: COMBINATIONS RULE

The number of combinations of r objects taken from n different objects is

$$
C_r^n = \frac{n!}{r!(n-r)!} \tag{4-8}
$$

Sometimes C_r^n is written as $C(n, r)$ or $\binom{n}{r}$. The symbol $n!$ is read "n factorial" and is defined as $n! = n \cdot (n-1) \cdot (n-2) \cdot \ \cdots \ \cdot 2 \cdot 1$. For example, $5! = 5 \cdot 4 \cdot 3 \cdot 2 \cdot 1 = 120$ and $2! = 2 \cdot 1 = 2$. By definition the term $0!$ is set equal to 1.

For the library example, $n = 5$ and $r = 2$; therefore

$$
C_2^5 = \frac{5!}{2!(5-2)!} = \frac{5 \cdot 4 \cdot 3!}{2 \cdot 1 \cdot 3!} = \frac{20}{2} = 10
$$

There are 10 combinations listed above. As you can imagine, the combinations rule can save a great deal of time in enumerating the elementary outcomes of an experiment.

Permutations Rule

Let us now consider a slightly different problem. Suppose the city has five sites for two new libraries, one which will be the main library and the other which will house special collections. The sites are labeled A to E, as before. Now, consider the outcome AB. This is interpreted as meaning site A is to be used for the main library and site B for special collections. Notice that outcome AB is not the same as outcome BA, since the two libraries are placed on different sites. In BA the main library is on site B, while in AB the main library is on site A. In this instance, the *order* of the elements in the set is important.

DEFINITION: PERMUTATION
A permutation is a distinct ordering of r objects.

The counting rule for permutations defines the number of different ordered arrangements of size r that can be drawn from a set of n objects.

RULE 3: PERMUTATIONS RULE
The number of permutations of r objects taken from n different objects is

$$P_r^n = \frac{n!}{(n-r)!} \tag{4-9}$$

Sometimes P_r^n is written as $P(n, r)$. In the library example, $n = 5$ and $r = 2$, so that

$$P_2^5 = \frac{5!}{(5-2)!} = 20$$

That is, there are twice as many permutations as combinations in a problem of this size. For each *combination AB* there is the reverse arrangement BA. This is not a general result. The formal relation between the number of combinations and the number of permutations is given by

---

$$C_r^n = \frac{1}{r!} P_r^n \tag{4-10}$$

Whenever $r = 2$, there are twice as many permutations as combinations.

Hypergeometric Rule

The final rule combines the combinations rule with the product rule.

RULE 4: HYPERGEOMETRIC RULE

A set of n objects consists of n_1 of type 1 and n_2 of type 2, where $n_1 + n_2 = n$. From this group of n objects, define an experiment in which r_1 of the first type and r_2 of the second type are to be chosen. The number of different groups that can be drawn is

$$C_{r_1}^{n_1} \cdot C_{r_2}^{n_2} \qquad (4\text{-}11)$$

As an example, consider the following counting problem. A survey is to be undertaken of neighborhoods in a metropolitan area. There are 10 suburban neighborhoods and 8 inner-city neighborhoods. A decision is made to survey three suburban and two inner-city neighborhoods. How many combinations of neighborhoods are possible for the survey? Clearly, the hypergeometric rule applies. There are two groups with $n_1 = 10$ and $n_2 = 8$. From these two groups $r_1 = 3$ and $r_2 = 2$ objects are to be drawn. From Equation (4-11), the number of combinations of neighborhoods that could be drawn is

$$C_3^{10} C_2^8 = \frac{10!}{3!(10-3)!} \frac{8!}{2!(8-2)!} = 120(28) = 3360$$

4.4. Basic Probability Theorems

Several useful theorems follow directly from the probability postulates outlined in Section 4.2 and the event relationships described in Section 4.1. Together with the counting rules presented in the previous section, these theorems can be used to determine event probabilities for quite complicated problems.

Addition Theorem

The addition theorem is used to obtain the probability that at least one of two events occurs.

ADDITION THEOREM

Let A and B be any two events defined over the sample space S. Then

$$P(A \cup B) = P(A) + P(B) - P(A \cap B) \qquad (4\text{-}12)$$

where $A \cup B$ represents the union and $A \cap B$ represents the intersection of events A and B.

Let us consider the problem of determining the probability of drawing a spade or an ace in a single draw from a deck of shuffled cards. As before, let event A be the selection of a spade and event B be the selection of an ace. For these two events we know

$$P(A) = \tfrac{13}{52} \qquad P(B) = \tfrac{4}{52} \qquad P(A \cup B) = \tfrac{1}{52}$$

Event A contains 13 outcomes; B contains 4 outcomes; and 1 outcome, the ace of spades, is common to both events A and B. Substituting these known values into Equation (4-12) yields $P(A \cup B) = \tfrac{13}{52} + \tfrac{4}{52} - \tfrac{1}{52} = \tfrac{16}{52} = .308$. It should be clear why $P(A \cap B)$ must be subtracted from $P(A) + P(B)$ to calculate $P(A \cup B)$. Since the elementary outcome the ace of spades is common to both events A and B, that outcome is counted twice when $P(A)$ and $P(B)$ are added. To avoid double counting, $P(A \cap B)$ is subtracted. This can be easily appreciated by closely examining Figure 4-2.

The addition theorem can be simplified when the two events A and B are known to be mutually exclusive. We know that $P(A \cap B) = 0$ for mutually exclusive events [from Equation (4-6)] so that the addition theorem reduces to

$$P(A \cup B) = P(A) + P(B) \qquad (4\text{-}13)$$

For events A and D of the card selection problem, there are no elementary outcomes in common, so the two events must be mutually exclusive. A spade and a diamond cannot be simultaneously drawn. Therefore, $P(A \cup D) = P(A) + P(D) = \tfrac{13}{52} + \tfrac{13}{52} = \tfrac{26}{52} = .50$. There is a 50 percent chance of drawing either a spade or a diamond.

The addition theorem for mutually exclusive events can be generalized for any number of events. For three events A, B, and C the addition rule is written as

$$P(A \cup B \cup C) = P(A) + P(B) + P(C) \qquad (4\text{-}14)$$

Generalizing the addition rule for more than two non-mutually-exclusive events is much more complicated.

Complementation Theorem

It is sometimes more difficult to find the probability of event A occurring than to find the probability of the complementary event $\bar{A}$ occurring. By using the complementation theorem, it is possible to derive $P(A)$ from $P(\bar{A})$.

COMPLEMENTATION THEOREM

For any event A and its complement $\bar{A}$ of a sample space S,

$$P(A) = 1 - P(\bar{A})$$

or $\qquad\qquad\qquad\qquad\qquad\qquad\qquad\qquad\qquad\qquad\qquad\qquad\qquad (4\text{-}15)$

$$P(\bar{A}) = 1 - P(A)$$

The complementation theorem follows directly from a few simple arguments. First, by definition, that $A \cup \bar{A} = S$. That is, the two events A and $\bar{A}$ exhaust the sample space. Therefore, $P(S) = P(A \cup \bar{A})$, and in turn $P(A) + P(\bar{A}) = 1$, by the addition rule for mutually exclusive events. The complementation theorem is a simple rearrangement of this last equation.

Let event E be defined as the selection of any card but a club in a single draw from a deck of cards. Event $\bar{E}$ is thus the selection of a club in a single draw. From (4-15), $P(E) = 1 - P(\bar{E}) = 1 - \frac{13}{52} = \frac{39}{52} = .75$.

Conditional Probabilities and Statistical Independence

To find the joint probability of two events occurring (the probability that both occur), it is necessary to introduce the notion of a *conditional probability*. A conditional probability is the probability of an event A occurring given that a second event B has occurred.

DEFINITION: CONDITIONAL PROBABILITY

If events A and B are defined over a sample space S, then the conditional probability of event A, given B, denoted $P(A \mid B)$, is defined as

$$P(A \mid B) = \frac{P(A \cap B)}{P(B)} \qquad (4\text{-}16)$$

The vertical line separating A and B in $P(A \mid B)$ denotes a conditional probability. The term $P(A \mid B)$ is read "the probability of A, given B."

To illustrate the concept of a conditional probability, let us consider a slightly different experiment. Suppose the experiment is to roll a perfectly balanced single die and record the number of dots appearing on the upward face. The sample space S for this experiment consists of six elementary outcomes. Define event A as "the upward face contains an odd number of dots" and event B as "the upward face contains less than or equal to three dots." These two events are illustrated within the Venn diagram of Figure 4-6. What is $P(A \mid B)$? It is the probability of the upward face containing an odd number of dots given that it has three or less dots. From Figure 4-6 we can see

$$P(A) = \tfrac{3}{6} = \tfrac{1}{2}$$

$$P(B) = \tfrac{3}{6} = \tfrac{1}{3}$$

$$P(A \cap B) = \tfrac{2}{6} = \tfrac{1}{3}$$

and from Equation (4-16), $P(A \mid B) = \frac{1}{3} / \frac{1}{2} = \frac{2}{3}$. The probability that the upward face of a die contains an odd number of dots is $\frac{2}{3}$, given that it must be less than 3. This result can also be directly inferred from the Venn diagram of Figure 4-6. If event A occurs, then one of the elementary outcomes in the left-hand column must be the result. Of these three elementary outcomes, two have less than or equal to three dots. Therefore, $P(A \mid B) = \frac{2}{3}$. Note that, in this case, $P(A) \neq P(A \mid B)$.

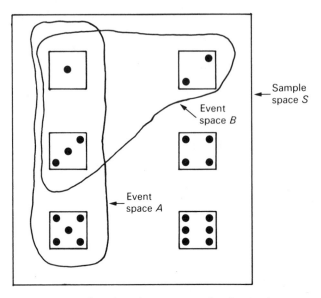

FIGURE 4-6. Sample and event spaces for die experiment.

Two events are *statistically independent* if the probability that one event occurs is unaffected by whether the other event occurs. This is formalized in the following definition.

DEFINITION: STATISTICALLY INDEPENDENT EVENTS
Two events A and B defined over a sample space S are said to be statistically independent if and only if

$$P(A \mid B) = P(A)$$

or equivalently,   (4-17)

$$P(B \mid A) = P(B)$$

If $P(A \mid B) \neq P(A)$, then events A and B are said to be dependent. For the two events A and B defined for the single roll of a die, we have already shown that $P(A \mid B) = \frac{2}{3}$ and $P(A) = \frac{1}{2}$, so that these two events are statistically dependent. That is, knowing that the die roll is less than or equal to 3 affects the probability that it is odd. For events A and B defined in the card selection experiment, $P(A) = \frac{13}{52}$, $P(B) = \frac{4}{52}$, and $P(A \cap B) = \frac{1}{52}$. From (4-16) it can be shown that $P(A \mid B) = \frac{1}{4} = \frac{13}{52} = P(A)$, so that these two events are independent. This implies that knowing that the outcome of the card experiment is an ace does not affect the probability of the card drawn being a spade.

Multiplication Theorem

The multiplication theorem is used to find the *joint* probability that two events occur. The multiplication theorem can be derived by rearranging Equation (4-16), the definition of a conditional probability.

MULTIPLICATION THEOREM

For any two events A and B defined over a sample space S,

$$P(A \cap B) = P(A \mid B) \cdot P(B)$$

or equivalently,   (4-18)

$$P(A \cap B) = P(B \mid A) \cdot P(A)$$

Here $P(A \cup B)$ is the probability that both events A and B occur. For events A and B of the card selection experiment,

$$P(A \cap B) = P(A \mid B) \cdot P(B) = \tfrac{1}{4}(\tfrac{1}{13}) = \tfrac{1}{52}$$

The probability of getting both an ace and a spade is $\tfrac{1}{52}$. The single elementary outcome satisfying both event definitions is the ace of spades. This can also be verified by examining the Venn diagram for this experiment, shown as Figure 4-2. For the two events of the experiment calling for the roll of a single die, $P(A \cap B) = P(A \mid B) P(B) = \tfrac{2}{3}(\tfrac{1}{2}) = \tfrac{1}{3}$. Two of the six elementary outcomes in this experiment both are odd and have less than three dots.

For independent events, the multiplication theorem can be simplified. Since $P(A \mid B) = P(A)$ for statistically independent events, Equation (4-18) can be reduced to

$$P(A \cap B) = P(A) \cdot P(B) \qquad\qquad (4\text{-}19)$$

The multiplication theorem for independent events, expressed as Equation (4-19), can be generalized to include n independent events. For three events, the joint probability is given by

$$P(A \cap B \cap C) = P(A) \cdot P(B) \cdot P(C) \qquad\qquad (4\text{-}20)$$

4.5. Illustrative Probability Problems

To solve probability problems, we frequently must resort to the event space–sample space rule (4-7), the counting rules presented in Section 4.3, and the compounding rules presented in Section 4.4. Sometimes two different methods can be used to solve the same problem. The following problems illustrate the use of these techniques in answering probability questions.

EXAMPLE 4-3. A local instrument supply firm received a shipment of 40 net pyradiometers and sold two to the local university for use in its climatology laboratories. A month after the instruments arrived at the university, the supply firm received a letter from the manufacturer stating that 8 of the 40 instruments sent to the local supplier were incorrectly calibrated. What is the probability that the university purchased two incorrectly calibrated instruments?

Solution. Let event A be defined as "both pyradiometers purchased by the university are incorrectly calibrated." It is possible to rewrite event A as the compound event $A_1 \cap A_2$, where event A_1 is "the first pyradiometer is incorrectly calibrated" and event A_2 is "the second pyradiometer is incorrectly calibrated." Thus $P(A)$ can be determined in two ways. First, $P(A)$ can be computed by using Equation (4-7) and determining the number of elementary outcomes in the sample space, n, and the event space, m, from counting rules. The number of elementary outcomes in the sample space is simply the number of ways of selecting two pyradiometers from the shipment of 40:

$$n = C_2^{40} = \frac{40!}{2!(40-2)!} = \frac{40 \cdot 39}{2} = 780$$

The set of pyradiometers from which the two purchased by the university were chosen can be divided into two types—correctly calibrated and incorrectly calibrated instruments. We can therefore use the hypergeometric rule to determine the number of ways of getting two incorrectly calibrated instruments in the university order:

$$m = C_2^8 C_0^{32} = \frac{8!}{2!(8-2)!} \frac{32!}{0!(32-0)!}$$

$$= \frac{8(7)}{2}(1) = 28$$

Here C_2^8 is the number of ways of drawing 2 defective instruments from the 8 instruments known to be incorrectly calibrated. This is multiplied by C_0^{32}, the number of ways of drawing 0 defective pyradiometers from the remaining 32. Therefore, $P(A) = m/n = \frac{28}{780} = .036$.

The second way of finding $P(A)$ is to use the multiplication rule on the compound event $A_1 \cap A_2$: $P(A_1 \cap A_2) = P(A_1) \cdot P(A_2 | A_1)$. Notice that these two events are not independent, and so the conditional probability must be used. If the first delivered instrument is not correctly calibrated, then this fact affects the probability that the second instrument is not correctly calibrated (since there will be one less instrument to choose from, and one less defective instrument). The required probability is therefore

$$P(A) = P(A_1) \cdot P(A_2 | A_1) = (\tfrac{8}{40})(\tfrac{7}{39}) = .036$$

Note that the probability of selecting the second incorrectly calibrated instrument is different from the first. Both this method and the method based on the event space–sample ratio give the same results. Thus there is often more than one way to approach most probability problems.

EXAMPLE 4-4. In a survey of landscape preferences, respondents were asked to express a preference for one of two landscape photographs. The sex of the respondent was recorded at the time of the interview. The results of the survey are summarized as follows:

	Prefer Landscape 1	Prefer Landscape 2	Total
Male	15	5	20
Female	25	5	30
Total	40	10	50

For example, it was found that 20 respondents were male, 40 preferred landscape photograph 1, and 25 were female *and* preferred landscape photograph 1. This table provides a convenient format for counting the elementary outcomes of an experiment in a sample space. The experiment consists of selecting one of the respondents from the survey at random and recording the sex and landscape preference.

Let event A be that the sex of the chosen respondent is male, and let event $\bar{A}$ be that the sex is female. Similarly, let event B be that the chosen subject expresses a preference for landscape 1, and event $\bar{B}$ is a preference for landscape 2. Determine $P(A)$, $P(B)$, $P(A \cap B)$, and $P(B \mid A)$.

Solution. There are 50 elementary outcomes in the experiment—the 50 individuals whose preferences and sex are cross-classified in the table. To find the event probabilities defined by using Equation (4-7), it is necessary to calculate the number of elementary outcomes in the appropriate event space, m. First, $P(A) = \frac{20}{50} = .4$ since 20 of the respondents are male. By the complementation theorem, $P(\bar{A})$ is calculated as $1 - .4 = .6$. Similarly, $P(B) = \frac{40}{50} = .8$, and $P(\bar{B}) = 1 - .8 = .2$.

The probability of the compound event $A \cap B$ refers to the probability that a randomly selected respondent will be a male *and* prefer landscape photograph 1. The number of elementary outcomes in this event space is $m = 15$, so $P(A \cap B) = \frac{15}{50} = .3$. The conditional probability $P(B \mid A)$ refers to the probability that the selected respondent prefers landscape 1, given that the respondent is a male. All the information necessary to compute this probability is contained in the first row of the table. There are 20 males, 15 of whom prefer landscape photograph 1. Therefore $P(B \mid A) = \frac{15}{20} = .75$. This conditional probability could also have been computed by using the definitional formula

$$P(B \mid A) = \frac{P(B \cap A)}{P(A)} = \frac{.3}{.4} = .75$$

Again, two approaches to solving a probability problem lead to an identical result. Inferential questions about the relationship between two variables cross-tabulated in this way are discussed extensively in Chapter 11.

EXAMPLE 4-5. The proportion of people living in various regions of a country and the proportion in each region owning their homes are given in the following table:

Region	Proportion of population	Proportion owning homes
East	.44	.20
West	.19	.40
North	.28	.80
South	.09	.60

What is the probability that a person randomly selected from the population will own his or her home?

Solution. Let $P(E)$, $P(W)$, $P(N)$, and $P(S)$ be the probability that the person selected at random is a resident of the east, west, north, and south, respectively. These values can be read directly from the table as .44, .19, .28, and .09, respectively. The probability that the randomly selected person lives in the east and owns her or his home is $P(E)P(0|E)$, where $P(0|E)$ is the probability the person owns a home given that he or she lives in the east. Thus $P(E)P(0|E) = .44(.20) = .088$. Similarly, in the west, the required probability is $P(W)P(0|W) = .19(.40) = .076$. In the north the required probability is $.28(.80) = .224$, and in the south it is $.09(.60) = .054$. The probability that the randomly selected individual owns a home is thus $.088 + .076 + .224 + .054 = .442$. These four probabilities can be added, since residence in each of the four regions is mutually exclusive.

4.6. Summary

When it is used in statistics, the term *probability* is a measure of the likelihood of an event occurring. It is a measure having a value between 0 and 1. The basic concepts of probability are founded on the notions of a statistical experiment. To compute event probabilities, it is usual to determine the elementary outcomes of the experiment. These elementary outcomes should constitute the set of mutually exclusive and collectively exhaustive outcomes of the experiment. Together, these outcomes represent the sample space of the experiment. If all elementary outcomes of the experiment are equally likely, the probability that an event occurs in the experiment can be determined by taking the ratio of the number of outcomes in the event space to that of the sample space. The enumeration of the elementary outcomes in either the event or the sample space is often made easier by the application of one or more counting rules.

The identification of the probabilities associated with complex events can be determined once the relationship between the individual events which constitute the complex or compound event are known. Whether events are mutually exclusive or are independent are the two most important considerations in deciding which probability laws should be applied to solve a particular problem. Frequently, there are several different ways to approach any given probability question. In the next chapter, these basic concepts of probability are related to the procedures of statistical inference.

PROBLEMS

1. Explain the meaning of the following terms:
   a. Statistical experiment or random trial
   b. Sample space
   c. Elementary outcome of an experiment
   d. Event
   e. Complementary event
   f. Event space
   g. Event intersection and event union
   h. Relative frequency definition of probability
   i. Conditional probability

2. Differentiate between an *objective* and a *subjective* interpretation of the concept of probability. Give an example of each. Why are objective interpretations preferred in statistical analysis?

3. Draw a Venn diagram and illustrate the event space for the following problems:
   a. The experiment is the selection of a single card from a deck of cards; event *A* is the selection of a jack, event *B* is the selection of a jack *or* a queen, and event *C* is the selection of the queen of hearts or king of diamonds.
   b. The experiment is defined as the roll of a single die with the outcome being the number of dots showing on the upward face; event *A* is the roll yields a 1 or a 2, event *B* is the roll yields an even number of dots, and event *C* is the roll yields an odd number of dots.

4. Differentiate between events which are (a) mutually exclusive and those which are not and (b) statistically independent and those which are not.

5. Residents of a city are asked to rank the desirability of five different neighborhoods *A*, *B*, *C*, *D*, and *E*. How many different ways can they be ranked, assuming there can be no ties?

6. A traveling salesperson beginning a trip at city *A* must visit (in order) cities *X*, *Y*, and *Z* before returning to city *A*. Several highway routes connect each pair of cities. There are four different ways of traveling between cities *A* and *X*, three routes between *X* and *Y*, five between *Y* and *Z*, and two between *Z* and *A*. By how many different routes can the salesperson complete the entire trip?

7. Construct an outcome tree outlining the elementary outcomes for three tosses of a coin. What is the probability that three consecutive heads occur, assuming the coin is fair?

8. The probability that a student selected at random from an introductory geography course at a certain university is male is .60. The probability that the student is unmarried is .80, and the probability that the student is both married and male is .05. If one student in the class is selected at random, calculate the probability that the student is (a) female, (b) married, (c) male or unmarried, and (d) female and unmarried. *Hint*: Set up an appropriate contingency table.

9. The license plates in a certain jurisdiction have six alphanumeric digits. The first digit is A, B, or C. The second digit is N, W, E, or S. The last four digits are restricted to the values 0, 1, . . . , 9. How many possible license plates are there?

10. How many license plates are possible in a six-digit license plate system if (a) all six digits are letters, (b) all six digits are numbers, and (c) there are three letters and three numbers?

11. Given $P(A) = \frac{1}{3}$, $P(B) = \frac{2}{3}$ and $P(A \cap B) = \frac{1}{3}$, (a) find $P(A \cup B)$, (b) find $P(B|A)$, and (c) draw a Venn diagram of the two events. (d) Are A and B independent?

12. Draw Venn diagrams for the following sets of events, and develop the addition rule [for example, Equations (4-12) and (4-14)] for the problem.
    a. Of the three events A, B, and C, only A and C are not mutually exclusive.
    b. Of the three events A, B, and C, only B and C are not mutually exclusive.
    c. All events A, B, and C are not mutually exclusive when they are taken together as a group of 3 or in any pair.

13. The probability that it rains on a given day in July is .10.
    a. Assuming independent trials, what is the probability that it does not rain for 3 days in a row?
    b. Again assuming independent trials, what is the probability that one day of rain is followed by 2 nonrainy days?
    c. Is the assumption of independent trials in this experiment reasonable?

14. A retail geographer surveys 200 shoppers after each has visited one of three shopping centers A_1, A_2, and A_3. He records whether each made a purchase, with $B_1 = $ yes and $B_2 = $ no. The survey results are as follows:

Center	B_1	B_2	Total
A_1	25	25	50
A_2	10	50	60
A_3	65	25	90
Total	100	100	200

    a. Find $P(A_3)$.
    b. Find $P(A_1 \cap B_2)$.
    c. Find $P(A_2 \cap B_1)$.
    d. Find $P[(A_1 \cup A_3)|B_2]$.
    e. Explain in your own words the meanings of parts (a) to (d).

15. Reg, Les, Emilio, and Gunnar take the bus together to work 75 percent of the time. The bus is on schedule 80 percent of the time, and they arrive at work on time. The remaining 25 percent of the time, Reg, Les, Emilio, and Gunnar take a car to work. When they take a car to work, 90 percent of the time they arrive on time. This morning, Reg, Les, Emilio, and Gunnar arrived at work on time. What is the probability they came by bus?

16. A certain geography journal contains one map for every 100 pages of text. Randomly selecting one issue of this journal, I examine the first six pages to see whether they include a map.
    a. What is the probability that there is at least one map in these six pages?
    b. What is the probability that exactly one map is included in these six pages?

17. Given events A and B where $P(A) = .20$ and $P(B) = .40$ and $P(B|A) = .50$, find (a) $P(A|B)$ and (b) $P(A \cup B)$.

18. A city planner is interested in determining the occupancy rates of all the rooming houses in some city. There are exactly 1000 rooming houses in the city, so she cannot possibly visit them all. She decides to estimate the occupancy rate by visiting 10 of these rooming houses. To be absolutely fair, she decides to select them randomly. She writes the address of each rooming house on a small piece of paper and puts them into a large jar. Blindfolded, she then reaches into the jar and picks 10 pieces of paper.

   a. What is the probability that any particular rooming house is selected for the sample? Assume that each time a rooming house is selected, it is not replaced into the jar.

   b. Suppose that, after each paper is selected from the jar, the address of the rooming house is noted and the paper is replaced into the jar. What is the probability that a particular rooming house is selected for the sample if it is drawn in this way?

   c. Given the sampling conditions of part (b), what is the probability that a rooming house is selected twice for the sample?

5

Random Variables and Probability Distributions

In Chapter 4, the basic concepts of probability were developed in considerable detail. Although most people are familiar with notions of probability in games of chance, calling heads or tails in a coin toss, and winning or losing a bet, the fundamental concepts of probability theory are frequently quite perplexing and difficult to comprehend on first exposure. Definitions of statistical experiments, statistical independence, and mutual exclusivity seem far removed from the context of these games of chance. Yet we know that probability theory lies at the core of statistical inference since all statistical judgments are necessarily probabilistic. The purpose of this chapter is to help clarify the link between statistical judgments, on the one hand, and probability theory, on the other.

Central to these arguments is the notion of a *random variable*. In Section 5.1 a random variable is formally defined and related by example to the process of population sampling. We also differentiate between the two major classes of random variables—those which are discrete and those which are continuous. Sections 5.2 and 5.3 are devoted to the specification of probability distribution models for these two classes of random variables. In Section 5.4, we extend our knowledge of random variables and probability functions to include bivariate random variables and bivariate probability functions. The various models are summarized and compared in Section 5.5.

5.1. Concept of a Random Variable

Suppose we consider any measurable characteristic of the units which comprise some typical population. Since this characteristic can assume one or more values in the population, we normally refer to it as a variable. To appreciate the concept of a *random variable*, let us consider a simple example. Assume that a population consists of eight different households and that the variable under consideration is the size, or total number, of individuals in the household. If we select one household *at random* from this population and note its size, then we could consider the variable called size of household to be a random variable. Thus, in Table 5-1, we treat the eight individual households as elementary outcomes from a random sampling experiment. The

TABLE 5-1
The Concept of a Random Variable

Sample space of elementary outcomes		Values of the random variable
Household 1: 3 members		$X(H_2) = 2$
Household 2: 2 members		$X(H_1)$
Household 3: 5 members	**Function**	$X(H_8)$ $\Big\} = 3$
Household 4: 5 members	$X = \{\text{size of household } H_i\}$	$X(H_5)$
Household 5: 4 members		$X(H_6)$ $\Big\} = 4$
Household 6: 4 members		$X(H_7)$
Household 7: 4 members		$X(H_3)$ $\Big\} = 5$
Household 8: 3 members		$X(H_4)$

values of the random variable called size of household are limited to 2, 3, 4, and 5. The sampling experiment defines one and only one numerical value for each household. If the first household is selected, then the value of the random variable is 3; if the second household is chosen, it is 2; and so on. The variable is random not because a household makes a random decision to include so many individuals, but because our sampling experiment selects a household randomly.

Let us define the random variable as X and denote any particular value that the variable may take as x. Then X is a random variable *because it is a numerically valued function that is defined over a sample space.* A random variable always has numerical values and probabilities associated with these values. The term *probability distribution*, or *probability function*, refers to a listing or graph of all values of a random variable and their corresponding probabilities. Probability distributions can be presented in three different ways: (1) in tabular form, such as Table 5-2; (2) in a graph, such as Figure 5-1; and (3) by using a mathematical formula, such as Equation (5-7).

The probabilities of Table 5-2 are relative frequency probabilities of the random variable called size of household for the population described in Table 5-1. The graph in Figure 5-1 is constructed from the same data. Henceforth, we use the notation $P(X)$ to denote the probability distribution of random variable X and

TABLE 5-2
Relative Frequency Probabilities for the Random Variable Called Household Size of Table 5-1

x_i	$P(x_i)$ or $P(X = x_i)$
2	$\frac{1}{8} = .125$
3	$\frac{2}{8} = .250$
4	$\frac{3}{8} = .375$
5	$\frac{2}{8} = .250$

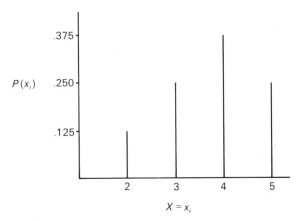

FIGURE 5-1. A probability mass function.

$P(X = x)$ or $P(x)$ to denote the probability that the random variable X takes on any particular value x. For example, $P(2)$ is the probability that X equals 2.

It is now important to distinguish between two types of random variables—those that are discrete and those that are continuous.

Discrete Random Variables

A random variable is said to be discrete if it can assume a finite number of values. Table 5-3 lists several different examples of discrete random variables. Size of household, sex, and many rating scales can all be considered discrete random variables if the particular value taken by the random variable is governed by a random sampling experiment. The outcome of a coin toss experiment, the roll of a six-sided die, and the

TABLE 5-3
Examples of Discrete Random Variables

Random variable X	Values of random variable
From random sampling	
Size of household	1, 2, 3, ...
Sex	0, 1; or 1, 2; or any two integers
Rating scale (strongly agree, agree, neutral, disagree, strongly disagree)	1, 2, 3, 4, 5 or $-2, -1, 0, 1, 2$
From simple experiments	
Number of heads in N tosses of a coin	1, 2, 3, ..., N
Number of dots on upward face of a die after a single roll	1, 2, 3, 4, 5, 6

selection of a card from a shuffled deck are also discrete random variables since they result from statistical experiments.

Note that in many instances the discrete outcomes of a random variable are qualitative. To be random variables, each must be assigned numerical values since a random variable is by definition a numerically valued function. For example, Table 5-3 indicates that we assign the numerical values male $= 0$ and female $= 1$ or male $= 1$ and female $= 2$ to the random variable called sex. We must make similar numerical assignments for rating scales.

The probability distribution of a discrete random variable is specified by a *probability mass function*. This function assigns probabilities to the values taken by the discrete random variable. Formally, we say that a probability mass function assigns probabilities to the k discrete values of a random variable X with the following two provisions:

$$0 \le P(x_i) \le 1 \qquad i = 1, 2, \ldots, k$$

$$\sum_i P(x_i) = 1$$

where x_i, $i = 1, 2, \ldots, k$, are the k different values of the discrete random variable. The first condition simply requires that all probabilities be nonnegative and less than 1. The second condition requires the sum of all probabilities be equal to 1. The probability mass function for the variable called household size illustrated in Figure 5-1 and Table 5-2 satisfies both conditions. Why is such a function called a probability mass function? It is termed a *mass function* because the probabilities are massed at discrete values of the random variable.

We can also portray probability mass functions in cumulative form in much the same way as frequency tables and histograms could be cumulated. For a discrete random variable with values ordered from lowest to highest as $x_1 < x_2 < x_3$, the cumulative mass function $F(x_i) = P(X \le x_i)$ is determined by summing the probabilities that $X = x_1$, $X = x_2, \ldots, X = x_i$. That is,

$$F(x_i) = \sum_{x = x_1}^{x = x_i} P(x) = P(x_1) + P(x_2) + \cdots + P(x_i) \qquad (5\text{-}1)$$

Consider again the discrete random variable called household size in Table 5-2. To cumulate these values, note $F(2) = P(X = 2) = .125$, $F(3) = P(X = 2) + P(X = 3) = .125 + .250 = .375$, and so on. The complete cumulative mass function is listed in Table 5-4 and illustrated in Figure 5-2.

TABLE 5-4
Cumulative Mass Function

x	$P(x_i)$	$P(X \le x_i) = F(x_i)$
2	.125	.125
3	.250	.375
4	.375	.750
5	.250	1.000

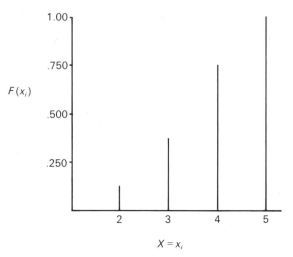

FIGURE 5-2. A cumulative probability mass function.

Just as an empirically derived variable has a mean value and variance, so does a random variable. The average of the values of a random variable is known as the *expected value* of the random variable. For a discrete random variable with values $x_1, x_2, \ldots, x_k$, the expected value of X, denoted $E(X)$, is defined as

$$E(X) = \sum_{i=1}^{k} x_i P(x_i) \qquad (5\text{-}2)$$

For the household size data we find the following:

x_i	$P(x_i)$	$x_i P(x_i)$
2	.125	.250
3	.250	.750
4	.375	1.500
5	.250	1.250
		3.750

Thus the mean or expected value of the household size is 3.75. Note that this need not be one of the original discrete values of X. Equation (5-2) is a "weighted" mean where the weights are defined as the relative frequencies or probabilities of each discrete value. The formula is not unlike that used to compute the grouped approximation to the mean in descriptive statistics, Equation (2-5).

The variability of a discrete random variable is measured by the *variance V(X)*.

The standard deviation is also defined as the square root of the variance. For a discrete random variable

$$V(X) = \sum_{i=1}^{k} [x_i - E(X)]^2 P(x_i) = \sum_{i=1}^{k} x_i^2 \, P(x_i) - [E(X)]^2 \qquad (5\text{-}3)$$

The second form of $V(X)$ is a more efficient computational formula. For the variable called household size,

x_i	$P(x_i)$	x_i^2	$x_i^2 \, P(x_i)$
2	.125	4	.500
3	.250	9	2.250
4	.375	16	6.000
5	.250	25	6.250
			15.000

and $V(X) = 15.000 - (3.75)^2 = 15.000 - 14.0625 = .9375$. The standard deviation is calculated as the square root of the variance, or $\sqrt{.9375} = .9682$.

Continuous Random Variables

A random variable is said to be *continuous* if it can assume all the real number values in some interval of the real number line. The number of different values that a continuous random variable can assume is therefore infinite. Table 5-5 lists several different variables which can be considered to be continuous random variables if they are selected by some random sampling procedure. For example, if annual rainfall

TABLE 5-5
Examples of Continuous Random Variables

Random variable X	Values of random variable
Annual rainfall	$X \geq 0$
Distance traveled to some facility	$X \geq 0$
Stream discharge	$X \geq 0$
pH Value of a water sample	$0 \leq X \leq 14$
Pseudo-continuous variables	
Population of city, region, census tract	$X \geq 0$
Interaction between two places	$X \geq 0$

totals are available from 1000 meteorological stations in North America, then annual rainfall can be considered a random variable if we select one of these stations at random. This is not to say that the amount of rainfall at any station is determined randomly! In addition, many discrete variables are often modeled as continuous variables. For example, the population of a city or region or census tract can be treated as if it were a continuous variable even though, in the strictest sense, it is a discrete variable. This is a common simplification for many discrete variables having extremely large ranges.

The probability distribution of a random continuous variable is represented by a *probability density function*. Let X be a continuous random variable defined over some interval of the real number line, say, from a to b. A probability density function of X, denoted $f(X)$, satisfies two conditions:

1. $f(x) \geq 0$ for $a \leq x \leq b$.
2. The area under $f(x)$ from $x = a$ to $x = b$ must be equal to 1.

These two conditions are formally identical to the two conditions defined for discrete random variables. In the discrete case, the probabilities are said to be *massed* at the discrete values of the random variable. In the continuous case, we say the probability is *spread densely* over the range of the random variable.

The probability density function for a random continuous variable is illustrated as Figure 5-3. Just as the heights of the vertical bars in a probability mass function of a discrete variable sum to 1, the area under a probability density function must also equal 1. It is possible to define random continuous variables over an infinite-length real number line, that is, from $-\infty$ to $+\infty$. If $f(x)$ is asymptotic to the x axis as it approaches $-\infty$ and $+\infty$, then $f(x)$ may still define a random continuous variable if the area under $f(x)$ is 1. Figure 5-4 illustrates such a probability density function. For other distributions the limits to the variable may be $[0, \infty]$. The most famous of all continuous distributions, the normal distribution, is defined for a continuous variable with potential values in the range $[-\infty, +\infty]$.

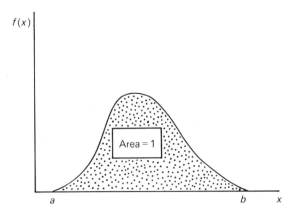

FIGURE 5-3. A probability density function.

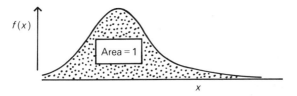

FIGURE 5-4. A probability density function defined over the entire real number line.

How can we compute probabilities by using these probability distributions? Suppose we are interested in computing $P(c \leq x \leq d)$ for some random variable X. If X is discrete, then this is simply a matter of summing the probabilities of X taking on any discrete value between (and including) c and d:

$$P(c \leq x \leq d) = \sum_{x=c}^{x=d} P(x) \qquad (5\text{-}4)$$

For a continuous random variable we use the convention that the probability that X takes on a value between c and d is equal to the area under the probability density function $f(x)$ between c and d. This is illustrated in Figure 5-5. Now, depending on the exact shape of the probability density function, this may not be an easy matter. However, it is not necessary to know advanced calculus to comprehend the concept. For computational purposes, tables of probabilities are provided at the end of this text; they can be used to determine these values for all the continuous-valued probability distributions encountered in this book.

A simple example illustrates the procedure for computing probabilities of a random continuous variable. Suppose the probability density function is given by

$$f(x) = \begin{cases} 0.5 & 0 \leq x \leq 2 \\ 0 & x < 0 \text{ or } x > 2 \end{cases} \qquad (5\text{-}5)$$

as illustrated in Figure 5-6. The total area under this curve can be calculated as the area of a rectangle with base $= 2$ and height $= 0.5$. The area under this probability density function is $0.5 \times 2 = 1$. Note also that $f(x) \geq 0$ for all x. To calculate

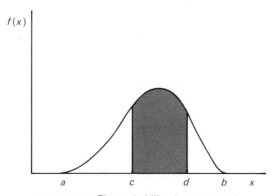

FIGURE 5-5. The probability that $c \leq x \leq d$.

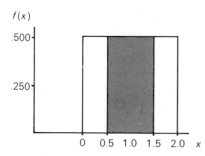

FIGURE 5-6. Computing a probability from a simple random continuous probability density function.

$P(0.5 \leq x \leq 1.50)$, simply find the area under the density function between these two limits. The base is $1.50 - 0.50 = 1.00$, and the height is 0.5, so the total area, and thus the probability, is $0.50 \times 1.0 = .50$. As illustrated in Figure 5-6, this is one-half the area under the probability density function. Unfortunately, it is not always possible to use such simple arguments from elementary plane geometry for more complex probability density functions.

Although there are many similarities between continuous and discrete random variables, there is one key difference. Consider the expression $P(X = r)$. For a discrete variable, this is simply the probability that the variable takes on the discrete value r. This probability can be determined directly from the probability mass function as the height of the bar graph for the value $X = r$. This operation is not possible for a continuous random variable. For a continuous variable, $P(X = r) = 0$. Why? Assume that the random variable is defined on the real number line from $X = a$ to $X = b$. If we were to let $P(X = r) = f(r)$, the height of the density function at $X = r$, then this would have to apply to all other values in the interval $[a, b]$. Since there are an infinite number of such values, say $r_1, r_2, \ldots, r_n$, the sum of the probabilities $f(r) + f(r_1) + f(r_2) + \cdots + f(r_n)$ would undoubtedly exceed 1.0. It is therefore not possible to interpret $f(r)$ directly as a probability. In fact, there is nothing in the definition of a probability density function that prohibits any single value of $f(r)$ from exceeding 1. The height of $f(r)$ is only constrained to be nonnegative. Consider the probability density function

$$f(x) = \begin{cases} 2 & 0 \leq x \leq 0.5 \\ 0 & x < 0 \text{ or } x > 0.5 \end{cases} \tag{5-6}$$

Clearly $f(0) = f(0.25) = f(0.5) = 2$. But the area under the probability density function, $P(0 \leq x \leq 0.5) = 0.5 \times 2 = 1$.

There is another explanation of why $P(X = r)$ for any continuous variable must equal zero. Let us suppose we calculate $P(c \leq x \leq d)$ and derive some nonnegative probability. As we move c and d closer together this probability must get smaller and smaller as the area under $f(x)$ declines. In the limit, $c = d$, and there is no area under the curve. The probability that a continuous random variable assumes a particular value must always be zero, regardless of the value or of the underlying probability distribution.

Computing the expected value $E(X)$ and variance $V(X)$ of a continuous random variable is not nearly so straightforward as the computation for discrete

TABLE 5-6
A Comparison of Continuous and Discrete Random Variables

	Discrete random variable	Continuous random variable
Number of values that can be assumed	Finite (or countably infinite), say k	Infinite, from a to b or $-\infty$ to $+\infty$
Graphical summary	Probability mass function as linear bar chart	Probability density function $f(x)$, area under curve
$P(X = a)$	$P(a)$	0
$P(c \leq X \leq d)$	$\displaystyle\sum_{x=c}^{d} P(x)$	Area under density function from $x = c$ to $x = d$, or $$\int_{c}^{d} f(x)\, dx$$
Conditions on $P(x)$		
(1)	$0 \leq P(x_i) \leq 1,$ $i = 1, 2, \ldots, k$	$f(x) \geq 0,\ a \leq x \leq b$ or $f(x) \geq 0,\ -\infty \leq x \leq +\infty$
(2)	$\displaystyle\sum_{i=1}^{k} P(x_i) = 1$	Area under $f(x) = 1$, that is, $$\int_{a}^{b} f(x)\, dx = 1 \text{ or } \int_{-\infty}^{\infty} f(x)\, dx = 1$$

variables. The derivation of these quantities requires a background in calculus well beyond the scope of this text. The expected value and variance of the continuous probability distributions described in Section 5.3 are presented without derivations. Students with a background in calculus may wish to consult Appendix 5A for a more complete specification of the derivation of $E(X)$ and $V(X)$ for continuous variables.

The properties of continuous and discrete random variables are summarized in Table 5-6. Both types of variables must satisfy an identical set of basic properties including (1) nonnegativity and (2) the sum of probabilities equal to 1. The major difference arises from the number of different values that the random variable can assume. For a discrete random variable this is always a finite number. By definition, a continuous random variable can assume an infinite number of different values. This distinction leads to several important differences in the specification of the properties of these two classes of random variables.

5.2. Discrete Probability Distribution Models

Discrete random variables can be represented by a number of different probability mass functions. If a complete listing of the *population* of interest is available, then the relative frequency distribution is the appropriate probability mass function. Figure 5-1 represents the probability mass distribution for the variable called household size, as distributed in the population listed in Table 5-1. For cases in which the available data are a sample from some larger population, this procedure is not feasible. Instead, it is usual to see how well the available data fit certain probability distribution models derived from specific statistical experiments. For discrete variables, three different probability distributions are often used in statistical analyses:

1. Discrete uniform distribution
2. Binomial distribution
3. Poisson distribution

It is not unusual to find close parallels between the assumptions underlying the statistical experiments governing these distributions and the nature of some empirical statistical problem. In Chapter 10, methods are presented for testing the fit of an empirically derived frequency distribution to any probability distribution model.

Discrete Uniform Distribution

One of the simplest probability distributions is the discrete uniform distribution. It occurs frequently in games of chance. Examine the probability of getting a head or a tail in the single toss of a fair coin. Each outcome has an equal probability of .5. In a single draw from a shuffled deck, the probability of drawing a card of any particular suit is equal to .25. Each of these experiments defines a discrete uniform random variable. In such a distribution, the probability of occurrence for each outcome or value of the random variable is equal. For two outcomes, each has a probability of .5; for three outcomes, each has a probability of .33; and so on.

The uniform distribution also has applications in statistical decision making in situations of complete uncertainty. Suppose, for example, someone is lost and arrives at an unmarked fork in the road. There is no reason to suspect that either road necessarily leads to a known location. Which road should be taken? In the absence of any other information, each road could be assigned an equal probability. A coin toss could be used to make the decision. Many other decision-making situations are similar to this example. All involve the notion of complete uncertainty.

To formalize this probability mass function, assign a single integer from $1, 2, \ldots, k$ to each of the k different outcomes that can be assumed by the random variable X. If X is a discrete uniform random variable, then

$$P(x) = \frac{1}{k} \qquad x = 1, 2, \ldots, k \qquad (5\text{-}7)$$

That is, each state or value of the random variable has an equal probability of occurrence. Figure 5-7(a) illustrates the general form of the probability mass function of a discrete uniform variable. Figure 5-7(b) and (c) illustrates the probability mass functions for the coin toss and card selection examples.

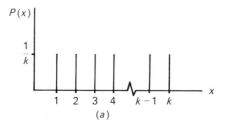

(a)

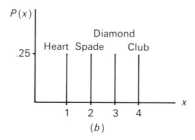

(b)

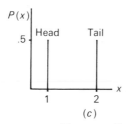

(c)

FIGURE 5-7. Discrete uniform random variables.

What are the mean and variance of a discrete uniform variable? Using (5-2), we can calculate the expected value as

$$E(X) = \sum_{x=1}^{k} xP(x) = \sum_{x=1}^{k} x \left(\frac{1}{k} \right) \qquad (5\text{-}8)$$

Now, since $1/k$ is a constant, it can be brought outside the summation (recall Rule 2 of Appendix 2A). The expression Σx is the sum of the first k integers and can be simplified to $k(k + 1)/2$. Therefore,

$$E(X) = \frac{1}{k} \left[\frac{k(k + 1)}{2} \right] = \frac{k + 1}{2} \qquad (5\text{-}9)$$

Although it is somewhat more difficult to derive, the variance of a discrete uniform variable is

$$V(X) = \frac{k^2 - 1}{12} \qquad (5\text{-}10)$$

As an example, consider again the outcome of a single coin toss. If we let $x = 1$ be the result of getting a head and $x = 2$ be the result of getting a tail, then $P(1) = P(2) = .5$ for the $k = 2$ potential values of the random variable. The mean or expected value is $(k + 1)/2 = \frac{3}{2} = 1.5$. Note once again that this average is not one of the two values that can be assumed by X. It is midway between each. From a decision-making point of view, this tells us that neither a head nor a tail is a preferable choice in a coin toss situation. Using (5-3), we find $V(X) = (k^2 - 1)/12 = .25$. This has no immediate interpretation.

Binomial Distribution

A second important distribution of a discrete random variable is the binomial distribution. It is applicable to variables having only two possible outcomes: A person is either a male or female, a person answers a question either yes or no, or the toss of a coin yields either a head or a tail. The simplest example of a binomial random variable is the operation of a coin toss. Suppose we toss a coin n times and record the number of heads occurring in the n tosses. If the tosses are all independent events and the random variable X is defined to be the number of heads occurring in the n tosses, then X is a binomial random variable.

A binomial random variable is produced by a statistical experiment known as a *Bernoulli process*, or a set of *Bernoulli trials*. A set of Bernoulli trials is defined by the following conditions:

1. There are n independent trials of the experiment.
2. The same pair of outcomes is possible on all trials.
3. The probability of each outcome is the same on all trials.
4. The random variable is defined to be the number of "successes" in the n trials.

Let us relate this definition to the example of a coin toss experiment. Our interest lies in the number of heads, or "successes," that occur in an experiment in which a coin is tossed n times. The probability of a head, or a success, is labeled π, and the probability of a tail, or failure, is $1 - \pi$. Each trial is represented by one toss of the coin. Table 5-7 lists the possible outcomes from this experiment in which a coin is tossed $n = 1, 2, 3$, or 4 times.

Consider the first row of the table. There are only two possible outcomes: Either we get a head (H) or a tail (T). Thus, we may get either zero heads or one head depending on the outcome. The probability of zero heads is $1 - \pi$, and the probability of one head is π. Since $\pi + (1 - \pi) = 1$, this is a legitimate probability distribution for a random variable. What if the coin is tossed twice? Now there are three different results: There can be zero heads, one head, or two heads. However, although there is only one way of getting zero heads or two heads, there are two ways of obtaining one head. To get two heads requires a head on each of the $n = 2$ tosses. This is listed as outcome {HH}. Similarly, only one sequence leads to two tails, {TT}. But, to obtain one head, two different sequences are possible: {HT} and {TH}.

For $n = 3$ tosses there are now eight possible outcomes. For example, there are

TABLE 5-7
Possible Outcomes of Coin Toss Experiment

$n = 1$			T		H	
$n = 2$		TT		TH		HH
				HT		
			TTH		THH	
$n = 3$		TTT	THT		HTH	HHH
			HTT		HHT	
				TTHH		
		TTTH		THTH	THHH	
$n = 4$	TTTT	TTHT		HTTH	HTHH	HHHH
		THTT		THHT	HHTH	
		HTTT		HTHT	HHHT	
				HHTT		

now three ways of obtaining one head. It can appear on the third, second, or first toss of the coin. These three outcomes are listed as {TTH}, {THT}, and {HTT}. Note that these three alternatives can be derived from the sequences for $n = 2$ which are listed immediately above. The sequence {TTH} results from the addition of a head to the sequence {TT}, and {THT} and {HTT} derive from the sequences {TH} and {HT} with the addition of a tail. Further, the three sequences with two heads—{THH}, {HTH}, and {HHT}—can be generated by the addition of the appropriate result to the sequences listed above as {TH}, {HT}, and {HH}. The entire list of outcomes in Table 5-7 is constructed in such a way that the sequences in any row can be generated from the immediately preceding row with the addition of the appropriate outcomes. The size of this triangle expands quite rapidly as the number of tosses increases.

One way of summarizing these results is to calculate the different number of sequences which lead to any outcome for a given number of tosses. For example, there are three ways of getting one head in $n = 3$ tosses. These numbers are termed the *binomial coefficients* and are displayed in Table 5-8 for $n = 1$ to $n = 6$ Bernoulli trials. This representation is known as *Pascal's triangle* in recognition of the role of the French mathematician Blaise Pascal in the development of these results. Note that any number in this table can be generated by adding the number of sequences above the entry in the immediately preceding row. The 6 in the fourth row is $3 + 3$, the sum of the two numbers directly above it. How are these binomial coefficients to be interpreted? Each entry represents *the number of different ways of obtaining exactly x heads in n tosses of a coin*. There are exactly 6 ways of obtaining 2 heads in 4 tosses of a coin. This is denoted as C_2^4 since there are this many combinations of heads and tails leading to the desired result. From Equation (4-8), C_2^4 is calculated as $4!/[2!(4 - 2)!] = 24/4 = 6$ combinations. These six combinations are listed in the third

TABLE 5-8
Pascal's Triangle of Binominal Coefficients

$n = 1$						1		1				
$n = 2$					1		2		1			
$n = 3$				1		3		3		1		
$n = 4$			1		4		6		4		1	
$n = 5$		1		5		10		10		5		1
$n = 6$	1		6		15		20		15		6	1

column of the $n = 4$ row of Table 5-7. Remember that the five numbers in the $n = 4$ row refer to the appearance of 0, 1, 2, 3, or 4 heads in the $n = 4$ tosses of the coin.

Now, consider the probability that any of these individual sequences occurs. We know that the probability of a head is π, the probability of a tail is $1 - \pi$, and the trials are independent. For example, what is the probability of each of the six sequences {TTHH}, {THTH}, {HTHT}, {THHT}, {HTHT}, and {HHTT}? Each contains $x = 2$ heads in $n = 4$ tosses. It is rather easy to show that the probabilities of these sequences are equal and are given by $\pi^2(1 - \pi)^2$. For the first sequence, the independence of trials allows us to rewrite this as $P(\text{TTHH}) = P(\text{T}) \cdot P(\text{H}) \cdot P(\text{H})$. Since $P(\text{T}) = 1 - \pi$ and $P(\text{H}) = \pi$, this simplifies to $(1 - \pi)(1 - \pi)(\pi)(\pi)$ or, with rearrangement, $\pi^2(1 - \pi)^2$. All sequences can be rearranged in this form.

TABLE 5-9
Binomial Distributions

Number of trials n	Number of heads	Probability of number of heads
1	0	$1 - \pi$
	1	π
2	0	$(1 - \pi)^2$
	1	$2\pi(1 - \pi)$
	2	π^2
3	0	$(1 - \pi)^3$
	1	$3\pi(1 - \pi)^2$
	2	$3\pi^2(1 - \pi)$
	3	π^3
4	0	$(1 - \pi)^4$
	1	$4\pi(1 - \pi)^3$
	2	$6\pi^2(1 - \pi)^2$
	3	$4\pi^3(1 - \pi)$
	4	π^4

Our ultimate concern is the probability of obtaining exactly x heads in n tosses of a coin. To continue the same example, what is the probability of obtaining exactly 2 heads in 4 tosses of a coin? Since there are 6 different sequences with exactly 2 heads, each with a probability $\pi^2(1 - \pi)^2$, the required probability is $6\pi^2(1 - \pi)^2$. The probability of obtaining x heads in n tosses of a coin for up to $n = 4$ trials is listed in Table 5-9. Notice the coefficients of the probabilities of the last column of Table 5-9. For example, for $n = 4$ and $x = 2$ the five coefficients are 1, 4, 6, 4, and 1. These are the entries of Pascal's triangle, Table 5-8, for the row $n = 4$. For this reason the entries in Pascal's triangle are known as the *binomial coefficients.* Also the exponent of π is the number of heads, and the exponent of $1 - \pi$ is the number of tails. In general, the probability of obtaining x heads in n tosses of a coin is equal to the product of the number of sequences of length n with x heads and the probability that the sequence contains x heads with $n - x$ tails.

Let us now formally define the probability mass function for a binomial

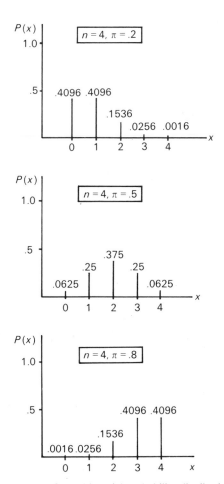

FIGURE 5-8. Some binomial probability distributions.

random variable. If X is a binomial random variable, then the probability mass function is given by

$$P(x) = C(n, x)\pi^x(1 - \pi)^{n - x} \qquad x = 0, 1, \ldots, n \qquad (5\text{-}11)$$

where n is the number of trials, x is the number of successes, and π is the probability of a success in each trial.

The specific form of the binomial distribution depends on the values of the two parameters π and n. Given values of π and n, we can easily compute the probability for any number of successes by using (5-11). If $\pi = .2$ and $n = 4$ and we wish to compute the probability of three successes, then by (5-11) $P(3) = C(4, 3)(.2)^3(.8)^1 = 4(.008)(.8) = .0256$. Figure 5-8 illustrates the probability mass functions for three members of the *family* of binomial probability distributions. If the coin we are tossing is a fair coin, then $\pi = .5$ and the distribution of successes is always symmetric. If $\pi < .5$, then the distribution is negatively skewed; and if $\pi > .5$, the distribution is positively skewed. When n is very small, the probability mass function for a binomial random variable, Equation (5-11), can be used directly to compute the required probabilities. But when n is large, it is more convenient to consult statistical tables which compile binomial probabilities for many different values of π and n. Statistical tables for a few members of the binomial family are included at the end of this text.

The mean and variance of the binomial distribution can be determined from (5-2) and (5-3) as

$$E(X) = \sum_{x=0}^{n} xC(n, x)\pi^x(1 - \pi)^{n - x} = n\pi \qquad (5\text{-}12)$$

$$V(X) = \sum_{x=0}^{n} [x - E(X)]^2 C(n, x)\pi^x(1 - \pi)^{n - x} = n\pi(1 - \pi) \qquad (5\text{-}13)$$

EXAMPLE 5-1. Suppose a student taking a course in statistics is required to take an examination consisting of 10 multiple-choice questions. The questions are designed in such a way that the probability of correctly guessing the answer to any question is .2. If the student can only guess at the correct answers, what is the expected number of correct answers? What is the standard deviation of the number of correct answers? Assuming a grade of 50 percent is required to pass the test, what is the probability that the student passes?

Solution. Let the binomial random variable X be the number of correct answers on the test. There are $n = 10$ trials, and it is known that $\pi = .2$. The appropriate probability mass function is

$$P(x) = C(10, x)(.2)^x(.8)^{10 - x} \qquad x = 0, 1, \ldots, 10 \qquad (5\text{-}14)$$

where x is the number of correct answers. The expected number of correct answers is obtained from (5-12):

$$E(X) = n\pi = 10(.2) = 2 \qquad (5\text{-}15)$$

And $V(X)$ is calculated from (5-13):

$$V(X) = n\pi(1 - \pi) = 10(.2)(.8) = 1.6 \qquad (5\text{-}16)$$

Therefore the standard deviation is $\sqrt{1.6} = 1.26$ correct answers. To determine the probability that the student passes the test, it is necessary to calculate $P(5) + P(6) + P(7) + P(8) + P(9) + P(10)$ since all these outcomes lead to passing grades:

$$P(5) = C(10, 5)(.2)^5(.8)^5 = \quad .026$$

$$P(6) = C(10, 6)(.2)^6(.8)^4 = \quad .006$$

$$P(7) = C(10, 7)(.2)^7(.8)^3 = \quad .001$$

$$P(8) = C(10, 8)(.2)^8(.8)^2 = \quad .000$$

$$P(9) = C(10, 9)(.2)^9(.8)^1 = \quad .000 \qquad \text{(to three places)}$$

$$P(10) = C(10, 10)(.2)^{10}(.8)^0 = \underline{.000}$$

$$.033$$

There is thus a 3.3 percent chance, or a probability of .033, that a student could pass the test just by guessing.

In our discussion of the binomial distribution, we assumed that the values of the two parameters π and n are known. With these values we can specify the appropriate probability mass function, calculate the expected value and variance, and solve many probability problems. What if they are not known? Then it is necessary to estimate these parameters from some sample data. We see how this problem in estimation is solved in Chapter 7. In Chapter 10 we see how to solve the related problem of testing the "fit" of the binomial distribution to sample data. Such methods are extremely useful for determining which particular probability mass function best describes empirical data that we wish to analyze.

Poisson Probability Distribution

The final discrete distribution discussed in this chapter is the *Poisson probability distribution*. As we see in Chapter 6, this distribution proves to be very important in the analysis of geographical *point patterns*. Point pattern analysis involves a variety of techniques that describe the spatial distribution of phenomena represented by points on the traditional dot map. Figure 5-9 illustrates a typical dot map. Each dot represents the location of one observation of some item of interest. These dots might represent settlements, individual dwelling units, households of some particular ethnic group, or specific geographic features such as drumlins, erratics, or members of a specific plant species. To derive the Poisson probability distribution and to develop the notion of a Poisson random variable, consider the analysis of such a dot map. Let us suppose that the dots represent drumlins and that these drumlins are *randomly distributed* over the area depicted on the dot map. Suppose it is also known that there is, on average, one drumlin per square mile.

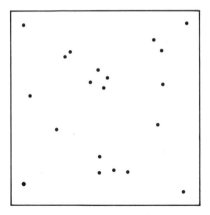

FIGURE 5-9. A typical dot map.

One way of characterizing this dot map is to record the number of dots appearing in quadrats of the map. *Quadrats* are simply areal subdivisions of a map. Usually they are regular in size and square. The region depicted in Figure 5-9 is divided into 100 quadrats. Initially, let us analyze this map as a Bernoulli process in which the examinations of quadrats are treated as "trials" of a statistical experiment. When examining a quadrat, we define a success as the existence of a drumlin and a failure as the absence of a drumlin. Since there is, on average, one drumlin per square mile, the expected number of drumlins in any quadrat is 1.

Now, suppose each 1-mi² quadrat is divided into four equal-sized 0.25-mi squares, as in Figure 5-10(*a*). Also let us make the additional assumption that no more than one drumlin can occur in any 0.25-mi quadrat. Let us examine these four 0.25-mi squares as a set of Bernoulli trials. Each quadrat either contains a drumlin or does not. Since the expected number of drumlins per square mile is 1, the probability that there is a drumlin in any of the four quadrats is $\frac{1}{4}$. Examining each of these quadrats as four trials of a Bernoulli process leads to five possible outcomes—there can be 0 drumlins, 1 drumlin, or 2, 3, or 4 drumlins in these four quadrats. Clearly this is a binomial random variable. The probability that the four quadrats contain exactly 2 drumlins is

$$P(2) = \frac{4!}{2!(4-2)!}\left(\frac{1}{4}\right)^2\left(\frac{3}{4}\right)^2 = .2109375 \tag{5-17}$$

This expression is derived by the application of Equation (5-1) with $\pi = \frac{1}{4}$, $1 - \pi = \frac{3}{4}$, $n = 4$, and $x = 2$.

To use Equation (5-11), we must make the assumption that the probability we might find 2 drumlins (or even more) in any 0.25-mi quadrat is 0. This violates the assumption that the distribution of points on the map is random. If the distribution is truly random, then there must be some probability that any quadrat could contain two, three, or even all dots of the map. To get a closer approximation, let us divide these four 0.25-mi quadrats into 16 quadrats of $\frac{1}{8}$-mi square, as in Figure 5-10(*b*). This time we assume that there can be at most 1 drumlin in each of these 16 quadrats. The

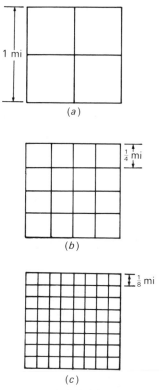

FIGURE 5-10. Subdividing quadrats of a map.

possible outcomes of the set of Bernoulli trials based on these 16 quadrats are
0, 1, 2, . . . , 16 drumlins. The probability of exactly 2 drumlins is

$$P(2) = \frac{16!}{2!(16-2)!}\left(\frac{1}{16}\right)^2\left(\frac{15}{16}\right)^{14} = .1899 \qquad (5\text{-}18)$$

since for this set of trials $\pi = \frac{1}{16}$, $1 - \pi = \frac{15}{16}$, $n = 16$, and $x = 2$.

The estimate of $P(2)$ given by (5-18) is more accurate than (5-17) since the
probability of more than 1 drumlin in any quadrat is smaller. This is because the
quadrats themselves are smaller. To get an even better estimate, let us divide our 16
quadrats into 64 quadrats, as in Figure 5-10(c). The probability that any 1-mi²
quadrat contains exactly 2 drumlins is given by

$$P(2) = \frac{64!}{2!(64-2)!}\left(\frac{1}{64}\right)^2\left(\frac{63}{64}\right)^{62} = .1854 \qquad (5\text{-}19)$$

Since 64 quadrats are a better basis for approximation than 16 quadrats, we
might ask what would happen if we examined this process as the number of quadrats
approached infinity and the probability that any quadrat contained a drumlin

approaches zero. Examining the sequence of our estimates $P(2)$ with increasing trials, or equivalently with increasingly smaller quadrats, we note the following:

n	4	16	64
$P(2)$	.2109	.1899	.1854

To determine the expression for $P(2)$ for an infinite number of trials requires an evaluation of the limit as n approaches infinity of the following expression:

$$P(2) = \frac{n!}{2!(n-2)!}\left(\frac{1}{n}\right)^2\left(\frac{n-1}{n}\right)^{n-2} \tag{5-20}$$

Although some calculus is required to derive the result, it can be shown that the limit converges to

$$P(2) = \frac{e^{-1}1^2}{2!} = .1839 \tag{5-21}$$

where e is the base of the natural logarithm, or approximately 2.71828. The limit of $P(2)$ thus converges to .1839, and the increasing accuracy of our estimate with an increasing number of quadrats is clearly illustrated.

This result can now be extended to determine the limits for $P(0)$, $P(1)$, $P(2)$, $P(3)$, and the probabilities for the rest of the positive integers. Let λ be the average number of drumlins per unit area. In the example considered above, $\lambda = 1$ drumlin per square mile. The random variable X is the number of drumlins to be found in a randomly selected quadrat, and X may take on the values $x = 0$, $x = 1$, or any other positive integer. The probability that a quadrat contains x drumlins is

$$P(x) = \frac{e^{-\lambda}\lambda^x}{x!} \qquad x = 0, 1, 2, \ldots \tag{5-22}$$

Equation (5-22) is the probability mass function for a Poisson random variable. There is only one parameter for this distribution—λ.

To generalize, the experiment generating a Poisson random variable is described in the following way. Let the random variable X be the number of occurrences $x = 0, 1, 2, \ldots$ of some specific event in a given continuous interval. The interval may be a time interval, a spatial interval (such as a quadrat), or even the length of some physical object such as a rope or a chain. The random variable X is termed a *Poisson random variable* if the experiment generating the values of X satisfies the following conditions:

1. The number of occurrences of the event in two mutually exclusive intervals is independent.
2. The probability of an occurrence of the event in a small interval is small and proportional to the length of the interval. That is, the event is rare.
3. The probability of two or more occurrences of the event in a small interval is near zero.

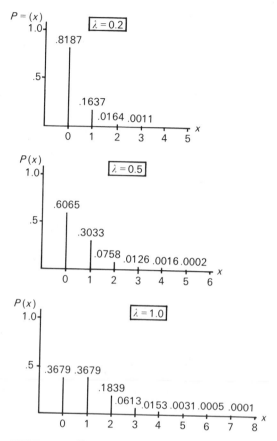

FIGURE 5-11. Three Poisson probability distributions.

For the point pattern problem, the event is the existence of a dot in the quadrat. Since the pattern is random, the number of dots in any two quadrats is independent. Finally the probability of two or more dots in one quadrat is extremely small.

There is a family of Poisson probability mass functions, each member of which is specified by selecting a particular value of λ. Figure 5-11 illustrates the probability mass functions for three members of this family. Note that all distributions are truncated and depict probabilities only for small values of x. But, as is indicated in Equation (5-22), the probability mass function of a Poisson random variable assigns probabilities to all positive integers. The probabilities for larger values of x are extremely small and have been omitted for clarity. A comparison of these three probability mass functions indicates that the probabilities of a Poisson random variable become increasingly spread over the potential values of X as λ increases. It is feasible to compute Poisson probabilities for only a relatively small number of cases. Compact tables for a few values of λ are provided at the end of the text. It is often easier to consult this table than to compute the required probabilities from (5-22).

The mean and variance of a Poisson random variable can be shown to be

$$E(X) = \sum_{x=0}^{\infty} xP(x) = \sum_{x=0}^{\infty} x\frac{e^{-\lambda}\lambda^x}{x!} = \lambda \tag{5-23}$$

$$V(X) = \sum_{x=0}^{\infty} [x - E(X)]^2 \frac{e^{-\lambda}\lambda^x}{x!} = \lambda \tag{5-24}$$

Thus a characteristic feature of a Poisson random variable is the equality of the mean and variance. This is a useful check for examining the applicability of the Poisson random variable to some sample data.

EXAMPLE 5-2. The number of arrivals to a wilderness park during the summer season is found to be Poisson-distributed with a mean of 2.5 camping groups per day. On a particular day during this summer period, what is the probability that no groups arrive at the park? What is the probability that between one and three groups arrive? What is the most likely number of arriving camping groups?

Solution. The appropriate probability mass function is

$$P(x) = \frac{e^{-2.5}(2.5)^x}{x!} \qquad x = 0, 1, 2, \ldots \tag{5-25}$$

where x is the number of camping groups arriving at the park per summer season day. To find the probability that no groups arrive, we simply compute $P(0)$ from (5-25):

$$P(0) = \frac{e^{-2.5}(2.5)^0}{0!} = .0821 \tag{5-26}$$

We see that there is an 8 percent chance that no groups arrive on a particular day. This computation can be checked by consulting the Poisson table at the end of the text for the value $x = 0$ and $\lambda = 2.5$. To determine the probability that between one and three groups arrive, $P(1 \le x \le 3)$, we can either consult the table or calculate $P(1) + P(2) + P(3)$ from (5-25). Using the table, we get $P(1 \le x \le 3) = P(1) + P(2) + P(3) = .2052 + .2565 + .2138 = .6755$.

Examining the Poisson table for $\lambda = 2.5$, we note

$$P(0) = .0821 \qquad P(2) = .2565$$

$$P(1) = .2052 \qquad P(3) = .2138$$

The mode is $x = 2$ since it occurs with the greatest probability. The most likely number of arriving camping groups is therefore 2.

The Poisson distribution is applied extensively in problems to model the distribution of the number of persons joining a *queue*, or line. Examples of this situation include arrivals at service stations or traffic facilities such as toll booths or ferries. These types of models are then used to make decisions about how many servicing units (for example, toll booths or ferries with a given capacity) to provide to keep the length of the queue manageable.

The experiment which leads to the value of a random variable in a given problem should always be compared to the experimental conditions generating a Poisson random variable. In the example of the wilderness park, the random variable X is the number of camping groups arriving in a 1-day interval. The possible values that X can assume are $x = 0, 1, 2, \ldots$, an infinite number. If we consider the 1-day interval to be composed of many small subintervals, each 10 min long, then the conditions for X to be a Poisson random variable appear to be satisfied. In a 10-min interval,

1. The number of arriving camping groups is independent of the number arriving in any other 10-min interval of the day.
2. The probability that a camping group arrives in a 10-min interval is small and proportional to the length of the interval.
3. The probability of more than one camping group arriving in a 10-min interval is extremely small.

The Poisson distribution appears to be a reasonable model for the random variable in this case. A specific member of the family of Poisson distributions is chosen by the estimation of λ from some sample data.

5.3. Continuous Probability Distribution Models

Unlike a discrete random variable, a continuous random variable can assume an infinite number of values. Although the distribution of a random continuous variable can also take on many forms, two distributions are presented in this text. First, the uniform continuous distribution is defined. It is the continuous version of the uniform discrete probability model described in Section 5.2. The second distribution, a bell-shaped curve discovered by Gauss and now known as the normal distribution, is the most important distribution in conventional statistical analysis.

Uniform Continuous Distribution

Just as we can define a uniform discrete distribution in which each value of X is equally probable, we can define a uniform continuous distribution in which all values of X are equally likely. Let the random variable X be defined over the range $a \leq x \leq b$. The probability density function for a uniform random variable is

$$f(x) = \begin{cases} \dfrac{1}{b-a} & a \leq x \leq b \\ 0 & \text{otherwise} \end{cases} \tag{5-27}$$

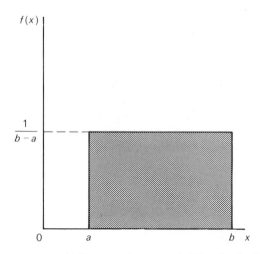

FIGURE 5-12. Uniform continuous probability distribution.

An example of a uniformly distributed random variable is shown in Figure 5-12. In this example both a and b are positive, but there is no restriction on the sign of a or b. The mean $E(X)$ and variance $V(X)$ of a uniform random variable are given by

$$E(X) = \frac{b + a}{2} \tag{5-28}$$

$$V(X) = \frac{(b - a)^2}{12} \tag{5-29}$$

It is not possible to determine the probability that a continuous random variable assumes a particular value of X, but it is possible to determine the probability that X assumes any value between c and d by finding the area under the probability density function from c to d. Suppose a uniform random variable is defined over the interval from 0 to 10. What is the probability that X assumes a value between 3 and

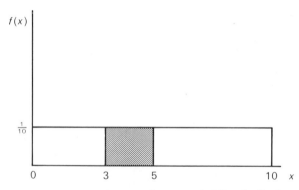

FIGURE 5-13. A uniform continuous probability distribution.

5? The density function is illustrated in Figure 5-13, where the area under the density function between 3 and 5 is shaded. To determine the required probability, it is necessary to calculate the ratio of the shaded area to the total area under the probability density function. By definition, the total area under the distribution is 1. The required probability is $(5 - 3)/(10 - 0) = 2/10 = .2$. In general, the probability that a uniform random variable takes on a value between c and d, where $a \leq c \leq d \leq b$ is $(d - c)/(b - a)$. Because of this formula it is not necessary to have extensive tables for the uniform random distribution. They can be calculated from this simple formula whenever necessary.

Normal Probability Distribution

For many continuous random variables the governing probability distribution is a specific bell-shaped curve known as the *normal distribution*. Originally developed by the German mathematician Gauss, this distribution is sometimes termed the *Gaussian distribution*. The basic shape of the normal distribution is a simple bell. Three different normal distributions are illustrated in Figure 5-14. Why is this distribution so common? One reason is that this basic bell shape can take on any of the forms illustrated in Figure 5-14. The center of the distribution can be located at any point along the real number line. The distribution can be very flat, or it can be very peaked. Notice that distribution A in Figure 5-14 is centered on the lowest value of X, and distribution C on the highest value of X. However, distribution C is the flattest and B the most peaked of the three.

What do all three distributions have in common? All randomly distributed normal variables follow a specific mathematical expression which defines the characteristic bell shape. Specifically, we say that a continuous random variable X is normally distributed if its probability density function is given by

$$f(x) = \frac{1}{\sigma\sqrt{2\pi}} \exp\left[-\frac{1}{2}\left(\frac{x - \mu}{\sigma}\right)^2\right] \tag{5-30}$$

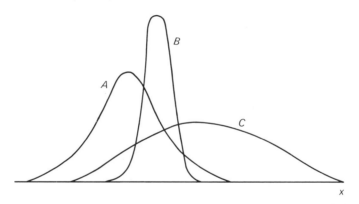

FIGURE 5-14. Three normal distributions.

where $-\infty < x$, $\mu < +\infty$, and $\sigma > 0$. The constants π and e are mathematical constants equal to 3.14149 and 2.71828, respectively. There are two parameters of the normal distribution, μ and σ. Although the derivation requires a knowledge of calculus beyond the scope of this text, it can be shown that the expected value and variance of a random normal variable are

$$E(X) = \mu \qquad (5\text{-}31)$$

$$V(X) = \sigma^2 \qquad (5\text{-}32)$$

The parameter μ is the mean of the distribution and locates the central value along the real number line. The variance σ^2 controls the dispersion of the values about this center. It should now be clear why the members of the normal family come in such a variety of shapes. Two *independent* parameters can be used to combine any degree of dispersion with a central value located anywhere along the real number line. Although there is a great diversity in the appearance of many random normal distributions, all are symmetric about μ. Thus, all normal variables are characterized by the equality of the mode, the median, and the mean. Not only is the highest point on the probability density function located at μ, but also it divides the distribution into two equal parts. Exactly .5000 of the area under the probability density function lies on either side of μ.

To compute the probability that a random normal variable assumes a value between two values of X, say c and d, we must calculate the area under the probability density function given by (5-30) from c to d. This area is illustrated in Figure 5-15. For the uniform continuous distribution, these areas could be determined from simple geometry. Unfortunately this is not possible for the normal distribution. What is the alternative? Of course, we could calculate and tabulate these areas for any two given values of μ and σ, but there are an infinite number of normal distributions— each with a unique combination of μ and σ. Fortunately, we can convert *any* normally distributed random variable to a single probability distribution known as the *standard normal distribution*. This is done by using the transformation

$$z = \frac{x - \mu}{\sigma} \qquad (5\text{-}33)$$

which converts a variable measured in units of x to a standard normal variable measured in units of z. When measured in units of x, a random normal variable has

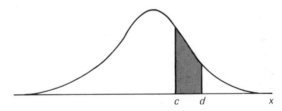

FIGURE 5-15. Probability that a normal random variable takes on a value in the interval $[c, d]$.

a mean of μ and a standard deviation of σ. When it is converted to a standard normal variable, it has a mean of 0 and a standard deviation (and therefore variance) of 1. The probability density function of the standard normal distribution is

$$f(z) = \frac{1}{\sqrt{2\pi}} e^{-\frac{1}{2}z^2} \tag{5-34}$$

For example, consider the two random normal variables X_1 and X_2, the first with mean $\mu_1 = 100$ and $\sigma_1 = 10$ and the second with mean $\mu_2 = 10$ and $\sigma_2 = 2$. Both can be converted to standard normal variables by taking the original measurements and coding them according to (5-33). Consider the first variable X, which is illustrated in Figure 5-16(a). The value $x = 110$ is converted to a standard z value by calculating $z_{110} = (110 - 100)/10 = 1.0$. In what units is z measured? It is always measured in *standard deviation units* of the variable being considered. The value

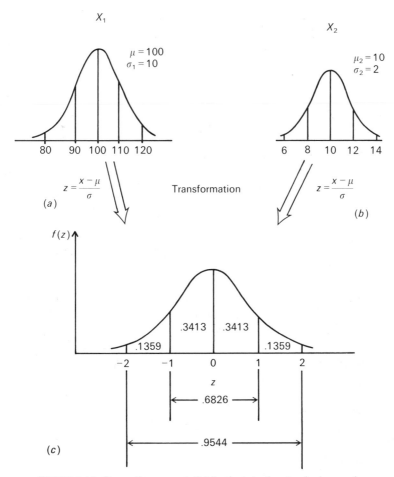

FIGURE 5-16. Converting normal distributions to the standard normal.

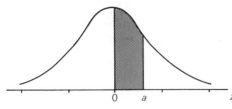

FIGURE 5-17. Areas under the normal in the z interval $[0, a]$.

$z = 1.0$ implies that $x = 110$ is 1 standard deviation above the mean $\mu_1 = 100$. Similarly, $x = 90$ is 1 standard deviation ($\sigma_1 = 10$) *below* the mean of $\mu_1 = 100$, so that $z_{90} = -1.0$. To check, we note that $z_{90} = (90 - 100)/10 = -1.0$.

Variable X_2 of Figure 5-16(b) can be transformed in the same way. Consider first a measurement of $x = 14$. Expressed as a *standard score* (or a *z score*), $z_{14} = (14 - 10)/2 = 2.0$. That is, the value $x = 14$ lies 2 standard deviation units above the mean of $\mu_2 = 10$. As another example, we know that $x = 4$ lies 3 standard deviations *below* the mean of 10. Using (5-33), we calculate $z_4 = (4 - 10)/2 = -3.0$. Any value of x can be converted to an equivalent value of z, and vice versa. For variable X_2, what value of x corresponds to $z = -2.0$? From (5-33) we know $(x - 10)/2 = -2.0$. Solving for x leads to $x = 6$. The value $x = 6$ lies 2 standard deviations below the mean of 10. Thus, we see that both X_1 illustrated in Figure 5-16(a) and X_2 illustrated in Figure 5-16(b) can be converted to the form of the standard normal distribution illustrated in Figure 5-16(c).

When this transformation is made, it is easy to see the characteristics shared by all random normal variables. As illustrated in Figure 5-16(c), exactly 68.26 percent, or .6826, of the values of a normal variable are *within* 1 standard deviation of the mean. For the standard normal this is between $z = 1.0$ and $z = -1.0$. For variable X_1, 68.26 percent of the area under the probability density function lies between 1 standard deviation below the mean, 90, and 1 standard deviation above the mean, 110. For variable X_2, the corresponding limits are $x = 8$ and $x = 12$. Also, 95.44 percent of the area under the probability density function of the standard normal lies within 2 standard deviations of the mean. For X_1 these limits are 80 and 120, and for X_2 the limits are 6 and 14. For all random normal variables the limits to the distribution are $+\infty$ and $-\infty$, but virtually 100 percent of the area under the curve (in fact, 99.72 percent) lies within 3 standard deviations of the mean.

To determine probabilities for the standard normal distribution, it is necessary to consult the standard normal table in the Appendix at the end of the text. This table contains areas under the probability density function (and therefore probabilities) of the form $P(0 \leq z \leq a)$, as in Figure 5-17. It is only necessary to provide the table for positive values of a since the distribution is symmetric about the mean $z = 0$. Thus, $P(a \leq z \leq 0)$, where $a < 0$ must equal $P(0 \leq z \leq -a)$, which can be found in the table. It is often necessary to consult the standard normal table in statistical inference problems. Use of this table is explained in the following set of problems.

EXAMPLE 5-3. Find the area under the standard normal curve (or the probability) for the following intervals:

1. Between $z = 0$ and $z = 2.20$
2. Between $z = 1.32$ and $z = 2.43$
3. Between $z = -1.21$ and $z = 1.87$
4. Greater than $z = 2.53$

First, $P(0 \leq z \leq 2.20)$ can be read directly from the standard normal table. Find the row in the table for $z = 2.2$, and then proceed to the first column for the hundredths digit of 0. The required probability can then be read directly as .4861.

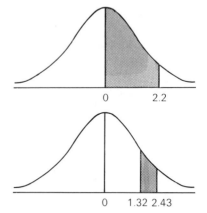

Second, as shown in the accompanying figure, neither endpoint of this interval is $z = 0$. To find the required area, we can *subtract* the area for $z = 0$ to $z = 1.32$ from the area for $z = 0$ to $z = 2.43$. From the table,

Area to 2.43 row for 2.4, column for .03	.4925
Area to 1.32 row for 1.3, column for .02	$-.4066$
	.0859

Third, this area is composed of two parts—the area between $z = -1.21$ and $z = 0.0$ and the area between $z = 0$ and $z = 1.87$. The symmetry implies that the area from $z = -1.21$ to $z = 0$ is equal to the area from $z = 0$ to $z = 1.21$. So,

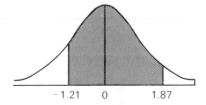

Area to -1.21 row for 1.2, column for .01	.3869
Area to 1.87 row for 1.8, column for .07	$+.4693$
	.8562

Fourth, this value cannot be read directly from the table. But since .5000, or one-half, of the distribution lies between $z = 0$ and $z = +\infty$, we can determine the area from $z = 0$ to $z = 2.53$ and then subtract that value from .5000. Hence

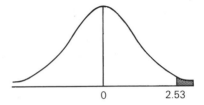

Area to $z = +\infty$	.5000
Area to $z = 2.53$	$-.4943$
	.0057

But what about random normal variables that are not standardized? To use the standard normal table, it is first necessary to compute standard scores from (5-33).

The area under the probability density function of a random normal variable X between the values c and d is exactly equal to the area under the standard normal between $(c - \mu)/\sigma$ and $(d - \mu)/\sigma$. A simple example clarifies this procedure.

EXAMPLE 5-4. The distance traveled in the journey to work in a metropolitan area averages 7.2 mi with a standard deviation of 1.4 mi. Assuming these distances are normally distributed, what is the probability that a randomly selected individual's journey to work is between 6 and 9 mi?

Solution. The distribution of distances is illustrated in the accompanying diagram. To show the correspondence between values of z and values of x, both are listed. To use the standard normal table, we must first convert the values of $x = 6$ and $x = 9$ to units of z. Note that $z = (6 - 7.2)/1.4 = -0.86$ and $z = (9 - 7.2)/1.4 = 1.29$. From the standard normal table

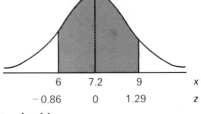

Area from $z = -0.86$ to $z = 0$	.3051
Area from $z = 0$ to $z = 1.29$	+ .4015
	.7066

Thus, the probability is .7066 that the journey to work is between 6 and 9 mi. In all problems requiring the computation of normal probabilities, it is useful to draw a quick sketch of the required area in the manner indicated above.

Normal Approximation to the Binomial Distribution

To this point we have made a clear distinction between probability distribution models for discrete random variables and for continuous random variables. Earlier in this chapter we noticed that in many cases it is useful to treat some discrete variables such as population as if they were continuous variables. Moreover, in some cases it is possible to approximate a discrete probability distribution model with a continuous model. The most important example is the use of the normal approximation to the binomial probability distribution. Because of the closeness of the approximation, the difficulty of working with the probability mass function of the binomial distribution, and the extensive tables required for the binomial, this proves to be an especially useful approximation. In Figure 5-18 an appropriate normal distribution is

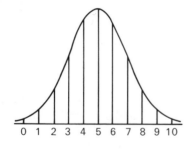

FIGURE 5-18. Normal and binomial distributions.

superimposed on the probability mass function for a binomial random variable with $\pi = .5$ and $n = 10$. Except for the fact that the normal distribution has limits of $(-\infty, +\infty)$, the normal curve appears to virtually duplicate the binomial mass function. The limits of the normal distribution are not a significant problem. Recall the small proportion of the normal curve that appears in the extreme tails of the distribution.

Continuing the example of $n = 10$ and $\pi = .5$, we note that the mean $E(X) = n\pi = 5$ and the variance $V(X) = n\pi(1 - \pi) = 2.5$, so the standard deviation is $\sqrt{2.5} = 1.58$. Therefore, it is possible to approximate this binomial distribution with a normal random variable having a mean $\mu = 5$ and a standard deviation of $\sigma = 1.58$. The normal density function of Figure 5-18 has been drawn by using these parameters. But how do we reconcile the fact that a random normal variable assigns probabilities continuously for values of X but the binomial distribution "masses" these probabilities at discrete integer values of X? The answer lies in the *correction for continuity*.

To illustrate, suppose we are interested in determining the probabilities that $P(X \geq 7)$ and $P(X \leq 6)$ for this variable. Using in Table 5-10, we note that $P(X \geq 7) = .117 + .044 + .010 + .001 = .172$. And $P(X \leq 6) = .828$ since the two probabilities must add to 1. This is easily checked in Table 5-10. Now, let us calculate these probabilities by using the normal distribution with $\mu = 5$ and $\sigma = 1.58$. First, we convert $X = 6$ and $X = 7$ to standard normal form. Thus $Z_6 = (6 - 5)/1.58 = .63$ and $Z_7 = (7 - 5)/1.58 = 1.27$. Using the normal tables, we find

$$P(Z \leq 6) = .5000 + .2357 = .7357$$

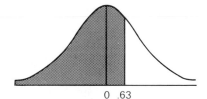

0   .63

$$P(Z > 7) = .5000 - .3980 = .1020$$

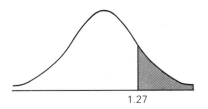

1.27

TABLE 5-10
Binomial Distribution for $\pi = .5$, $n = 10$

x	$P(x)$	x	$P(x)$
0	.001	6	.205
1	.010	7	.117
2	.044	8	.044
3	.117	9	.010
4	.205	10	.001
5	.246		

This is not a good approximation to the probabilities calculated from the binomial. In fact, the sum $.7357 + .1020 = .8377$ does not equal 1. Why? We have failed to account for the area between 6 and 7! Although there is no probability assigned to this area in the binomial, the normal distribution assigns probability to this interval. To correct this, split the area into two parts, putting one-half with $P(X \leq 6)$ and one-half with $P(X \geq 7)$. To approximate $P(X \leq 6)$ by using a normal probability density function, we must calculate $P(X \leq 6.5)$. Similarly, to approximate $P(X \geq 7)$, calculate $P(X \geq 6.5)$. The .5 difference is known as the *correction for continuity*. With this correction, $Z_{6.5} = (6.5 - 5)/1.58 = 0.95$. From the standard normal table this leads to $P(X \leq 6.5) = .5000 + .3289 = .8289$ and $P(X \geq 6.5) = .5000 - .3289 = .1711$. These estimates are excellent approximations to the true values of $P(X \geq 7) = .172$ and $P(X \leq 6) = .828$.

The obvious question is, Under what conditions is the approximation valid? Remember, the normal distribution is always symmetric, but the binomial is not! The general rule is that the binomial is sufficiently symmetric whenever $n\pi$ and $n(1 - \pi)$ are greater than or equal to 5. In the previous example, $n\pi = 5$ so that the approximation is close. Whenever $n\pi$ or $n\pi(1 - \pi)$ is less than 5, the binomial distribution is too skewed to be closely approximated by the normal distribution. Examples of asymmetric binomial distributions are seen in Figure 5-8. The normal approximation to the binomial proves to be important in several instances in statistical inference problems. Instead of using the cumbersome binomial distribution, we use the normal approximation to it, recognizing the conditions under which our approximation is sufficiently close. This procedure is illustrated in Chapter 9.

5.4. Bivariate Random Variables

When two or more random variables are jointly involved in a statistical experiment, their outcomes are generated by a *multivariate probability function*. *Bivariate probability functions* are a class of multivariate functions in which two random variables are jointly involved in the outcome of a statistical experiment. In turn, bivariate probability functions can be classed as either *discrete* or *continuous*. The concepts of bivariate random variables are most easily explained in relation to discrete distributions without resorting to the mathematical complications inherent in continuous models. The relevant features of continuous models can then be explained by analogy to the discrete case.

Bivariate Probability Functions

The *joint* or *bivariate probability mass function* for two discrete variables is a function that assigns probabilities to joint values of X and Y such that two conditions are satisfied:

1. The probability that random variable X takes on the value $X = x$ *and* that random variable Y takes on the value $Y = y$, denoted $P(x, y)$, is nonnegative and less than or equal to 1. That is, $0 \leq P(x, y) \leq 1$ for all (x, y) pairs of the discrete random variables X and Y.

2. The sum of the probabilities $P(x, y)$ taken over all discrete values of X and Y is 1. That is, $\Sigma_x \Sigma_y P(x, y) = 1$.

To understand these two conditions, consider the relationship between the size of a household and the number of cars owned by the household. Suppose 100 households are surveyed, and the number of members of the household and the number of cars owned by that household are determined. The responses of the households are classified in Table 5-11. Of the 100 survey households, 10 have 2 members and 0 cars available, 8 have 2 members and 1 car available, and so on. When households are categorized by size, 23 have 2 members, 26 have 3 members, 27 have 4 members, and 24 have 5 members. Similarly, the totals by number of cars owned are given in the last row of the table.

These data can be converted to a bivariate probability distribution by dividing each entry in Table 5-11 by the number of households in the survey, $n = 100$. The bivariate probability function $P(X, Y)$ for these two variables is given in Table 5-12. In the last column of the table is the *marginal probability distribution* of random variable X, the probability distribution of random variable X taken alone. The marginal probability distribution of X can be calculated from the relation $P(x) = \Sigma_y P(x, y)$. Note that, for example, the probability of a randomly selected household having exactly 3 members is $P(X = 3) = .07 + .10 + .06 + .03 = .26$. Similarly, the marginal probability distribution of Y can be determined from $P(y) = \Sigma_x P(x, y)$. The probability that a household has no cars is obtained by summing the first column: $P(Y = 0) = .10 + .07 + .04 + .01 = .22$. It is thus possible to construct the marginal probability distributions $P(X)$ and $P(Y)$ from a joint or bivariate probability function $P(X, Y)$.

The joint probability distribution $P(X, Y)$ can also be used to determine the *conditional probability functions* for variables X and Y. What is the conditional probability distribution of Y, given that X takes on some specific value $X = x$? For example, what is the conditional probability distribution of car ownership, given that

TABLE 5-11
Car Ownership and Household Size

Household size \ Cars owned	0	1	2	3	Total
2	10	8	3	2	23
3	7	10	6	3	26
4	4	5	12	6	27
5	1	2	6	15	24
Total	22	25	27	26	100

TABLE 5-12
Probabilities of Car Ownership and Household Size

y x	0	1	2	3	$P(x)$
2	.10	.08	.03	.02	.23
3	.07	.10	.06	.03	.26
4	.04	.05	.12	.06	.27
5	.01	.02	.06	.15	.24
$P(y)$	.22	.25	.27	.26	1.00

a household contains three members? The conditional probability distribution of Y given $X = x$ is calculated from the relation

$$P(y \mid x) = \frac{P(x, y)}{P(x)} \qquad (5\text{-}35)$$

and the conditional distribution of X given $Y = y$ is

$$P(x \mid y) = \frac{P(x, y)}{P(y)} \qquad (5\text{-}36)$$

The conditional probability distribution of car ownership for three-person house-holds can be computed from Equation (5-35) with the data of Table 5-12. The conditional probability function $P(y \mid x = 3)$ is

$$P(y = 0 \mid x = 3) = \frac{.07}{.26} = .269$$

$$P(y = 1 \mid x = 3) = \frac{.10}{.26} = .385$$

$$P(y = 2 \mid x = 3) = \frac{.06}{.26} = .231$$

$$P(y = 3 \mid x = 3) = \frac{.03}{.26} = .115$$

Note that this is a valid probability distribution since it sums to 1. To obtain this distribution, we simply divide each entry in the row for $x = 3$ by the row sum $P(3)$ given in the last column. It is not possible to use the entries in the row as they stand. Why? Because they do not sum to 1. The division by .26 standardizes the distribution

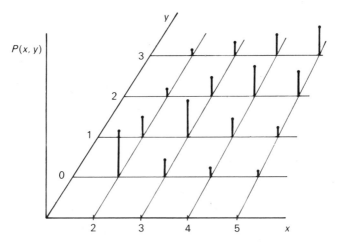

FIGURE 5-19. Graph of bivariate probability distribution of Table 5-12.

so that it sums to 1. Similarly, the conditional probability distribution of household size given that the household has three cars is

$$P(x = 2 \mid y = 3) = \frac{.02}{.26} = .077$$

$$P(x = 3 \mid y = 3) = \frac{.03}{.26} = .115$$

$$P(x = 4 \mid y = 3) = \frac{.06}{.26} = .231$$

$$P(x = 5 \mid y = 3) = \frac{.15}{.26} = .577$$

It is also possible to illustrate a bivariate probability function by using a simple three-dimensional graph. The vertical axis is used for $P(x, y)$, and the two random variables are depicted in the XY plane. Where there are only a few discrete values for the variables, such illustrations give a visual impression of the joint distribution of the two random variables involved. Figure 5-19 illustrates the bivariate probability distribution for the data in Table 5-12. It is apparent, for example, that larger households tend to own more cars than smaller households. In fact, there appears to be a direct relation between the size of the household and the number of cars owned. How can the strength of this relation be measured? The answer lies in a measure known as *covariance*.

Covariance of Two Random Variables

For a single random variable it is possible to calculate the mean or expected value. For two random variables, a useful statistical measure is the *covariance*. The covariance of two random variables is defined as

$$C(X, Y) = E\{[X - E(X)][Y - E(Y)]\} = \sum_{x,y} P(x, y)[X - E(X)][y - E(Y)] \quad (5\text{-}37)$$

or

$$C(X, Y) = E(X, Y) - E(X)E(Y) = \sum_x \sum_y xyP(x, y) - E(X)E(Y) \quad (5\text{-}38)$$

Here $C(X, Y)$ is the covariance, and E is the expectation operator. Recall from Equation (5-2) that $E(X) = \sum_{i=1}^{k} x_i P(x_i)$. The covariance is a direct statistical measure of the degree to which two random variables X and Y tend to vary together. Whenever large values of X tend to be associated with large values of Y, and small values of X with small values of Y, then $C(X, Y)$ has a large positive value. Whenever large values of X are associated with small values of Y and small values of X with large values of Y, then $C(X, Y)$ is a large negative number. Whenever there is no pattern, $C(X, Y) = 0$ or is close to zero.

Consider the data of Table 5-12. Although Equation (5-38) is the most efficient way to calculate the covariance, Equation (5-37) gives the identical result. This equivalence is illustrated by the results of the computations using each formula, summarized in Tables 5-13 and 5-14. The calculated value of $C(X, Y) = .6336$

TABLE 5-13
Computation of Covariance for Data of Table 5-12 by Using Equation (5-37)

(x, y)	$P(x, y)$	$x - E(X)$	$y - E(Y)$	$[x - E(X)][y - E(Y)]$	$P(x, y)[x - E(X)][y - E(Y)]$
2, 0	.10	−1.52	−1.57	2.3864	.2386
2, 1	.08	−1.52	−0.57	0.8664	.0693
2, 2	.03	−1.52	0.43	−0.6536	−.0196
2, 3	.02	−1.52	1.43	−2.1736	−.0435
3, 0	.07	−0.52	−1.57	0.8162	.0571
3, 1	.10	−0.52	−0.57	0.2964	.0296
3, 2	.06	−0.52	0.43	−0.2236	−.0134
3, 3	.03	−0.52	1.43	−0.7436	−.0223
4, 0	.04	0.48	−1.57	−0.7536	−.0301
4, 1	.05	0.48	−0.57	−0.2736	−.0137
4, 2	.12	0.48	0.43	0.2064	.0248
4, 3	.06	0.48	1.43	0.6864	.0412
5, 0	.01	1.48	−1.57	−2.3236	−.0232
5, 1	.02	1.48	−0.57	−0.8436	−.0169
5, 2	.06	1.48	0.43	0.6364	.0382
5, 3	.15	1.48	1.43	2.1164	.3175
					$C(X, Y) = .6336$

TABLE 5-14
Computation of Covariance for Data of Table 5-12 by Using Equation (5-38)

x	P(x)	xP(x)	y	P(y)	yP(y)	x, y	P(x, y)	xyP(x, y)
2	.23	0.46	0	.22	0	2, 0	.10	0
3	.26	0.78	1	.25	0.25	2, 1	.08	0.16
4	.27	1.08	2	.27	0.54	2, 2	.03	0.12
5	.24	1.20	3	.26	0.78	2, 3	.02	0.12
		$E(X) = 3.52$			$E(Y) = 1.57$	3, 0	.07	0
						3, 1	.10	0.30
						3, 2	.06	0.36
						3, 3	.03	0.27
						4, 0	.04	0
						4, 1	.05	0.20
						4, 2	.12	0.96
						4, 3	.06	0.72
						5, 0	.01	0
						5, 1	.02	0.10
						5, 2	.06	0.60
						5, 3	.15	2.25
								6.16

$$C(X, Y) = 6.16 - (3.52)(1.57) = .6336$$

indicates that the two variables tend to covary, or "run together." Since $C(X, Y)$ is positive, household size and car ownership are positively correlated. Unfortunately, the covariance is not an easily interpretable measure of correlation. The numerical value of $C(X, Y)$ is completely dependent on the magnitudes of the two random variables X and Y. If either of these two variables is measured in different units, then $C(X, Y)$ must change. As we shall see in Chapter 11, Pearson's product-moment correlation coefficient standardizes the covariance and overcomes this problem.

First, however, we must introduce the notion of independence of two random variables X and Y. Two variables are said to be *independent* if

$$P(x, y) = P(x)P(y) \qquad (5\text{-}39)$$

for all possible combinations X and Y. To show that two variables are *dependent*, it is sufficient to show that Equation (5-39) does not hold for *any* one combination of X and Y. Clearly, household size and car ownership are not independent but covary. To see this, examine the relation given by (5-39) for the values $x = 3$ and $y = 1$. From Table 5-12

$$P(x = 3) = .26$$

$$P(y = 1) = .25$$

$$P(x = 3, y = 1) = .10$$

so that $P(x = 3)P(y = 1) = (.26)(.25) = .065 \neq P(x = 3, y = 1) = .10$.

Independence and covariance are closely related. If two random variables X and Y are independent, then their covariance must equal zero. However, it is not necessarily true that the two variables are independent if the covariance is zero. It is possible for two variables to have a covariance $C(X, Y) = 0$, yet still not satisfy the demanding requirements of Equation (5-39).

Bivariate Normal Random Variables

For *continuous* bivariate random variables, the joint probability distribution is specified by using a density function $f(X, Y)$. The most important distribution is the *bivariate normal distribution*. Figure 5-20 contains a graphical representation of a bivariate normal distribution. For every pair of (x, y) values there is a density $f(x, y)$, represented by the height of the surface at that point. The surface is continuous, and probability corresponds to the volume under the surface. If two random variables are jointly normally distributed, then

1. The *marginal* distributions of both X and Y are univariate normal.
2. *Any* *conditional* distribution of X or Y is also univariate normal.

The implication of item 2 is that the cross section of a "slice" through the bivariate normal for a given value of X, say $X = x_a$, has the characteristic shape of the normal curve. A slice at $Y = y_a$ has the same characteristic. Both cases are illustrated in Figure 5-20.

There are five parameters to the bivariate normal density function: μ_X and σ_X, μ_Y and σ_Y, and ρ_{XY} (pronounced "row sub XY"). The first four parameters are the means and standard deviations of the marginal distributions for X and Y. The final

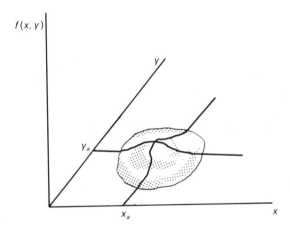

FIGURE 5-20. Bivariate normal probability density function.

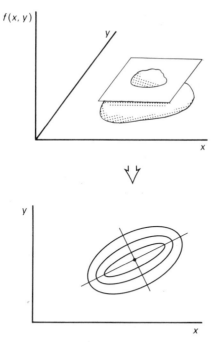

FIGURE 5-21. Three- and two-dimensional representations of bivariate normal distribution.

parameter ρ_{XY} is the *correlation coefficient* between random variables X and Y and is defined as

$$\rho_{xy} = \frac{C(X,\,Y)}{\sigma_X \sigma_Y} \tag{5-40}$$

The correlation coefficient is a pure number, and it can take on any value between -1 and $+1$ inclusive. If variables X and Y are independent, then $C(X,\,Y) = 0$ and therefore $\rho_{XY} = 0$. If the two variables are perfectly positively correlated, then $\rho_{XY} = 1$. If $\rho_{XY} = -1$, then the perfect linear relation is a negative or inverse one.

It is common to portray a bivariate normal distribution by using a contour diagram. A contour diagram is created by taking horizontal slices through the bivariate normal distribution, as in Figure 5-21. A contour is composed of all $(x,\,y)$ pairs having a constant density $f(x,\,y)$. The contour curves of all bivariate normal distributions are elliptical, except where $\rho_{XY} = 0$ and $\sigma_X = \sigma_Y$. In this single situation the contours will appear as concentric circles. Several examples of bivariate normal distributions are shown in Figure 5-22. In (a) note that $\rho_{XY} = 0$ and the two standard deviations are equal, so all contours are circular. In (b), X and Y are independent since $\rho_{XY} = 0$, but the contours are elliptical since $\sigma_X > \sigma_Y$. In both (c) and (d), variables X and Y are positively related and $\rho_{XY} > 0$. The principal axis of each ellipse has a positive slope, indicating that the surface tends to run along a line with positive slope. Similarly, when X and Y are negatively related and $\rho_{XY} < 0$, the principal axis

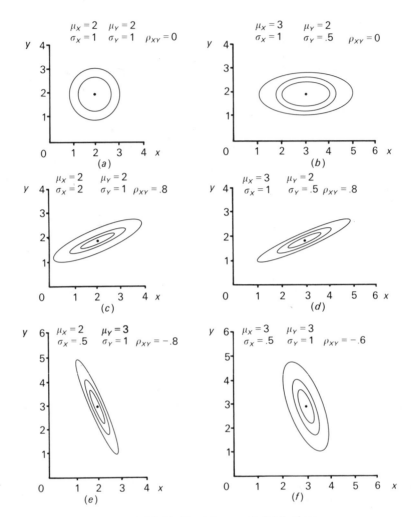

FIGURE 5-22. Bivariate normal distributions.

has a negative slope. This case is illustrated in (e) and (f). In all cases in Figure 5-22, the centers of the ellipses are at coordinates (μ_X, μ_Y). The role of the parameters is clearly illustrated by these eight cases. The two means μ_X and μ_Y control the location of the center of the surface, and the two standard deviations and the correlation coefficient control the shape.

5.5. Summary

In this chapter the notion of a random variable has been formally introduced. We recognize that there are two different types of random variables, continuous and discrete. Five important probability distribution models for random variables were

TABLE 5-15
Summary of Five Probability Distributions

Distribution	Continuous (C) or discrete (D)	Parameters	Mean	Variance
Discrete uniform	D	k	$\dfrac{k+1}{2}$	$\dfrac{k^2-1}{12}$
Binomial	D	n, π	$n\pi$	$n\pi(1-\pi)$
Poisson	D	λ	λ	λ
Continuous uniform	C	a, b	$\dfrac{a+b}{2}$	$\dfrac{(a-b)^2}{12}$
Normal	C	μ, σ	μ	σ^2

introduced in Sections 5.2 and 5.3. The parameters, mean, and variance of these five distributions are summarized in Table 5-15. In addition, Section 5.4 developed the ideas underlying bivariate probability distributions and introduced the bivariate normal distribution. The material presented in Section 5.6 proves useful when we begin to study the relationships between variables in Chapters 11, 12, and 13.

Appendix 5A

There is a direct correspondence between the area under a probability density function between any two limits c and d and the probability that a continuous random variable X assumes a value between these two limits. Students familiar with calculus will recognize that the area under $f(x)$ between c and d is given by

$$\int_c^d f(x)\, dx$$

where the symbol $\int$ represents the integration operator. It is directly analogous to the summation operator Σ used for discrete variables. The requirement that the sum of the probabilities assumable by a random variable equal 1 can be stated as

$$\int_a^b f(x)\, dx = 1$$

where a and b are the limits of the values assumable by the random variable. For continuous random variables defined over the infinite length of the real number line, the appropriate equation is

$$\int_{-\infty}^{+\infty} f(x)\, dx = 1$$

A second important property is that the probability of a continuous random variable assuming a particular value, say $X = c$, is always zero. That is,

$$\int_c^c f(x)\, dx = 0$$

since any definite integral over the limits c to c must be zero. The expected value of a continuous random variable is

$$E(X) = \int_a^b x f(x)\, dx$$

and the variance $V(X)$ is

$$V(X) = \int_a^b [x - E(X)]^2 f(x)\, dx = \int_a^b x^2 f(x)\, dx - [E(X)]^2$$

As an example, let us consider the uniform continuous random variable defined by

$$f(x) = \begin{cases} \frac{1}{10} & 0 \le x \le 10 \\ 0 & \text{otherwise} \end{cases}$$

First, to show that the area under $f(x)$ equals 1, we note

$$\int_0^{10} \tfrac{1}{10}\, dx = \tfrac{1}{10}(10 - 0) = 1$$

To calculate the expected value, we write

$$E(X) = \int_0^{10} x \left(\frac{1}{10}\right) dx = \int_0^{10} x\, dx$$

$$= \frac{1}{10} \frac{x^2}{2} \Big|_0^{10}$$

$$= \frac{1}{10} \left(\frac{10^2}{2} - \frac{0^2}{2}\right)$$

$$= 5$$

By solving the general case

$$E(X) = \int_a^b x \left(\frac{1}{b - a}\right) dx$$

it is possible to derive the general formula $E(X) = (b + a)/2$, given as (5-28). The variance is

$$V(X) = \int_0^{10} x^2 \left(\frac{1}{10}\right) dx - [E(X)]^2$$

$$= \frac{1}{10} \frac{x^3}{3} \Big|_0^{10} - 5^2$$

$$= \frac{1}{10} \left(\frac{1000}{3}\right) - 25$$

$$= 8.33$$

By solving the general case

$$V(X) = \int_a^b x^2 \left(\frac{1}{b-a}\right) dx - [E(X)]^2$$

the variance of a uniform continuous random variable can be shown to be $V(X) = (b - a)^2/12$. This is given in the text as (5-29).

FURTHER READING

Introductory statistics textbooks for geographers seldom discuss the concept of random variables in detail, nor do they give an extended treatment of probability distribution models. Introductory statistics textbooks for business and economics students and some texts for applied statistics courses usually give a more complete presentation of these topics. Three representative examples are Neter, Wasserman, and Whitmore (1982); Pfaffenberger and Patterson (1981); and Winkler and Hays (1975). The development of the Poisson probability distribution using the analogy to a dot map only hints at the widespread use of probability models in this situation. Additional material on quadrat analysis and a related technique known as nearest-neighbor analysis is provided in the problems below. Extended treatment of these topics can be found in Getis and Boots (1978), Taylor (1977), and Unwin (1981).

A. Getis and B. Boots, *Models of Spatial Processes* (Cambridge, England: Cambridge University Press, 1978).
J. Neter, W. Wasserman, and G. Whitmore, *Applied Statistics* (Boston: Allyn and Bacon, 1982).
R. Pfaffenberger and J. Patterson, *Statistical Methods for Business and Economics* (Homewood, Ill.: Richard Irwin, 1981).
P. Taylor, *Quantitative Methods in Geography* (Boston: Houghton Mifflin, 1977).
D. Unwin, *Introductory Spatial Analysis* (London: Methuen, 1981).
R. Winkler and W. L. Hays, *Statistics: Probability, Inference and Decision Making*, 2d ed. (New York: Holt, 1975).

PROBLEMS

1. Explain the meaning of the following terms:
   a. Random variable
   b. Probability distribution
   c. Expected value of a random variable
   d. Variance of a random variable
   e. Bernoulli trial
   f. Binomial coefficients
   g. Pascal's triangle
   h. Point pattern analysis
   i. Standard normal distribution
   j. Standard score
   k. Correction for continuity
   l. Bivariate probability function
   m. Covariance
   n. Correlation coefficient

2. Differentiate between the following:
   a. A discrete and a continuous random variable
   b. A probability mass function and a probability density function
   c. A normal distribution and the standard normal distribution
   d. A marginal probability distribution and a conditional probability distribution
   e. A probability mass function and a cumulative mass function

3. Explain what is meant when we say two random variables X and Y are (a) dependent and (b) independent.

4. Let X be a random variable with the following probability distribution:

x	$P(x)$
0	.25
1	.50
2	.15
3	.10

   a. Verify that this is a valid probability distribution model.
   b. Determine $E(X)$.
   c. Determine $V(X)$.
   d. What is the mode of X?

5. A census of all households in a town is undertaken, and the number of trips made by members of each household on a given day is recorded. The trip frequencies follow this probability distribution:

Trips per day x	Probability of making x trips $P(x)$
0	.13
1	.14
2	.21
3	.18
4	.10
5	.08
6	.07
7	.05
8	.03
9	.01

   a. Portray this distribution as a graph.
   b. Graph the cumulative mass function.
   c. Find $E(X)$ and $V(X)$.
   d. Use Chebyshev's theorem to determine the proportion of values of x contained within 2 standard deviations of the mean. Compare this proportion with the actual proportion computed directly from this table.

6. Graph the probability mass functions for the following discrete probability distribution models:
   a. Uniform: $k = 3$, 5, and 10
   b. Binomial: $n = 5$, $\pi = .20$; $n = 5$, $\pi = .50$; $n = 5$, $\pi = .70$
   c. Poisson: $\lambda = .20$, .40, and .80

7. The number of persons X in a camping party arriving at a wilderness park is a uniform discrete random variable. The maximum number in a party is 9.
   a. Determine $P(3)$ and $P(7)$.
   b. Find $P(X < 6)$.
   c. Determine $P(X \geq 6)$.
   d. What is the expected number of persons in a camping party?
   e. What is $V(X)$?

8. Suppose the number of persons in a camping party is a Poisson random variable with a mean of $\lambda = 3$. Solve (a) to (e) of Problem 7.

9. The maximum temperature reached on any day can be classified as above freezing (a success!) or below freezing (a failure?). In a certain city of eastern North America, January weather statistics indicate the probability a January day will be above freezing is .30. Use the binomial distribution to determine these probabilities:
   a. Exactly 2 of the next 7 January days will be above freezing.
   b. More than 5 of the next 7 days will be above freezing.
   c. There will be at least 1 day above freezing in the next 7 days.
   d. All seven days in the next week will be above freezing.
   e. Is this a reasonable application of the binomial distribution? Why or why not?

The remaining problems involve *quadrat analysis*. In addition, material is introduced and problems are defined for a closely related technique known as *nearest-neighbor analysis*.

10. A grid of squares has been placed over a map, and the number of points (say, houses) falling in each square is counted. The number in each quadrat of the 25-cell map illustrated below represents this frequency.

0	1	1	1	0
0	2	2	1	2
1	0	0	2	0
1	1	1	2	3
0	3	0	1	0

   a. Construct a frequency distribution of points per quadrat.
   b. Determine the mean number of points per quadrat.
   c. What would the frequency distribution of points be if the point pattern were random (i.e., given by the Poisson distribution) with mean given by the value calculated in part (b)?
   d. Compare the frequency distribution generated in part (a) with the frequency distribution predicted by the Poisson distribution for these 25 quadrats. Does the Poisson distribution appear to give a reasonable fit?

11. Complete parts (a) to (d) of Problem 10 for the following map.

1	0	0	0	0	0	0	2	0	1
0	0	0	0	0	0	0	1	0	0
1	0	0	1	1	0	0	0	2	0
0	0	0	0	0	0	0	0	0	1
0	1	0	0	0	0	1	1	0	0
0	0	0	0	0	0	0	1	2	0
0	2	0	1	0	1	0	1	0	0
1	0	0	0	1	1	0	0	0	0
0	0	0	0	2	0	0	0	0	0
0	1	0	0	0	0	0	1	0	1

12. One problem in quadrat analysis is that the results are highly dependent on the size of the quadrats chosen. Conclusions drawn from a study based on a certain quadrat size may be contradicted by conclusions of a second study of the same data based on a different quadrat size. This phenomenon might be termed the *scale problem*, by analogy to the scale problem identified in descriptive geographical statistics of Chapter 4. To examine the impact of quadrat size, complete parts (a) to (d) of Problem 11, using the same base map, but with the quadrats grouped into 4s, forming 25 larger square quadrats.

13. A second problem in quadrat analysis is that markedly different point patterns can give rise to identical frequency distributions of points by quadrats. This is because the frequency distributions cannot indicate whether quadrats with zero points are located close to or away from other quadrats with zero points. In Chapter 3 we termed this the *pattern problem*. To see the significance of this problem, perform the following experiment. Construct two 5×5 grids of 25 quadrats each. Into each grid place a distribution of points with the following frequency distribution:

Number of points	Number of quadrats
0	12
1	9
2	3
3	1

In the first map, place the points in the grid so that the distribution appears random. In the second map, place the points so that the map appears clustered.

14. Quadrat analysis also suffers from the effects of the boundary problem and the modifiable areal units problem identified in Chapter 3. Generate dot maps that expose the significance of each of these problems.

15. In this chapter, we have shown that an independent random spatial process can be modeled by the Poisson probability distribution. If a map has a random distribution of points, the Poisson distribution should provide a good fit to the frequency distribution of points per quadrat. However, there are many potential point distributions aside from a random

one. In fact, a random distribution can be considered to be an intermediate pattern, a mix of two other types: dispersed and clustered. Consider these five point patterns:

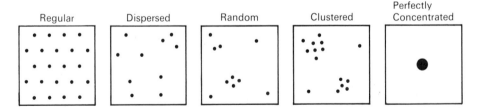

First, there are dispersed patterns. These are distributions of points more or less evenly spread over the map. The limiting case is termed *regular* and describes a pattern in which each point is equidistant from its neighbors. This is known as a *triangular lattice* of points. *Clustered* patterns have one or more groups of points in clusters and large areas of the map without any points. In the limiting case, all points share one location, and we have a perfectly clustered pattern. A random pattern has elements of both clustered and dispersed patterns. One way of classifying point patterns as clustered, dispersed, or random is to examine the variance/mean ratio of the observed frequency distribution of points per quadrat. A random pattern will have a variance/mean ratio of exactly 1, since it can be described by the Poisson distribution for which $V(X) = E(X) = \lambda$. Hence, $V(X)/E(X) = \lambda/\lambda = 1$. Dispersed patterns will have small variance/mean ratios, and clustered patterns will have variance/mean ratios larger than 1.

a. Briefly explain why the variance/mean ratio of dispersed patterns will be low, and conversely why clustered patterns will have high variance/mean ratios.

b. Classify the following five frequency distributions of quadrat counts as clustered, random, or dispersed based on their variance/mean ratios:

Number of points in a quadrat, x	Number of quadrats				
	Map A	Map B	Map C	Map D	Map E
0	18	1	15	61	10
1	6	5	25	30	20
2	1	12	25	7	40
3	0	5	30	1	20
4	0	1	4	1	5
5	0	1	1	0	5
	25	25	100	100	100

Hint: Use Equations (2-5) and (2-20) to estimate the mean and variance.

16. One alternative to quadrat analysis that avoids some of the difficulties described in Problems 12 to 14 is termed *nearest-neighbor analysis*. By using this technique it is also possible to classify point patterns. The method is quite straightforward. The distance d_{ij} between each pair of points i and j in a given point pattern is calculated by using Pythagoras's theorem. For each point $i = 1, 2, \ldots, n$, the closest point is determined, that is, $\min_j d_{ij}$. The mean, or average, of all these nearest-neighbor distances is denoted $\bar{d}_a$. Unfortunately, this measure cannot be used to compare maps since it depends especially on the size of the area depicted on the map. What is needed is some standard against which $\bar{d}_a$ can be

compared. The obvious standard is the expected average distance between nearest neighbors in a random point pattern. It turns out that the expected mean nearest-neighbor distance for a random point pattern is given by

$$\bar{d}_e = \frac{1}{2\sqrt{n/A}}$$

where n is the number of points in the pattern and A is the area of the study region. Therefore, the ratio $R = \bar{d}_a/\bar{d}_e$ is always equal to 1 if the point pattern being analyzed is random. Where there is a clustered point pattern, distances to nearest neighbors will be small, $\bar{d}_a$ is less than $\bar{d}_e$, and thus $R < 1$. For a perfectly regular point pattern, R attains the maximum possible value of 2.149. Thus, for dispersed point patterns $1 < R \leq 2.149$. The minimum value of R is zero, and it occurs when the point pattern is perfectly concentrated and all points share the same location. In this case $\bar{d}_a = 0$ and $R = 0$. So for clustered point patterns $0 < R < 1$. Coordinates for a set of $n = 20$ points of five maps A, B, C, D, and E are given below. The coordinates are based on a 10×10 km grid laid over an area of $A = 100$ km^2.

	Map coordinates				
Point	A	B	C	D	E
1	(2, 1)	(0.5, 0.5)	(2, 1)	(1, 3)	(2, 4)
2	(2, 3)	(4, 0.25)	(2, 3)	(1, 5)	(2, 4.5)
3	(2, 5)	(7, 0.75)	(1, 2)	(1, 9)	(2, 5)
4	(2, 7)	(9.5, 0.5)	(3, 2)	(2, 1)	(2, 5.5)
5	(2, 9)	(5, 2)	(5, 4)	(2, 4)	(2, 6)
6	(4, 1)	(2, 2.25)	(1, 5.5)	(3, 2)	(1.5, 5)
7	(4, 3)	(4, 4)	(4, 5.5)	(3, 6)	(1.5, 1.5)
8	(4, 5)	(5, 5)	(3, 6)	(3, 10)	(3,5)
9	(4, 7)	(8, 4)	(3, 7)	(4, 8)	(3, 5.5)
10	(4, 9)	(8.5, 3.5)	(3.5, 6)	(5, 1)	(1.5, 6)
11	(6, 1)	(8.5, 4.5)	(6, 5)	(5, 4)	(7, 4)
12	(6, 3)	(2, 6.5)	(6, 5.5)	(6, 6)	(7, 4.5)
13	(6, 5)	(0.5, 9.5)	(6, 6)	(6, 9)	(7, 5)
14	(6, 7)	(3, 8)	(6, 6.5)	(7, 2)	(7, 6)
15	(6, 9)	(5, 9.5)	(9, 2)	(7, 4)	(6.5, 4)
16	(8, 1)	(9.5, 9.5)	(8, 9)	(8, 7)	(6.5, 6)
17	(8, 3)	(4, 7.5)	(8, 8)	(8, 9)	(8, 4)
18	(8, 5)	(4.5, 7)	(8, 7)	(9, 3)	(8, 5)
19	(8, 7)	(4.5, 8)	(9, 8.5)	(9, 8)	(8, 6)
20	(8, 9)	(5, 7.5)	(7, 8.5)	(10, 5)	(7, 5.5)

a. Draw maps of the five point patterns. Use graph paper.
b. Calculate distance between each point and its nearest neighbor.
c. Determine $\bar{d}_a$ for each map, the actual mean distance to the nearest neighbor.
d. Find $\bar{d}_e$ for a map with $n = 20$ points and $A = 100$ km^2.
e. Calculate R for each map. Classify each map as clustered, dispersed, or nearly random.
f. Suppose the actual area on the map from which the coordinates are taken is actually 200 km^2. Does this alter any of your conclusions?

6

Sampling

In Chapter 1, statistical methodology was conveniently divided into *descriptive statistics* and *inferential statistics*. In inferential statistics, a descriptive characteristic of a sample is linked with probability theory so that a researcher can generalize the results of a study of a few individuals to some larger group. This idea was made more explicit in Chapter 5, where the notion of a *random variable* and its *probability distribution* were defined. At the core of inferential statistics is the distinction between a *population* and a *sample*. From a statistical universe or population, a small subset of individuals is selected for detailed study. This sample is used to estimate the value of some population characteristic or to answer a question about a particular characteristic of the population. However, to make such inferences, the sample must be collected in a specific way. It is not possible to make statistically reliable inferences from *any* sample. Whereas street corner interviews, for example, tend to make interesting news items, they may not reflect the views of the population they are supposed to represent.

Ideally, it would be best to have a sample that is a good representation of the population from which it has been drawn. Quality inferences are made by using quality samples. Unfortunately, unless we know everything about the population, say from a census, we have no way of knowing if we do have a representative sample. The very act of sampling thus introduces some uncertainty into our inferences, simply because the sample may *not* be representative of the population. This is known as *sampling error*. Suppose, for example, we wished to sample the students at a university and determine the number of hours the average student spends studying in any given week. By mere chance, we may just select a sample which includes more industrious students than average and thus overestimate the amount of time spent studying. We might be led to believe that the average student spent more time studying than is, in fact, the case. Our only safeguard against such sampling error is to select a larger sample. The larger the sample, the more likely it includes a true cross section of the population, that is, the more likely it is *representative* of the population.

Besides sampling error, there is another reason why our sample may not be representative of the population. This could occur if the way in which the sample is collected is itself biased. This is known as *sampling bias*. In the example of university student study habits, a sample would surely be biased if it were selected by interviews of students leaving the university library late in the evening!

DEFINITION: SAMPLING BIAS
Sampling bias occurs when the procedures used to select the sample tend to favor the inclusion of individuals of the population with certain population characteristics.

Sampling bias can usually be avoided, or at least minimized, by selecting an appropriate sampling plan. Errors in recording, editing, and processing sample data can likewise be limited by various checks. When data are collected through mail questionnaires, a form of bias due to *nonresponse* often occurs. The respondents to the questionnaire may not be representative of the overall population. Many studies have found these respondents to be more highly educated, wealthier, and more interested in the subject of the questionnaire than members of the population at large. Since the quality of the inferences from a sample depends so much on the sample itself, clearly any researcher must carefully select a sampling plan capable of minimizing, or at least controlling to acceptable limits, both sampling error and sampling bias.

Sampling techniques with this characteristic are the focus of this chapter. First, the advantages of sampling are enumerated in Section 6.1. Why do we favor a sample over a complete census of a population? In Section 6.2 an extremely useful four-step procedure for sampling is outlined. The tasks defined in these steps are encountered in every sampling problem. Various types of samples are identified in Section 6.3. Only specific types of samples can be used to generalize to a population—with a *known* degree of risk. The most commonly used type, the *simple random sample*, is explained in Section 6.4. It is then compared to a few other sample designs. In Section 6.5 the concept of a *sampling distribution* is introduced. The sampling distribution of sample statistics such as the mean $\bar{X}$ and proportion p is central to both the estimation and the hypothesis testing procedures of statistical inference. Finally, the issues in geographic sampling are presented in Section 6.6.

6.1. Why Do We Sample?

Seldom must we collect information from all members of a population in order to make reliable statements about the characteristics or attributes of that population. Often a sample constituting only a small percentage of the total population is sufficient for such inferences. There are several reasons for choosing a *sample* rather than *census* of an entire population.

1. Usually it is not *necessary* to take a complete census. Valid, reliable generalizations about the characteristics of a population can be made with a sample of modest size—if the sample is properly taken. The uncertainty inherent in generalizing from the few to the many not only is within acceptable limits, but sometimes is even less than the uncertainties which arise when we try to precisely control the enormous amount of data generated from a complete enumeration of an extremely large population. It is simply far easier to check the data of a small sample than those of a large population.

2. The time, cost, and effort of collecting data from a sample are usually

substantially less than are required to collect the same information from a larger population. The workforce or available financial resources usually constrain a researcher from taking a full census.

3. The population of interest may be infinite, and therefore sampling is the only alternative. We could, for example, consider the population to be the water temperature at a certain depth at a given reach of a particular stream. There are an infinite number of times when we could record the water temperature—even in a small time interval. Since space itself can be treated as a continuous variable, there are an infinite number of places in any area where a set of sample measurements could be taken. This issue is explored more fully in Section 6.6.

4. The act of sampling may be destructive. To estimate the mean lifetime of light bulbs, for example, any light bulb in a sample must be tested until it is no longer of use. A census of the light bulbs produced by a manufacturer would destroy the entire production!

5. The population may be only hypothetical. In the case of the light bulb manufacturer, the real population of interest is the set of bulbs that *will* be produced by the manufacturer in the future. At the time any sample is taken, this population is not observable.

6. The population may be empirically definable, but not practically available to a researcher. Not all slopes in a study region may be accessible to a geomorphologist interested in studying the dynamics of freeze-thaw weathering. Even an experienced climber may find only a few suitable sites for study.

7. Information from a population census can be quickly outdated. Given the volatility of political polls, it would certainly be unwise to determine the party supported by each member of a population. *Repeated* censuses of this type would be impossible, and sufficient accuracy can be obtained by using only a small proportion of voters. Repeated polls of this type are a usual occurrence now.

8. When the topic of the study requires an in-depth study of individuals in the population, only a small sample may be possible. By restricting attention to only a few individuals, extremely comprehensive information can be collected. A study of the residential mobility of residents of a large city might require detailed questions concerning the history of moves, characteristics of the current and past residences, motivations for each move, the search process used to locate new residences, and characteristics of the household itself. Probing for sufficient detail in all these areas precludes the possibility of a complete census of the population.

Such a task would certainly be beyond the resources of most institutions. For a variety of reasons, then, many research questions must be answered through the use of a small sample from a population. Providing the sample is collected properly, valid conclusions about key characteristics of the population can be drawn, with only a surprisingly small degree of uncertainty.

6.2. Steps in the Sampling Process

Having decided that no suitable data exist to answer some research question, and concluding that a sample is the only feasible method of collecting the necessary data,

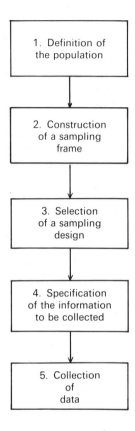

FIGURE 6-1. Steps in the sampling process.

the researcher must specify a sampling plan. Rushing out to collect the data as quickly as possible is often the *worst* thing that could be done. It is far better to follow the simple five-step sampling procedure illustrated in Figure 6-1. Many potential problems not anticipated by the researcher can be addressed and successfully solved *before* a considerable effort has been put into the actual task of data collection. In literally hundreds, if not thousands, of studies, insufficient time and care in devising a sampling plan have led to the collection of large data sets with only limited possibilities for statistical inference. Let us consider each of these steps in turn.

Definition of the Population

The first step is to define the population. What at first glance might appear to be a trivial task often proves to be an extremely difficult chore. It is easy to conceive of a statistical population as a collection of individual elements which may be individual people, objects, or even locations. However, to actually identify which individuals should be included in the population and which should be excluded is not so simple. To see some of the potential issues and difficulties, consider the problem of defining the population for a study of the elderly in some particular city. A number of practical questions immediately surface:

1. How will we distinguish the elderly from the nonelderly, by age? If so, what age? Age 60? Age 65?
2. Or, will the elderly be defined by occupational categories? Should we restrict ourselves to retired individuals? Or should we restrict the population to individuals over 65 years of age *and* retired?
3. Will we include all elderly, or those living independently, that is, not in some long-term care home?
4. Is the study concerned with elderly individuals or elderly households? What about mixed households with both elderly and nonelderly members?

As you can see, even if we can conceptualize the population of interest, arriving at a strict, operationally useful definition may require considerable thought and difficult choices. In the study of the elderly, it is still necessary to define a time frame and a geographical limit to the study region.

Construction of a Sampling Frame

Once we have chosen the specific definition to be used in identifying the individuals of a population, it is necessary to construct a *sampling frame*.

DEFINITION: SAMPLING FRAME

A sampling frame (also called a *population frame*) is an ordered list of the individuals of a population.

There are two key properties of the sampling frame. First, it must include all individuals in the population; that is, it must be exhaustive. Second, each individual element of the population must appear once and only once on the list. Obtaining a sampling frame for a particular population may itself be a time-consuming task. It is usually easy to compile a list of all current students at a university in a given academic year from existing academic records or transcripts. But what if the population of interest is not regularly monitored in any way? Where, for example, could a list of all the elderly residents of a city be obtained? It may be possible to extract a fairly complete list of the elderly by examining the list of recipients of social security or old-age assistance from a government agency. But would the list include *all* the elderly? What about noncitizens or residents otherwise ineligible for this type of aid?

As a second example, consider the use of telephone surveys for evaluation of voter preferences for political parties. Although the population of interest is all eligible voters, the population actually sampled is composed of those residents with telephones—or, more accurately, the set of individuals who answer these phones. There is clearly a great deal of overlap between these two groups, but they are not exactly the same. Restricting ourselves to those with listed telephone numbers may exclude some relatively wealthy residents with unlisted numbers as well as some poorer households without telephones. So it is useful to distinguish the *target* population from the *sampled* population.

DEFINITION: TARGET AND SAMPLED POPULATIONS

The target population is the set of all individuals relevant to a particular study. The sampled population consists of all the individuals listed in the sampling frame.

Obviously, it is desirable to have the sampled and target populations as nearly identical as possible. When they do differ, it is extremely important to know the particular way(s) in which they do differ, since this is a form of sampling bias. It is sometimes necessary to qualify the inferences made by using a sampled population which differs in significant ways from the target population. This is equivalent to recognizing the limitations imposed on the study by the sampling frame.

Selection of a Sample Design

Next we must decide how we are going to select individuals from the sampling frame to include in the sample.

DEFINITION: SAMPLE DESIGN

A sample design is a procedure used to select individuals from the sampling frame for the sample.

There are several ways that this could be done. We could select a sample of size n simply by taking the first n individuals listed in the sample frame. Or we could select the last n individuals or every kth individual on the list until we get n members for the sample. There are many types of samples and sample designs. Because of the importance of this step, it is described in depth in Sections 6.3 and 6.4. At this point, it is sufficient to note that a *random* sample is an extremely useful design in statistical analysis. Individuals to be included in the sample are chosen by using some procedure incorporating chance. The mechanical devices used in many lotteries are one example. An urn is filled with identical balls, one for each member of the sampled population. The sample is chosen by selecting balls from a well-mixed urn, one at a time.

The important characteristic of this type of sample is that we are able to assess the probability that each individual in the population is included in the sample. In this case, each individual has an equal chance of being included. A number of variations of this design are explained in Section 6.4.

Specify the Information to Be Collected

This step can usually be accomplished at *any* point prior to the commencement of data collection. The particular format used to collect data must be rigorously defined and pretested by using a pilot sample.

In a field study, the pretest can be used to check instruments, data loggers, and all
other logistics. For surveys—mail, telephone, or personal interview—the pretest can
sometimes reveal deficiencies for any of the following reasons: difficulty in locating
population members; dealing with an abnormally high percentage of refusals or
incomplete questionnaires; problems in questionnaire wording, sequence, or format;
unanticipated responses; or inadequately trained interviewers.

Collection of Data

Once all the problems indicated in the pretest have been successfully solved, the
ultimate task of data collection can begin. At this stage, careful tabulation and
editing are particularly important if we wish to minimize nonsampling error.

6.3. Types of Samples

In this section, we expand on the ideas discussed in the third step of the sampling
process, the selection of a sampling design. Sampling designs can be conveniently
divided into two classes: probability samples and nonprobability samples. Simple
random sampling is one type of probability sample.

Because we know only the probability that an individual is included in the sample, an
element of chance, or uncertainty, is introduced into any inferences made from the
sample. Expressed simply, the random sample may be quite unrepresentative of the
population it is supposed to reflect. The advantage of a probability sample is that we
can determine the *probable* accuracy of our results. The four principal types of
probability samples are shown in Figure 6-2. Because of their importance in statistical inference, these samples are discussed in depth in the following section.

Nonprobability samples may also be excellent or poor representations of the
population. The difficulty is that whether it is a good or bad sample can never be
determined. Four types of nonprobability samples are sometimes used to collect
sample data.

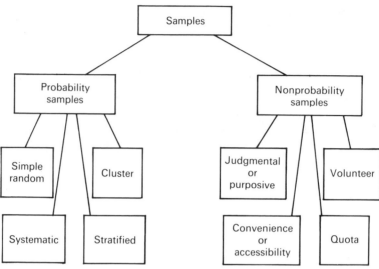

FIGURE 6-2. Types of samples.

DEFINITION: JUDGMENTAL, OR PURPOSIVE, SAMPLE

A judgmental, or purposive, sample is one in which personal judgment is used to decide which individuals of a population are to be included in the sample. These are individuals whom the investigator feels can best serve the purpose of the sample.

Obviously, a very skillful investigator with considerable knowledge of a population can sometimes generate extremely useful data from such a sample. If these deliberate choices turn out to be poor, however, poor inferences are made. The risk is not known. Purposive samples are sometimes selected in pretest, or pilot, samples. A range of respondents, including both "typical" and "unusual" individuals, is chosen. This sample is used to get an idea of what the full range of questions in the survey should be. In addition, it is not unusual to use these interviews to preview the types of answers likely to be given by the respondents in the actual survey. Many times, this information can be used to significantly improve the survey instrument. It would be completely erroneous to utilize such a sample to draw conclusions about the whole population.

DEFINITION: CONVENIENCE, OR ACCESSIBILITY, SAMPLE

A convenience, or accessibility, sample is one in which only *convenient*, or *accessible*, members of the population are selected.

Since they include only easily identifiable members of a population, convenience and accessibility samples are subject to sampling bias. These individuals are almost always special or different from other population members in some way. Street corner interviews rarely reflect overall opinion; only certain individuals will allow themselves to be interviewed by the media.

There are two other types of nonprobability samples. Before probability samples were widely accepted in empirical surveys, a pseudoscientific approach known as *quota sampling* was common.

DEFINITION: QUOTA SAMPLING

A quota sampling is an attempt to obtain a representative sample by instructing interviewers to obtain data from given subgroups of the population.

The basic idea behind quota sampling is sound. Suppose, for example, that we were attempting a voters' poll and knew voter preference varied by whether the respondent was male or female, resided in an urban or rural area, and was young or old. If we knew the breakdowns of the population among these categories, we would instruct one of our interviewers to obtain a specific number of young male voters in some given rural area. Similar instructions could be given to other interviewers for other categories. Some judgment is involved since the interviewer must select the actual respondents in that category. A more sophisticated form of quota sampling known as *stratified* random sampling is discussed in the following section.

Another type of sample which fails to take adequate safeguards against sampling bias is the *volunteer sample*. Individuals who volunteer to take part in a survey self-select themselves from the population; they are rarely representative. Usually, these individuals are more motivated and have a higher interest in the topic than the general population. Since the respondents to mail questionnaires sometimes have these characteristics, this type of survey is often subject to nonresponse bias.

The sampled population units in a probability sample are always chosen according to some probability plan. This minimizes the sampling bias. Even though the resulting sample need not be representative of the population, we can still determine the probable accuracy of our results. For these reasons, a probability sample is the preferred way in which a sample should be drawn from a population.

6.4. Simple Random Sampling and Related Probability Designs

In this section, we develop the notion of one kind of probability sample, and we illustrate its operationalization by using the random number table. Three other probability sampling techniques—stratified, systematic, and cluster sampling—are also discussed and compared to simple random sampling.

In most social science applications of simple random sampling, the population from which we are sampling is finite. Nevertheless, it is useful to distinguish between two definitions for simple random sampling, one that applies when the population is finite in size and the other that applies when the population is infinite.

DEFINITION: SIMPLE RANDOM SAMPLE AND FINITE POPULATION

A simple random sample from a finite population is one in which each possible sample of a given size n has an equal probability of being selected.

To illustrate this definition, consider a population consisting of the four elements A, B, C, and D. We wish to draw a sample of size $n = 2$ from this population according to the definition given. First, how many samples of size $n = 2$ are there? As illustrated in Figure 6-3, there are $C_2^4 = 6$ samples of size $n = 2$, and each must have an equal chance of being selected. Notice that the samples AA, BB, CC, and DD are not included in this list. When each unit of the population is selected for the sample, it is not replaced, so it can be drawn a second time. We restrict ourselves to such sampling, known as *sampling without replacement*. In fact, by not putting the unit first drawn back into the population, we must be aware that sampling without replacement does not yield a *statistically independent* sample. For such a sample, we would require that the act of choosing any observation from our sample not affect the probability of choosing any other element of the population. To develop a statistically independent sample, we would have to put each unit chosen back into the population, so that it could be drawn again. In most instances, the difference between these two samples is insignificant. The procedure based on sampling without replacement is thus used in this case.

It is relatively easy to select a simple random sample. For most populations and samples of modest size, it is not feasible to develop a list of all possible samples, such as in Figure 6-3. For a population of size $N = 10,000$ and a sample of $n = 100$, there are an impossibly large number of samples to list. Instead, we select one element for the sample at a time from the sampling frame. To illustrate this procedure, we consider the problem of selecting a sample of size $n = 10$ from a population with $N = 87$ members. The procedure described for this case can be easily generalized to larger samples from larger populations.

To perform this procedure, we number each individual in the sampling frame

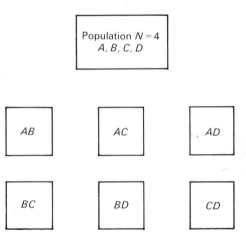

FIGURE 6-3. Samples of size $n = 2$.

consecutively from 1 to 87. Using some type of randomization device, we select the first person from the population. There are several ways that this can be done. One way is to take a set of 87 identical balls, numbered from 1 to 87, and select one from an urn in which the balls have been thoroughly mixed. This is often done mechanically in lottery draws. We then note the individual on the list corresponding to this ball, and we include that element in the sample. Nine more balls are drawn to complete the sample of $n = 10$ elements. The selected balls are not replaced prior to the next draw. This somewhat cumbersome procedure can be replaced by another randomization device—the *random number table*.

DEFINITION: RANDOM NUMBER TABLE

A random number table is a table of digits developed in such a way that each $0, 1, 2, \ldots, 9$ has an equal probability of occurring for each digit in the table.

A portion of a random number table generated by a computer is shown in Table 6-1. In an extended version of this table we would expect each of the 10 digits 0, 1, $2, \ldots, 9$ to appear with equal frequency. Because it is completely random, the table can be used in many ways. It is possible to start with any digit in the table and proceed in any direction. Because each single digit is random, each pair or set of three, four, or five numbers is also a random number.

The next step in the random sampling procedure is to associate one or more random numbers with each member of the population listed in the sampling frame. For any population with less than $N = 100$ members, it is possible to associate one two-digit random number with each of these individuals. The random number pair 01 refers to the first individual listed in the sampling frame, the pair 02 refers to the second individual, and so on. The random number pair 00 is assigned to the last, or $N = 100$th, individual of the population. Any random number we draw thus leads us to a particular individual in the population. For a population with up to $N = 100$ members, a three-digit random number can be associated with each individual in the sampling frame. For a population of $N = 200$ members, we can associate five three-digit random numbers with each individual. The digits 001, 002, 003, 004, and 005 all identify the first individual in the sample.

TABLE 6-1
200 Random Digits

61	56	24	90	10	28	59	01	45	47
28	61	34	93	93	02	83	73	99	29
42	40	63	36	82	63	15	01	28	36
68	02	63	83	41	24	30	10	18	50
41	05	14	37	40	64	10	37	57	18
04	59	34	82	76	56	80	00	53	36
69	16	43	43	46	04	35	64	32	05
74	90	09	63	92	36	59	16	05	88
25	29	81	66	99	22	21	86	79	64
43	12	55	80	46	31	98	69	22	22

To select the sample, choose a place in the random number table to begin—for example, by looking away from the page and placing a finger somewhere in the table. Suppose this technique leads to the entry in the fourth row and fourth column, 83. The 83d individual in the sampling frame is the first element chosen for the sample. Proceeding by columns (rows or even diagonals are also possible), we see that the following individuals will be selected for the first seven members of the sample:

Sample No.	Sampling frame identification
1	83
2	37
3	82
4	43
5	63
6	66
7	80

Thus, moving to the next column, we see the eighth member of the sample is identified as individual 10 on the sampling frame. The next random number of the list is 93. Since this is larger than 87, it is discarded and sampling is continued. The last two individuals are then selected by using the random numbers 82 and 41.

This simple procedure can be followed in all cases where random sampling from a finite population is required. Many computers, including even the smallest microcomputers, contain random number generators. In some cases it is simply a matter of entering the required sample size and the population size; the random number generator returns the list of individuals to be selected from the sampling frame.

Whenever the population from which we are sampling is infinite, it is necessary to modify the previous definition of simple random sampling. There are an infinite number of potential samples for any given sample size n.

DEFINITION: SIMPLE RANDOM SAMPLE IN AN INFINITE POPULATION
From an infinite population, a simple random sample is selected in such a way that all observations chosen are statistically independent.

As an example, consider the problem of selecting a simple random sample from the light bulbs produced on a single assembly line by a manufacturer. Conceptually at least, the manufacturer could continue to produce an infinite number of light bulbs. However, as the production process is operating, the machine may gradually wear out, and so we can no longer assume that the sample light bulbs are statistically independent. Since most populations in the social sciences can be considered to be finite, the definition of simple random sampling for finite populations is applicable. It is still possible to collect random samples from the infinite populations sometimes encountered in geographical research. Despite the fact there are an infinite number of time intervals in 1 day (or even a shorter period), it is still possible to sample randomly *during* this period. Similarly, legitimate random samples can be taken from maps (or in the field), even though space is continuous and there are an infinite number of places to collect the sample. This topic is explored in detail in Section 6.6.

Systematic Sampling

The *systematic random sample* is a variation on simple random sampling. This method is generally easier to operationalize than simple random sampling in terms of both time and cost.

DEFINITION: SYSTEMATIC RANDOM SAMPLE

To draw a systematic random sample, every k th element of the sampling frame is chosen, beginning with a randomly chosen point.

Consider a population with $N = 200$ elements from which we wish to draw a random sample of size $n = 10$. This sample can be drawn by selecting every $k = N/n = \frac{200}{10} = 20$th individual listed on the sampling frame. The next thing to be decided is *where* on the sampling frame selection is to begin. The easiest procedure is to draw a single random number between 01 and 20. The possible samples include the following:

$$
\begin{array}{ccccc}
01 & 21 & 41 & \cdots & 181 \\
02 & 22 & 42 & \cdots & 182 \\
03 & 23 & 43 & \cdots & 183 \\
\cdots\cdots\cdots\cdots\cdots\cdots\cdots \\
19 & 39 & 59 & \cdots & 199 \\
20 & 40 & 60 & \cdots & 200
\end{array}
$$

Why is a systematic random sample *not* a simple random sample? Notice there are not C_{10}^{200} different samples that could be selected—only 20. For example, it is impossible to collect a sample which includes individuals 2, 3, and 27. Only a small fraction of the potential samples of size n may be drawn.

The systematic sample is likely to be as good as a random sample provided that the arrangement of the individuals in the sampling frame is random. Where there are regularities or periodicities in the sampling frame, these will be picked up in the systematic sample. To see the potential problems this may create, consider a sampling frame which includes all the apartments in some high-rise block. Suppose there are 10 floors with 20 apartment units per floor. As is usual in such cases, apartment units above and below any given apartment are identical in floor plan, size, and other features. A systematic sample of size $n = 10$ would probably include 10 identical units. Further, we might expect them to be occupied by roughly similar sized households. Our systematic sample might include 10 households, each living in a two-bedroom unit. This would hardly be a representative sample if the building contained, one-, two-, and three-bedroom units. As we shall see in Section 6.6, this potential for bias also exists in spatial sampling. Before a systematic sample is drawn, it is always necessary to check for randomness in the sampling frame itself. If the list has been collected in an orderly manner, then one must be sure that such an ordering will not lead to a biased sample through systematic sampling. Many obvious periodicities exist in data ordered in time sequence or by regular spatial intervals.

Stratified Sampling

One of the drawbacks of simple random sampling is that there is always the possibility that the individuals selected for the sample may be unrepresentative of the population. Unlike with simple random sampling, the composition of a stratified random sample is not left entirely to chance.

DEFINITION: STRATIFIED RANDOM SAMPLE
A stratified random sample is obtained by forming classes, or strata, in the population and then selecting a simple random sample from each.

The basic idea behind stratified sampling is illustrated in Figure 6-4. The population is divided into a set of strata before sampling takes place. The divisions must be both exhaustive and mutually exclusive. That is, each member of the population must appear in one and only one class. Simple random samples are drawn independent of class. Population units may be stratified on the basis of one or more characteristics. Social surveys often stratify the population by age, income, and location of residence.

Why do we stratify? If it is properly organized, stratified sampling can use the additional control over the sampling process to reduce sampling error. Put simply, stratified sampling can decrease the likelihood of obtaining an unrepresentative sample. To reduce sampling error, we must define *homogeneous* strata, that is, strata consisting of individuals who are very much alike in terms of the principal characteristic of the study. Suppose we are interested in the trip-making behavior of the

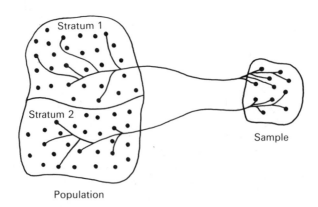

FIGURE 6-4. Stratified random sampling from a population with two strata.

households in a metropolitan area. Knowing that households with two or more cars make more total trips, make different types of trips, and are less likely to use public transport means that they are less variable than the population as a whole. By defining subpopulations or substrata with less *internal* variability, we effectively reduce sampling error. A stratified sample is preferred whenever two conditions are met: First, we must know something about the relationship between certain characteristics of the population and the problem being analyzed. Second, we must actually be able to achieve stratification, that is, to identify population members in each class. Sometimes we may not be able to develop a sampling frame for this purpose. Where, for example, could we obtain a list of households according to the number of cars owned? Various post hoc procedures of classifying households after they are sampled are usually less than satisfactory. Because of the advantages of stratification, it should be undertaken whenever possible. Virtually all political polls employ stratified sampling techniques, particularly by geographic area.

Cluster Sampling

The notion of cluster sampling is illustrated in Figure 6-5. First, the population is divided into mutually exclusive and exhaustive classes, as in the initial step in stratified sampling. These classes are usually defined on the basis of convenience, rather than according to some variable thought to be important to the problem under study. Next, certain clusters are selected for detailed study, usually by some random procedure. However, a random sample is *not* drawn from *each* cluster. Cluster sampling is then completed by either taking a complete census of those clusters selected for sampling or selecting a random sample from these clusters. These units are then combined and constitute the complete cluster sample. Ideally, each cluster should be internally heterogeneous. This is in direct contrast to the situation in stratified sampling where it is desired to define the strata to be as internally homogeneous as possible.

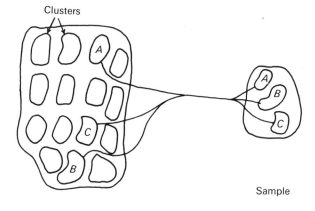

FIGURE 6-5. Cluster sampling.

Cluster sampling can often give very poor results. The sampling error from a cluster is usually higher than the sampling error from a simple random sample or a stratified sample. A cluster sample is effective only when the classes or clusters defined for the problem are each representative of the population as a whole. If they are not representative, then the clusters selected for the sample are likely to be biased and create considerable sampling error. Randomly or arbitrarily defined clusters are unlikely to be representative. Obviously, great care should be taken in specifying the clusters into which any population is to be divided.

Why are cluster samples taken? Usually, the reason is the efficiencies in time and cost. These efficiencies are particularly important whenever the population to be sampled is geographically dispersed over a large area. Suppose, for example, that the problem under study dictates that a personal interview be taken from a sample of such a population. If we were to select a random or systematic sample, this would require a considerable amount of traveling and hence would be a very costly collection procedure. But by restricting our sampling to a small number of clusters, each covering only a small area, these costs are significantly reduced.

Choice of Sampling Design

In terms of efficiency alone, stratified samples are best and cluster samples are worst. Random samples usually lie somewhere between these two extremes. That is, for a given level of precision, stratified samples require the fewest respondents and cluster samples require the most respondents. Unfortunately, sampling efficiency alone does not dictate the type of design chosen in any study. The selection of a stratified design presumes that the relevant information necessary to define the strata and construct the sampling frame is available. Also, the time and cost of collecting the data must be considered. If the sample requires a survey of some human population, then another concern is whether the information is to be collected through a personal interview, telephone interview, or a mail questionnaire.

Finally, many sampling designs used in practice are hybrids which combine elements of several different designs. A sample of housing in a metropolitan area might use stratification to ensure adequate representation of black, Oriental, and other ethnic groups. Within these strata, cluster sampling might be used to select particular neighborhoods and/or blocks within these neighborhoods. A simple random or systematic random sample might be used to select the individual households for the sample.

6.5. Sampling Distributions

In Chapter 2, we introduced several numerical measures of data sets, such as the mean, median, standard deviation, and variance. At that time, we made a distinction between numerical measures calculated for a data set which could be considered a *population* and measures for a data set which is a *sample*. Let us now clarify the distinction between such numerical measures. If the measure is computed from a population data set, it is termed the *value of the population parameter*; if it is calculated from a sample data set, it is called the *value of the sample statistic*. For a finite population, the *population mean parameter* and the *population standard deviation parameter* are defined in the following way.

DEFINITION: POPULATION MEAN AND STANDARD DEVIATION PARAMETERS (FINITE POPULATION)

Let $x_1, x_2, \ldots, x_N$ be the values of some population characteristic X in a finite population of N elements. The value of the population mean parameter, denoted μ, is

$$\mu = \frac{\sum_{i=1}^{N} x_i}{N} \tag{6-1}$$

and the value of the population standard deviation parameter, denoted σ, is

$$\sigma = \sqrt{\frac{\sum_{i=1}^{N}(x_i - \mu)^2}{N}} \tag{6-2}$$

There is only *one* population mean parameter and one population standard deviation parameter in any population. All population *parameters* are denoted by Greek letters.

The equivalent numerical measures in the sample, the values of the *sample statistics* for the mean standard deviation, are defined as follows.

DEFINITION: SAMPLE MEAN AND STANDARD DEVIATION STATISTICS

Let $x_1, x_2, \ldots, x_n$ be a set of values of n elements taken from a population. The value of the sample mean statistic $\bar{x}$ is defined as

$$\bar{x} = \frac{\sum_{i=1}^{n} x_i}{n} \tag{6-3}$$

and the value of the sample standard deviation statistic s is defined by

$$s = \sqrt{\frac{\sum_{i=1}^{n}(x_i - \bar{x})^2}{n-1}} \tag{6-4}$$

There are two principal differences between these two formulas. First, sample statistics are calculated by using the n values of a sample, and population parameters are calculated by using the N values of the population, $n < N$. Second, the divisor within the radical of σ is N, but for s it is $n - 1$. However, there is another more important difference. Whereas there is only one value for a population parameter such as μ, there are many different possible values for $\bar{x}$. Why? Each different sample of size n drawn from a finite population of N elements can have different value for $\bar{x}$. In fact, *if we choose a random sample from the population, then the value of the sample mean must itself be a random variable.*

In Chapter 4, we distinguished a random variable by an uppercase letter X and a specific value of this random variable by the corresponding lowercase symbol x. This distinction is also useful in the sampling context. When we refer to a random sample of size n drawn from a population *before* a value has been drawn, the sample is designated $X_1, X_2, \ldots, X_n$. *After* a single sample has been drawn, the *values* of these random variables are known. These n numbers constitute the sample $x_1, x_2, \ldots, x_n$ is a set of random variables. They are random variables since the value of any of these variables depends on which member of the population is drawn while sampling. As we noted in Section 6.4, randomness is introduced into sampling through the use of a random number table.

Sample Statistics

Now, the sample mean is also a random variable since its value is not known before the sample is actually drawn. Prior to taking the sample, the sample mean is a random variable

$$\bar{X} = \frac{\Sigma_{i=1}^{n} X_i}{n} = \frac{X_1 + X_2 + \cdots + X_n}{n} \tag{6-5}$$

Formally, we say that $\bar{X}$ is a *sample statistic*.

DEFINITION: SAMPLE STATISTIC

A sample statistic is a random variable based on the sample random variables $X_1, X_2, \ldots, X_n$. The *value* of this random variable can be determined once the values of the observations in a specific random sample have been drawn from the population.

Just as there are many numerical measures of a data set, so there are many sample statistics. In fact, each measure has an associated sample statistic. Thus, for example, the sample mean statistic $\bar{X}$ given by Equation (6-5) is the random variable, and its value is given by (6-3). This notation which differentiates $\bar{X}$ from $\bar{x}$ is necessary so that we appreciate that the statistic $\bar{X}$ will vary from sample to sample for a fixed sample size n.

Since $\bar{X}$ is known to be a random variable, the obvious question is: What is the distribution of this random variable?

Sampling Distributions

DEFINITION: SAMPLING DISTRIBUTION OF A STATISTIC

The sampling distribution of a statistic can be developed by taking all possible samples of size n from a population, calculating the value of the statistic for each sample, and drawing the distribution of these values.

To illustrate this concept, let us generate the sampling distribution of the sample mean $\bar{X}$ for the following population consisting of five elements:

Element	Values of X
A	$x_1 = 6$
B	$x_2 = 6$
C	$x_3 = 5$
D	$x_4 = 4$
E	$x_5 = 4$

Now first let us calculate the population mean and standard deviation, using formulas (6-1) and (6-2):

$$\mu_X = \frac{\Sigma_{i=1}^{5} x_i}{5} = \frac{6+6+5+4+4}{5} = 5 \tag{6-6}$$

$$\sigma_X = \sqrt{\frac{\Sigma_{i=1}^{5}(x_i - \mu_X)^2}{5}}$$

$$= \sqrt{\frac{(6-5)^2 + (6-5)^2 + (5-5)^2 + (4-5)^2 + (4-5)^2}{5}}$$

$$= \sqrt{\tfrac{4}{5}} = .894 \tag{6-7}$$

Notice the divisor in each case is equal to 5, the number of elements in the population.

Suppose a simple random sample of size $n = 3$ is to be taken from this population. How many samples are possible? There are $C_3^5 = 10$ different samples of size 3. These possibilities are listed in Table 6-2, along with the value of the statistic $\bar{X}$. The sampling distribution of the statistic $\bar{X}$, the distribution of sample means, is

TABLE 6-2
Possible Samples of Size $n = 3$ from Population $N = 5$

Elements in sample	Values of X	Mean $\bar{x}$
ABC	6, 6, 5	5.67
ABD	6, 6, 4	5.33
ABE	6, 6, 4	5.33
ACD	6, 5, 4	5
ACE	6, 5, 4	5
ADE	6, 4, 4	4.67
BCD	6, 5, 4	5
BCE	6, 5, 4	5
BDE	6, 4, 4	4.67
CDE	5, 4, 4	4.33

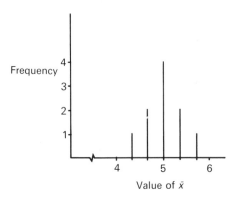

Value of $\bar{x}$ **FIGURE 6-6.** Sampling distribution of $\bar{X}$.

illustrated in Figure 6-6. Computing the mean and standard deviation of this distribution yields

$$\mu_{\bar{X}} = \frac{\Sigma_{i=1}^{10} \bar{x}_i}{10} = \frac{5.67 + 5.33 + \cdots + 4.33}{10} = 5 \tag{6-8}$$

and

$$\sigma_{\bar{X}} = \sqrt{\frac{\Sigma_{i=1}^{10}(\bar{x}_i - \mu_{\bar{X}})^2}{10}}$$

$$= \sqrt{\frac{(5.67 - 5)^2 + (5.33 - 5)^2 + \cdots + (4.33 - 5)^2}{10}} = .365 \tag{6-9}$$

The notation used to denote the mean of $\bar{X}$ is $\mu_{\bar{X}}$. Note that this is *not* the same as μ_X,
even though $\mu_{\bar{X}} = \mu_X$. Notice that the divisor for μ_X in (6-5) is 5, the number of
elements in the population, but it is 10 for $\mu_{\bar{X}}$ in (6-8), the number of samples.
However, we would expect the sample average to be distributed about the population
mean. Indeed, one of the important properties of the sample mean statistic is that its
mean is equal to the population mean.

To compute the sample standard deviation, the mean of each sample of size 3
(given in the last column of Table 6-2) is subtracted from the mean of $\bar{x}$ and squared.
The sum is then divided by 10, the number of samples of size 3 that can be drawn
from the population with $N = 5$ elements. The fact that $\sigma_{\bar{X}} < \sigma_X$ should not be
surprising. We should expect less variability in the distribution of the means $\bar{X}$ than
in the population itself. Why? Because when we take a sample of items from the
population, we tend to select samples in which large values of the variable are
averaged with small values, and the result lies close to the population mean μ. To
distinguish σ_X from $\sigma_{\bar{X}}$, it is common to call σ_X the population standard deviation and
$\sigma_{\bar{X}}$ the *standard error of the sampling distribution of means.* By convention, the standard deviation of a sampling distribution is known as its standard error.

Inferential statistics would be a time-consuming process if it were necessary to
manually generate sampling distributions in this way. Fortunately, this is not the
case. A set of theorems can be used to specify the nature of the sampling distributions

of most sample statistics under quite general conditions. We illustrate the principal theorem involved, using as an example the sampling distribution of the sample mean $\bar{X}$.

Central Limit Theorem

DEFINITION: CENTRAL LIMIT THEOREM

Let $X_1, X_2, \ldots, X_n$ be a simple random sample of size n. This random sample is assumed to be drawn from a population random variable which is normally distributed with mean μ and standard deviation σ. Then the sampling distribution of $\bar{X}$ (1) is normally distributed, (2) has a mean of μ, and (3) has a standard deviation of $\sigma_{\bar{X}} = \sigma/\sqrt{n}$.

The implications of the central limit theorem are illustrated in Figure 6-7. There are two qualifications on this result. First, the population must be infinite. Second, the population random variable must be normally distributed. We discuss the consequences of these assumptions later.

First, let us make two important observations from this theorem. The most important observation is the *inverse* relation between sample size and the variability of the sample statistic $\bar{X}$. As the sample size n increases, the variability of $\bar{X}$ decreases since $\sigma_{\bar{X}} = \sigma/\sqrt{n}$ falls. Suppose the population standard deviation is $\sigma = 100$. Then the following relations always hold:

n	1	4	16	25	100
$\sigma_{\bar{X}}$	100	50	25	20	10

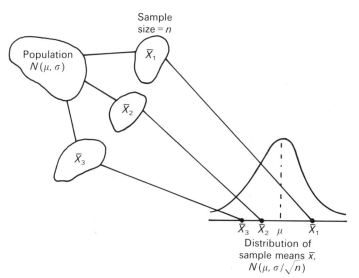

FIGURE 6-7. Central limit theorem and the distribution of sample means.

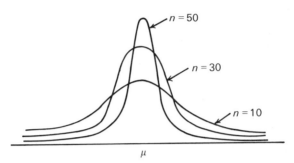

FIGURE 6-8. Sampling distributions of $\overline{X}$ with increasing n.

For $n = 1$, $\sigma_{\overline{X}} = 100$. This is intuitively appealing. If we were to randomly select one element of the population, we would expect our choice to exactly mirror the underlying population. Our second observation concerns the limit of $\sigma_{\overline{X}}$ as n increases. If the population is infinite, then n may be increased without bound. The limit of $\sigma/\sqrt{n}$ as n approaches infinity is zero. What does this imply? This tells us that as n increases, the sampling distribution of $\overline{X}$ converges to the mean $\mu_{\overline{X}} = \mu$, the value of the population mean. This is also intuitively obvious. We would expect the value of $\overline{X}$ to be closer to μ as the sample size n becomes larger. This feature is illustrated in Figure 6-8.

It is necessary to distinguish between situations where the random variable X is normally distributed and where it is not. Where X is *not* normally distributed, this theorem also holds but the sampling distribution of $\overline{X}$ is only approximately normal. This approximation improves with increasing sample size. This feature can be illustrated by a simple example. Suppose the population characteristic of interest is the time between the arrival of recreation parties at a wilderness park. Let us make two different assumptions regarding the distribution of the population random variable X, the time between arrivals:

1. Variable X is normally distributed with mean $\mu = 8$ h and standard deviation $\sigma = 4$ h.
2. Variable X is uniformly distributed over the interval from 0 to 16 h, with mean $\mu = 8$ h.

These two possibilities are illustrated in Figure 6-9.

Hypothetical samples were drawn from each population with the help of a computer. One hundred random samples were drawn from each population for sample sizes $n = 5$, $n = 50$, and $n = 100$. The frequency distributions of the sample means are illustrated in Figure 6-10. First, consider the case in which X is normally distributed. Except for the impact of sampling error, each of the sampling distributions of the sample mean statistic $\overline{X}$ is normally distributed. We could generate increasingly normal frequency distributions simply by taking more than 100 repeated samples at each of the three samples. Nevertheless, the sampling distribution of $\overline{X}$ for $n = 100$ is almost perfectly normal. Where the population random variable is

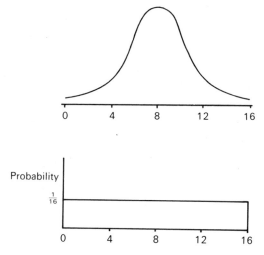

FIGURE 6-9. Alternative population distributions for time between arrivals.

uniform, the sampling distributions are only approximately normal. The distribution gradually converges to normality and is near normal by $n = 100$. The fact that the sampling distributions are normal, no matter what the form of the underlying population random variable, is quite remarkable.

To this point, we have used the central limit theorem to specify the sampling distribution of $\bar{X}$ for infinite populations. There is one significant modification to this

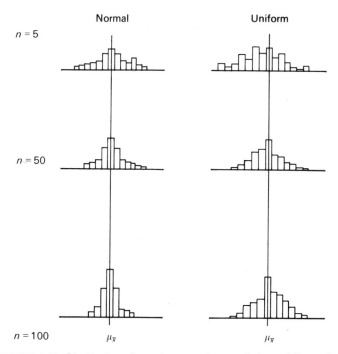

FIGURE 6-10. Distribution of sample means for populations of Figure 6-9.

theorem when the underlying population is finite. The standard deviation of the sampling distribution of $\bar{X}$ reduces to

$$\sigma_{\bar{X}} = \frac{\sigma}{\sqrt{n}} \sqrt{\frac{N-n}{N-1}} \tag{6-10}$$

The square root term is called the *finite-population correction factor*. It is necessary to correct this standard deviation because each time an element is selected from the finite population to be placed in the sample, the variability of the population is reduced. For an infinite population, such a correction is unnecessary. The finite-population correction factor compensates for this difference by adjusting the standard deviation of $\bar{X}$. If n is relatively small in relation to N, the value of the correction factor can safely be ignored. For example, suppose we had a sample of $n = 50$ from a population of $N = 100,000$. The correction factor is $\sqrt{(100,000 - 50)/(100,000 - 1)} = 0.9995$. Notice it is close to 1. When can the correction factor be ignored? A number of different alternative rules of thumb have been suggested. The simplest rule is to ignore the correction factor when we have a large sample size ($n \geq 30$) and the sample constitutes less than about 5 percent of the population size N. We can use Equation (6-10) to generate the identical result of (6-9). By (6-10),

$$\sigma_{\bar{X}} = \frac{.894}{\sqrt{3}} \sqrt{\frac{5-3}{5-1}} = .365$$

This is the same result generated by a complete enumeration of all possible samples.

The central limit theorem is not limited to the statistic $\bar{X}$. It applies to many other sample statistics including the sample variance S^2 and the sample proportion P. The specification of the sampling distributions for these statistics is left to Chapter 7.

The real value of this theorem is that it produces a method of deducing results concerning the outcome of a sample, based on only a knowledge of the population mean and standard deviation. In particular, the theorem allows us to determine the *probability* that the sample mean statistic $\bar{X}$ is larger than a given value, is smaller than a given value, or even falls within a given interval. This is best illustrated by a simple numerical example. Let us return to the example in which X is the time between arrivals at a wilderness park. We suppose that the time between arrivals is normally distributed with $\mu = 8$ with $\sigma = 4$ h. What is the probability that in a sample of $n = 36$ arrival intervals the mean time between arrivals is greater than 10 h?

Since the random variable X = time between arrivals is normally distributed, the central limit theorem is applicable. The statistic $\bar{X}$ is normally distributed with $\sigma_{\bar{X}} = 4/\sqrt{36} = .667$ h. This sampling distribution is shown in Figure 6-11. We are interested in determining the probability that $\bar{X}$ is greater than 10 h, the shaded area of the sampling distribution. To find this probability, we must first standardize the variable $\bar{X}$, using

$$z = \frac{\bar{x} - \mu_{\bar{X}}}{\sigma_{\bar{X}}} = \frac{10 - 8}{.667} = 3.0$$

Then, using the normal tables, we see the desired probability is $.5000 - .49865 = .00135$.

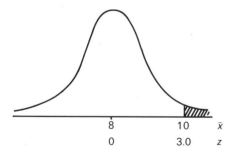

FIGURE 6-11. Sampling distribution of arrival time intervals for $\mu = 8$, $\sigma = 4$, and $n = 36$.

In Chapter 7 we use this theorem in another way. We take a characteristic of a sample such as $\bar{X}$ and then determine the probability that it comes from a population with hypothetical population characteristics.

6.6. Geographic Sampling

There are at least two instances in which geographers must take spatially distributed samples. First, there is the case where the geographer must go into the field and sample some areally distributed phenomenon. A biogeographer may wish to study the distribution of a particular plant species or even the distribution of seeds from a particular plant or set of plants. As another example, field sampling may be required for the compilation of a soil map. Second, sometimes areal sampling is necessary in direct sampling of locations on a map. This is particularly important when we are sampling from maps over which some phenomenon is continuously distributed. Land-use maps of cities, soil maps, and vegetation maps are typical examples of this form of sampling. For phenomena *discretely* distributed on a map, conventional sampling procedures can be used. Consider the problem of selecting a set of farms distributed on a map of some rural area. It is sufficient to compile a list of all farms in the area and then sample from this sampling frame in the usual way. We may even sample from all parts of the map by dividing the list into strata based on location. Such a procedure is basically independent of the map itself. For the complex patterns that exist on most continuously variable maps, such a procedure is totally inappropriate. Instead, these maps are sampled by using traverse samples (or lines), quadrats, or points. These three alternatives are illustrated in Figure 6-12.

Before examining the sampling designs used with each of these techniques, we specify the sampling frame implicit in spatial sampling. This is equivalent to the ordered list of the elements in a population used in conventional nonspatial sampling. For sampling from a map, the Cartesian coordinate system is sufficient to identify any point on the map and hence any element of the population. This coordinate system meets all the requirements of a sampling frame. It is ordered, with a unique address for each element of the population. The (x, y) coordinate pairs are mutually exclusive since no two points on the map can possibly have the same coordinates.

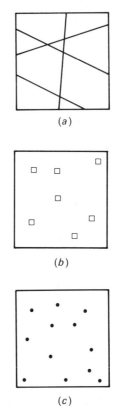

(a)

(b)

(c)

FIGURE 6-12. (a) Traverse samples; (b) quadrat samples; (c) point samples.

Finally, it is exhaustive of all elements in the population since it covers the entire map.

Consider the coordinate system used to sample from the area illustrated in Figure 6-13. The two axes follow the usual north-south and east-west orientations, with the origin placed at the lower left, or southwest, corner of the map. The axes are then subdivided into units at any desired degree of accuracy. Two possibilities are shown in Figure 6-13. The simplest one-digit grid uses the numbers 0 to 9 to identify particular points on both the x and y axes. Then random numbers can be used to locate any point on the map. The first random number drawn is used to define the x coordinate, and the second number drawn the y coordinate. If one-digit random numbers are used, then only the 100 points shown as grid intersections can actually be sampled. This *available population* is quite different from the *target*, or total, *population*, which is the entire map. If two-digit random numbers are used for the grid on each axis, then there are $100 \times 100 = 10,000$ possible points to be included in this sample. The grid used to define the sampling frame should be sufficiently fine that features with only small areas are included in the available population. Depending on the map in question, it may be necessary to use two-, three-, or even four-digit grid references so that the set of available points adequately covers the target map.

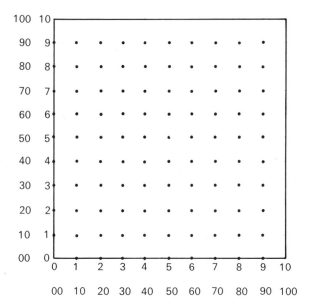

FIGURE 6-13. Coordinate systems and areal sampling.

Quadrat Sampling and Sampling Traverses

One of the more commonly used sampling methods is quadrat sampling. Quadrat sampling differs slightly from the quadrat systems used to analyze point patterns. Variations on three different sampling designs are illustrated in Figure 6-14. First, consider the random quadrat sampling design in (*a*). Such a design can be developed by using three simple rules:

1. Select a quadrat size. There is some choice in size, but larger quadrats sometimes become difficult to analyze. If one is forced to calculate, say, the proportion of the quadrat in various land use categories, then this task itself becomes extremely time-consuming for a large quadrat.
2. Draw two random numbers. Use the first random number to locate the x coordinate of the quadrat centroid and the second to locate the y coordinate of the centroid.
3. Select as many quadrats as necessary to reach a sample size deemed sufficient.

As a variation of this algorithm, it is also possible to randomly orient the quadrats by selecting a third random number to control the orientation of the quadrats. This alternative is shown in Figure 6-14(*b*). For very complex maps, the irregular distribution of the map pattern *within* a quadrat may make analysis very difficult. In such a case, point sampling may be a desirable alternative.

A systematic sampling design is illustrated in Figure 6-14(*c*). A pair of random numbers are used to locate one quadrat on the map, and the remaining quadrats are

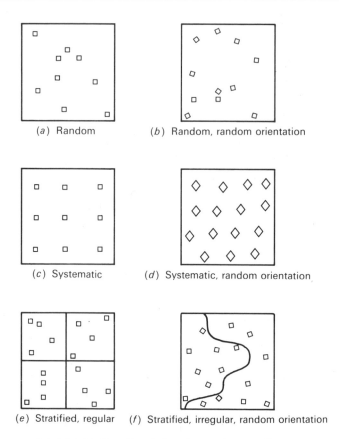

FIGURE 6-14. Sampling designs for quadrat sampling.

placed according to a prespecified spacing interval which can yield a sufficiently large sample. In Figure 6-14(*d*) a systematic sample with a random orientation is illustrated. In the stratified sampling designs of Figure 6-14(*e*) and (*f*), the number of quadrats in any stratum is proportional to the area of the map in that stratum. These areally defined strata may be based on either regularly or irregularly shaped portions of the map. In both designs the quadrats are randomly located within each stratum.

Four designs for sampling traverses are illustrated in Figure 6-15. To locate these traverses, the sampling frame must be slightly modified. As illustrated in Figure 6-15(*a*), the grid references are defined in a clockwise manner from the lower left corner. Again, there is some choice of precision depending on the number of digits used in the grid referencing system. Let us suppose we wish to sample L mi of linear traverse. For the random sampling design, this is accomplished in the following way. First, a single two-digit random number is drawn. This number specifies a point on the map boundary which can be used as one end of the first traverse. A second number is used to locate a point on the boundary for the other end of the traverse. A line connecting these two point determines the first traverse. Next, we determine the length of this line. Let this length be l_1 mi. We now have l_1 mi in our sample and

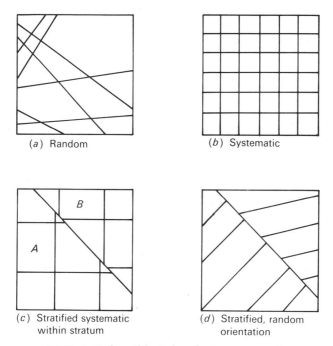

(a) Random   (b) Systematic

(c) Stratified systematic within stratum   (d) Stratified, random orientation

FIGURE 6-15. Sampling designs for traverse sampling.

require only $L - l_1$ mi to complete the sample. Two more random numbers are used to locate a second traverse, the length of the sample traverse is updated to $l_1 + l_2$ mi, and the remaining length to be sampled is reduced to $L - l_1 - l_2$ mi. This process is continued until *exactly l* mi, or alternatively until *at least L* mi of traverse, has been selected for sampling.

Let us consider the sequence of traverses generated in a study attempting to estimate the proportion of a 100-mi^2 map that is woodland, given a traverse of 200 mi. Figure 6-16 illustrates the sequence of traverses used to generate the sample. The first traverse is 15 mi long, of which 7 mi passes through woodland (W) areas and 8 mi through nonwoodland (NW) areas. These figures are recorded in the first three columns. A cumulative, or running, total of the traverse length, in woodland and in nonwoodland areas, is provided in the last three columns. Notice that to complete the sample of *exactly* 200 mi, the last traverse is truncated, and only 7 of its 10 mi are included in the sample. Alternatively a sample of $L = 203$ mi could be used by accepting the complete length of the final traverse in the sample. The estimated proportion of woodland on the map is $\frac{78}{200} = .39$.

The systematic design of Figure 6-15(b) is based on a regular pattern of perpendicular traverses with a prespecified interval. Suppose a sample of $L = 250$ traverse miles is to be taken from a 25-mi^2 area. There will be five traverses in each of the north-south and east-west directions, each 25 mi long. A spacing of 5 mi between traverses is used. A single random number can be used to locate the east-west traverses and another to locate the north-south ones. Where there are periodicities on

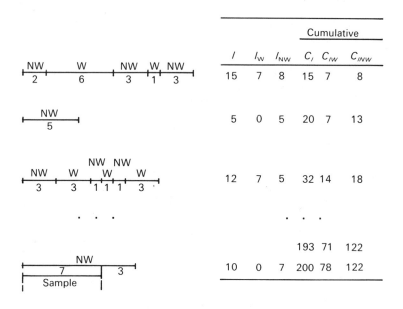

			Cumulative		
l	l_w	l_{NW}	C_l	C_{lW}	C_{lNW}
15	7	8	15	7	8
5	0	5	20	7	13
12	7	5	32	14	18
			193	71	122
10	0	7	200	78	122

$$\text{Proportion of woodland} = \frac{78}{200} = .39$$

FIGURE 6-16. Estimating the proportion of woodland on a map by using traverses.

the map, particularly linear or near-linear features, this sampling procedure can lead to biased estimates of certain map characteristics. Streets, highways, stream networks, eskers, and other similar map features can lead to significant sampling error when maps are sampled by using a systematic design.

It is possible to randomly select the orientation of traverses in a systematic sample, although this is slightly more difficult to operationalize. Additional modifications are also required in sampling from an irregularly shaped area. Nevertheless, there are significant time efficiencies to be realized from systematic sampling.

An areally stratified sampling design for an irregular map is illustrated in Figure 6-15(c). The areas or strata may be defined contour intervals, geomorphic or geologic features, or even municipal boundaries in some urban area. Two strata, A and B, are depicted on the map of Figure 6-15(c), in the proportion 60:40. Sixty percent of the traverse length should thus be drawn from area A and 40 percent from B. As usual, random numbers are drawn to locate individual traverses. As each traverse is selected and enumerated, a running total is kept of the total traverse length in each of the strata. Once the length of the traverses in area A reaches 60 percent of the desired sample traverse length, only the parts of subsequent traverses in stratum B are utilized to complete the sample. If the traverses first exhaust the desired length in area B, then subsequent traverse lengths in B are ignored and only lengths in A are included in the sample. This procedure can be extended to handle any number of areas or strata, of regular or irregular shape. For regular shapes, though, it may be easier to sample each stratum individually. Linear traverses for areally stratified maps can also be taken in a systematic manner and with a randomly chosen orientation, as in Figure 6-15(*d*).

Point Sampling

Perhaps the simplest form of spatial sampling is point sampling. To select a *simple random point sample* from a map is really not much more difficult than selecting a random sample from a conventional sampling frame. To identify each point in the sample, two random numbers are necessary. One is used for the *X* coordinate and the other for the *Y* coordinate. Two random numbers are drawn for each point in the sample until *n* points have been located. Figure 6-17(*a*) illustrates a typical random point pattern. There are usually a few areas on the map where no points are sampled and a few areas where there are point concentrations. Since the design is purely random, there is always the possibility that a nonrepresentative sample could be drawn.

To ensure that all parts of the map are sampled and there are no overconcentrations of sampled points, a *systematic random point sample* can be drawn. A typical systematic sample is illustrated in Figure 6-17(*b*). The 16 points for this particular map are regularly spaced at equal intervals in terms of both the *X* and *Y* coordinates. This design is exceptionally easy to operationalize. The interval is selected so that the number of points which fall on the map equals the desired sample size. Once the location of a single point in the design is known, the location of all other points is determined using the spacing interval. In Figure 6-17 (*b*) the location of the point in the southwest corner of the map is randomly chosen within the small area delimited by the dashed lines. Only two random numbers are required for a systematic point sample of any size.

Because this design selects points at regular intervals, the systematic design can lead to a biased estimate of the proportion of a map covered by some phenomenon. Consider the land-use map of a large city. There are many structural regularities in the land-use map of the typical North American city. Streets tend to follow a rectangular grid with a standard distance between street intersections. Successive points taken at regular intervals might all lie on streets. Or more of the points may lie on a street. In either case, the estimate of the land use in a city devoted to streets would be biased. The bias admitted with a systematic sample is usually small, except for cases in which the map has a systematic linear or near-linear trend. Most often, the ease with which this sample can be drawn outweighs the slight bias introduced by systematization. Systematic sampling is widely used in studies of soils and vegetation sampling.

When it is used in spatial sampling, *stratified random sampling* has the same advantage as systematic sampling—more or less complete coverage of the map. But, as illustrated in Figure 6-17(*c*), the procedure is less likely to pick up any regularities in the map. Stratified designs have smaller sampling errors than either simple random or systematic designs. A *cluster sample* can also be taken from a map that has been areally stratified. As shown in Figure 6-17(*d*), the usual procedure is to randomly select certain areas on the map and then intensively sample randomly within these

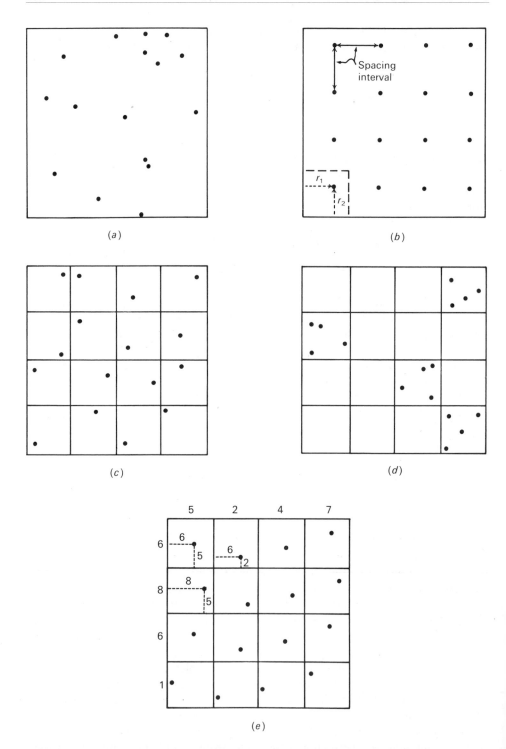

FIGURE 6-17. (a) A simple random point sample; (b) a systematic areal sample; (c) an areally stratified random sample; (d) a cluster sample; (e) a stratified systematic unaligned areal sample.

clusters. The advantage of sampling ease, particularly in field sampling, must be weighed against the possible disadvantage of omitting large areas of the map.

A hybrid design incorporating elements of these simpler designs is the *stratified systematic unaligned sample* shown in Figure 6-17(*e*). It is stratified, since the map is divided into cells prior to sampling. It is systematic since it is unnecessary to draw two random numbers for each point in the sample. However, unlike the case of a simple systematic sample, the pattern produced by this procedure is not aligned. To develop this axis, we define a simple (x, y) grid within each cell. A single random number is drawn for each row and each column. The row random number defines the X coordinate within each cell in that row, and the column random number locates the Y coordinate for all cells in that column. The locations of the points in the three most northwesterly cells are illustrated in Figure 6-17(*e*).

The choice of sampling design in spatial sampling depends very much on the nature of the map. Each spatial design shares the basic characteristics of its nonspatial counterpart. If the underlying map has systematic features, then a systematic design is a poor choice. If there is *no* obvious pattern on the map, then all designs will give fairly similar results. Typically, stratified samples are more precise than random samples, which in turn are more precise than cluster samples. An experiment to test the precision of all sample designs found the stratified systematic unaligned sample design to be decidedly superior to all other designs.

6.7. Summary

Samples can be drawn from a population in a number of ways. They are usually taken to minimize both the time and the cost of collecting data. Four steps usually precede the actual collection of data: specification of the population, delineation of a sampling frame, selection of a sampling design, and pretest of the data collection procedures. For purposes of statistical inference, a probability sample is the required sampling procedure. Although a probability sample still can lead to the selection of an unrepresentative sample, it is *always* possible to specify the probable accuracy of the results. Simple random samples are the most commonly used method of data collection, although other variations sometimes prove necessary and desirable in practice. *For the remainder of this text, all inferential procedures are based on the assumption that the sample has been collected in this manner.*

Sample statistics are random variables. The value of any sample statistic varies from sample to sample for all samples of a fixed size drawn from the same population. For sufficiently large samples, the sampling distribution of many statistics such as $\bar{X}$ or P is known to be normal or approximately normal. By knowing the characteristics of sampling distributions, it is possible to answer probability questions about one sample. That is, we can determine the likely sampling error in our sampling experiment. In Chapter 7, we show how these sampling distributions can be used to either *estimate* population characteristics or make judgments concerning their hypothesized value.

In Section 6.6, we briefly examined the nature and problems of geographic sampling. An exceedingly rich variety of sample designs can be devised based on linear, point, or quadrat sampling. All methods find considerable application in the geographical literature.

FURTHER READING

Textbooks devoted to advanced topics in sampling theory include Mendenhall, Ott, and Schaeffer (1971) and Cochran (1977). Many textbooks of this type, and in particular the text by Cochran, require considerable background. A more practical approach to sampling questions is taken by Sudman (1976). There are literally hundreds of journal articles on practical techniques of sampling. Two very useful journals to consult are the *Journal of Marketing Research* and *Public Opinion Quarterly*. For an analysis of sampling techniques in geography, see Dixon and Leach (1978) and Berry and Baker (1968).

B. J. L. Berry and A. Baker, "Geographic Sampling," in B. J. L. Berry and D. F. Marble (eds.), *Spatial Analysis* (Englewood Cliffs, N.J.: Prentice-Hall, 1968, pp. 91–100).

W. Cochran, *Sampling Techniques* (New York: Wiley, 1977).

C. J. Dixon and B. Leach, *Sampling Methods for Geographical Research* (Norwich, England: Geo Abstracts, 1978).

W. Mendenhall, L. Ott, and R. L. Schaeffer, *Elementary Survey Sampling* (Belmont, Calif.: Duxbury, 1971).

S. Sudman, *Applied Sampling* (New York: Academic, 1976).

PROBLEMS

1. Explain the meaning of the following terms:

   a. Sample
   b. Population
   c. Sampling error
   d. Sampling bias
   e. Nonresponse bias
   f. Sampling frame
   g. Target population
   h. Sampled population
   i. Sample design
   j. Pilot sample, or pretest
   k. Random number table
   l. Simple random sample
   m. Systematic sample
   n. Stratified random sample
   o. Sample statistic
   p. Sampling distribution
   q. Finite-population correction factor
   r. Standard error of a sampling distribution
   s. Stratified systematic unaligned point sample

2. Explain why, in general, it is advisable to sample rather than to attempt a complete census of a population.

3. Give three examples of finite populations and three examples of infinite populations.

4. Differentiate among (a) probability samples, (b) quota samples, (c) convenience samples, (d) judgmental samples, and (e) volunteer samples.

5. Why does a simple random sample from an infinite population differ from one from a finite population?

6. In stratified sampling, it is desirable to define strata that are internally homogeneous but different from other strata. In cluster sampling, both internal heterogeneity and between-cluster homogeneity are desirable. Why?

7. Why is a sample statistic a random variable?

8. Select a hypothetical population of 100 members. Draw the frequency distribution of some variable of interest. Illustrate the operation of the central limit theorem by drawing repeated samples of sizes $n = 4$ and $n = 10$ and drawing the sampling distributions of sample means. How do these sampling distributions illustrate the implications of the central limit theorem?

9. Distinguish among (a) the population mean of a variable μ, (b) the sample mean $\bar{x}$, and (c) the mean of the sampling distribution of the random variable $\bar{X}$, or $\mu_{\bar{x}}$.

10. A population consists of the following seven observations:

Observation	A	B	C	D	E	F	G
Value	4	3	5	7	4	8	11

   a. Find the population mean and standard deviation.
   b. Create a table listing all samples of size $n = 2$. For each sample, find the sample mean $\bar{x}$. Draw a histogram of the sampling distribution.
   c. Find the mean and standard deviation of the sampling distribution of means generated in part (b).
   d. Repeat parts (b) and (c) for samples of size $n = 3$.

11. The map below classifies the land use in a square city into four categories: residential, commercial, industrial, and open space. The percentage of the city land use in each class is as follows: residential (R), 69 percent; commercial (C), 8 percent; industrial (I), 10 percent; open space (O), 13 percent. Estimate the proportion of the map in each land-use category, using a point sample of 25 observations. Draw one sample each, using the following designs: simple random, stratified random, systematic, and systematic stratified unaligned.

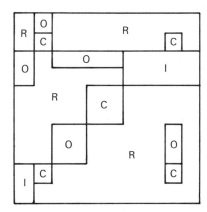

7

Parametric Statistical Inference: Estimation

The principal objective of statistical inference is to make inferences about a population based on a subset of that population, the sample. To be perfectly correct, statistical inference includes all procedures based on using a *sample statistic* to draw inferences about a *population parameter*. In Chapter 1, we distinguished between two ways of drawing inferences about population parameters. The first way is to *estimate* the value of a population parameter by using the information in our sample. This process is illustrated in Figure 7-1 for the population parameter μ. From some given statistical population, we draw a sample of size n. Usually, the value of the population mean μ is unknown. As we found in Chapter 6, the sample statistic used to make inferences about μ is $\bar{X}$. Once the sample is taken, the values of the sample random variables $X_1, X_2, \ldots, X_n$ are known to be $x_1, x_2, \ldots, x_n$, and the *value* of the sample statistic is used to estimate the population parameter μ.

The second way of drawing inferences about the value of a population parameter such as μ is called *hypothesis testing*. When we test a hypothesis, we make some reasonable assumption about the value of μ (or some other population parameter) and then use the sample information to decide whether this hypothesis is supported by the data. For example, we may be interested in a population parameter π, the proportion of residents of a city who moved in the last year. In particular, we may wish to determine whether the city has some abnormally high rate of residential mobility, say greater than $\pi = .20$. A sample of residents is taken, and the proportion of movers within the last year is determined as p. We then use this value of p to decide whether $\pi > .20$ or $\pi \leq .20$ is the more reasonable conclusion.

Both *statistical estimation* and *statistical hypothesis testing* utilize the theoretical relationship between samples and populations expressed by the central limit theorem. It should not be surprising that these two methods of statistical inference are intimately related. This relationship is fully explained in Chapter 8. In this chapter, we focus on statistical estimation. Section 7.1 explains the distinction between the two main categories of statistical estimation, *point* and *interval* estimation. This section also develops criteria differentiating *good* estimators from bad ones. In Section 7.2, point estimators for the population parameters μ, π, and σ are considered. Section 7.3 discusses interval estimation of the first two parameters and introduces the notion of a *confidence interval*. Finally, in Section 7.4 we see how the notions of

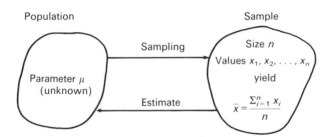

FIGURE 7-1. Estimating the value of the population parameter μ.

statistical estimation can be applied "in reverse" to determine the sample size that should be used in a statistical experiment. The key question in many statistical problems faced by geographers is often, "Just how large a sample is needed to be able to say, with a certain level of confidence . . . ?" This question can often be answered by using arguments developed from the process of statistical estimation.

7.1. Statistical Estimation Procedures

The procedures used in statistical estimation can be conveniently divided into two major classes: point estimation and interval estimation.

DEFINITION: POINT ESTIMATION
In problems of point estimation, a *single* number is calculated from the sample and is used as the best estimate of some unknown population parameter.

This number is called a *point estimate* since the population parameter is estimated by one point on the real number line. We might say, therefore, that based on our sample of some size, the best estimate of the proportion of movers in a city is $\pi = .26$.

DEFINITION: INTERVAL ESTIMATION
In interval estimation, *two points* on the real number line are used to define a range or interval within which the population parameter is thought to lie, with a certain probability.

We might say, for example, that we are 95 percent certain that the proportion of people who moved in the city last year is in the interval [.20, .32]. Let us consider each type of estimation in more detail.

Point Estimation

The process of point estimation of any population parameter θ is illustrated in Figure 7-2. The population parameter can be *any* population parameter such as the mean,

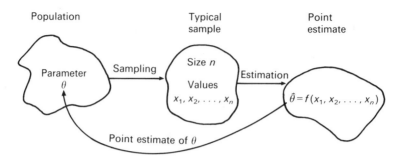

FIGURE 7-2. Point estimation of a population parameter θ.

variance, standard deviation, or proportion. The random sample is composed of n random variables $X_1, X_2, \ldots, X_n$. The *estimator* of the population parameter θ is some function of the sample random variables.

DEFINITION: STATISTICAL ESTIMATOR AND STATISTICAL ESTIMATE
A statistical estimator is a function of the n random variables $X_1, X_2, \ldots, X_n$ in a sample. An estimator is, therefore, also a random variable. Once the random sample is taken, the values of the sample random variables are known and are $x_1, x_2, \ldots, x_n$. The *value* of the estimator can thus be calculated. It is known as the *statistical* *estimate* of the population parameter θ.

As an example, let us consider the population mean parameter. There are several possible point estimators for μ. The sample mean statistic $\bar{X}$ is the obvious choice, but why should it be chosen over the sample median statistic or even the sample mode statistic? All are possible estimators and simple functions of the sample values $x_1, x_2, \ldots, x_n$.

There are two principal criteria in selecting the best sample statistic as the point estimator of a population parameter: It should be both *unbiased* and *efficient*. These criteria are characteristics of the sampling distribution of an estimator. As we noted in Chapter 6, an estimator such as $\bar{X}$ is a random variable with a sampling distribution specified by the central limit theorem. To explain the concepts of efficiency and bias, let us utilize the example of the sampling distribution of $\bar{X}$. Recall from Chapter 6 that the distribution of sample means from a normal population has a mean of μ and a variance of σ^2/n, where μ and σ are the population mean and standard deviation and n is the sample size.

DEFINITION: UNBIASED ESTIMATOR
An estimator of a population parameter θ is said to be unbiased if the expected value of the estimator is equal to the population parameter.

Note that the expected value or mean of the sampling distribution of $\bar{X}$ is μ, so $\bar{X}$ is an unbiased estimator of μ. The distinction between a biased and an unbiased

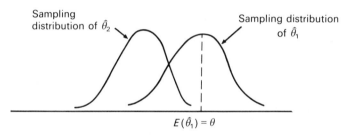

Sampling distribution of $\hat{\theta}_2$

Sampling distribution of $\hat{\theta}_1$

$E(\hat{\theta}_1) = \theta$

FIGURE 7-3. Sampling distributions for a biased and an unbiased estimator of θ.

estimator is shown clearly in Figure 7-3. The sampling distributions of two possible estimators of some population parameter θ are shown. Note that the sampling distribution of $\hat{\theta}_1$, the first estimator, is centered on θ. Since the expected value of this estimator is θ, it is an *unbiased* estimator. The sampling distribution of $\hat{\theta}_2$, another competing estimator, is not centered on θ and is, therefore, a *biased* estimator. Obviously, all other things being equal, $\hat{\theta}_1$ is preferred to $\hat{\theta}_2$ as an estimator of θ.

Notice, however, one important feature about the estimators $\hat{\theta}_1$ and $\hat{\theta}_2$. It is possible for a particular sample value of $\hat{\theta}_2$ to be closer to θ than a particular sample value of $\hat{\theta}_1$. However, by comparing the two sampling distributions, we can say that it is *more likely* that we will draw a sample in which the value of $\hat{\theta}_1$ is closer to θ than is the value of $\hat{\theta}_2$. That is not to say $\hat{\theta}_1$ is *always* better than $\hat{\theta}_2$. We may draw a sample for which the value estimated for $\hat{\theta}_1$ is worse than $\hat{\theta}_2$ (farther from θ), but the chances of this happening are small.

The fact that an estimator is unbiased is not sufficient in itself to qualify that estimator as the "best" or even a "good" estimator. Compare the two estimators $\hat{\theta}_3$ and $\hat{\theta}_4$ of Figure 7-4. Both are unbiased, but $\hat{\theta}_3$ is certainly the preferred estimator. Why? The possible values of this sample statistic are more concentrated about the population parameter θ.

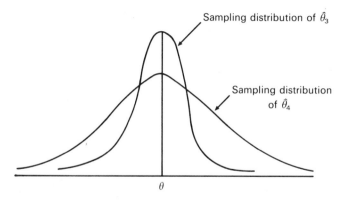

Sampling distribution of $\hat{\theta}_3$

Sampling distribution of $\hat{\theta}_4$

θ

FIGURE 7-4. Sampling distributions for two unbiased estimators of θ.

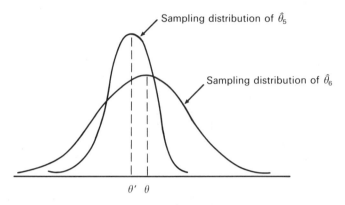

FIGURE 7-5. Difficulties in choosing a potential estimator.

DEFINITION: EFFICIENCY OF AN ESTIMATOR
The efficiency of an estimator refers to the degree of variability of its sampling distribution. Among all possible unbiased estimators, the goal is usually to select the most efficient one—the one whose sampling distribution has the minimum variance.

The best estimator for a population parameter is usually termed an *unbiased minimum variance estimator*. This recognizes the two factors of importance: unbiasedness and efficiency. In this chapter, unless otherwise specified, the estimator given has unbiased minimum variance properties.

Note that these two properties need not run together. Sometimes, we may wish to choose an efficient biased estimator over an inefficient unbiased one. This situation is depicted in Figure 7-5. Even though it is biased, estimator $\hat{\theta}_5$ would probably be preferred to $\hat{\theta}_6$. The degree of bias introduced by using this estimator is easily compensated for by its efficiency. It is also possible, at least theoretically, to envisage situations where an unbiased inefficient estimator is preferred to a biased efficient one. Sometimes, the situation is very unclear. Fortunately, for most of the commonly used parameters, there is a clearly superior choice.

Interval Estimation

The interval estimation procedure attempts to rectify the single most important drawback to point estimation. Although we may know that the value of some point estimator gives us the best guess as to the true value of the population parameter, we have no idea how good it really is. In statistical estimation, we obviously want to know how much difference is likely between the point estimate from a sample of size n and the actual value of the population parameter. The basic idea in interval estimation is illustrated in Figure 7-6. Sampling from a population, we utilize the values of our sample $x_1, x_2, \ldots, x_n$, to define *two* points on the real line which together define

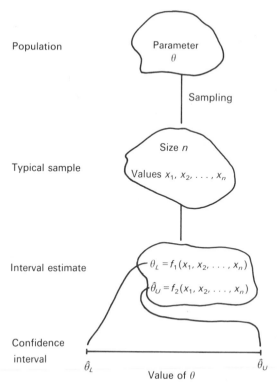

FIGURE 7-6. Interval estimation of a population parameter θ.

an interval. Two functions are used to define a lower limit for the population parameter $\hat{\theta}_L$ and an upper limit $\hat{\theta}_U$. For a given sample size n, we say that the interval contains the value of the population parameter, with a certain degree of confidence. For this reason, the interval estimator is often termed a *confidence interval*. A typical conclusion might be, "Based on a sample of 100 light bulbs, we estimate the true mean lifetime of these light bulbs to be between 1000 and 1200 h, with 95 percent confidence." The word *confidence*, when it is used in this context, has a particular meaning. This is thoroughly explained in Section 7.3. The usefulness of our inference is based on the *probability* that the specified confidence interval contains the actual value of the population parameter θ. In the example cited above, we would attach a probability of .95 to the interval [1000, 1200] based on a sample size of $n = 100$. It is possible that the interval [1000, 1200] does not contain the true mean lifetime for this brand of light bulb. Just as a result of random sampling, it is possible that our sample is not representative of the population of light bulbs. This probability is known to be .05.

Just as there are good point estimators, so are there good interval estimators. The principal requirement of a good interval estimator is that the specified interval $[\hat{\theta}_L, \hat{\theta}_U]$ be as short as possible for that sample size. Interval estimators for π and μ are specified in Section 7.3.

TABLE 7-1
Point Estimators for μ, π, and σ^2

Population parameters	Point estimator	Formula for point estimate
μ	$\bar{X}$	$\bar{x} = \dfrac{\sum_{i=1}^{n} x_i}{n}$
π	P	$p = \dfrac{x}{n}$ $\begin{array}{l} x = \text{number of successes,} \\ n = \text{number of trials} \end{array}$
σ^2	S^2	$s^2 = \dfrac{\sum_{i=1}^{n} (x_i - \bar{x})^2}{n-1}$

7.2. Point Estimation

The point estimators for the population parameters μ, π, and σ^2 are given in Table 7-1. Let us consider each in turn.

Population Mean μ

Clearly the best estimator of the population mean μ is the sample statistic $\bar{X}$. It is the unbiased minimum variance estimator. Although we could also use the sample median and sample mode to estimate μ, they are decidedly inferior estimators. The problem with the sample mode, as we found in Chapter 2, is that it is not necessarily unique. Moreover, it is a *biased* estimator. The sample median is a better alternative; it is an unbiased estimator. However, its sampling distribution is much more variable than the statistic $\bar{X}$.

Population Proportion π

The best estimator of the population proportion π is the sample statistic $P = X/n$, where X is the number of successes in n trials of an experiment. The population parameter π is the single parameter of a random vaiable having a binomial distribution. To estimate π, then, we simply calculate the observed number of successes x in a sample of n trials and estimate π by x/n. Suppose, for example, we are interested in the proportion of residents in a metropolitan area who take public transit on the journey to work. A sample of 100 residents finds 26 took public transit. Our best estimate of π is $p = \frac{26}{100} = .26$.

Population Variance σ^2 and Standard Deviation σ

For the population variance σ^2, the best estimator is S^2. The value of S^2, given the n specific values in the sample of $x_1, x_2, \ldots, x_n$, is

$$s^2 = \frac{\sum_{i=1}^{n} (x_i - \bar{x})^2}{n-1}$$

Here S^2 is the unbiased minimum variance estimator of σ^2. Thus if we took many samples of size n from a population, the values of S^2 computed for each sample would average to σ^2. Why is the denominator $n - 1$ and not n? If the divisor of n were used, the statistic would *not* have an expected value of σ^2 and would thus be a biased estimator. The formal proof is quite complicated [see, for example, Pindyck and Rubinfeld (1981)] and not critical for our purposes. However, it is important to remember that the statistic S^2 with $n - 1$ in the denominator is the unbiased estimator of σ^2.

The estimator for σ, the population standard deviation, is the square root of S^2, or S. Although it is most commonly used, it is a biased estimator of σ, even though the degree of bias is very small. However, the use of S is much more convenient than alternative estimators, and it is the one used in this text.

7.3. Interval Estimation

There are two important elements in a statistical inference: the inference itself and a measure of how good the inference is. In point estimation, we select a *value* that is used as an estimate of the population parameter in question. Even if the estimator is known to have unbiased minimum variance properties, this tells us nothing about how good the value from our sample may be. How far is this value from the true population parameter? What would be useful is some measure of the probability that the population parameter of interest lies within a certain range. Specifying these ranges, and the probabilities attached to them, is the goal of interval estimation. In this section we show how to develop interval estimates for two population parameters, μ and π, based on a knowledge of the sampling distributions of the estimators $\bar{X}$ and P.

Interval Estimator for μ

Let us review the nature of the sampling distribution of $\bar{X}$. Whenever the sampling distribution of the statistic $\bar{X}$ is normally distributed, we can make statements about the probability that a value of $\bar{X}$ falls into any interval, based on our knowledge of the normal distribution. Consider the sampling distribution shown in Figure 7-7. Since the area under the standard normal curve between -1.96 and $+1.96$ standard

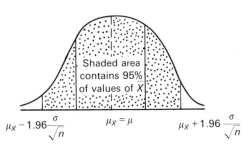

FIGURE 7-7. Sampling distribution of $\bar{X}$.

deviations about the mean encompasses 95 percent of the values of $\bar{X}$, it is possible to say

$$P(\mu - 1.96\sigma_{\bar{x}} \leq \bar{X} \leq \mu + 1.96\sigma_{\bar{x}}) = .95$$

or

$$P\left(\mu - 1.96\frac{\sigma}{\sqrt{n}} \leq \bar{X} \leq \mu + 1.96\frac{\sigma}{\sqrt{n}}\right) = .95 \tag{7-1}$$

This probability statement can be reexpressed in terms of the unknown μ as

$$P\left(\bar{X} - 1.96\frac{\sigma}{\sqrt{n}} \leq \mu \leq \bar{X} + 1.96\frac{\sigma}{\sqrt{n}}\right) = .95 \tag{7-2}$$

How can this probability statement be interpreted? Let us suppose repeated samples of size n are taken from a population with a mean of μ and standard deviation σ. See Figure 7-8. Using the data in each of these samples, we compile a *value* for $\bar{X}$, using

$$\bar{x} = \frac{x_1 + x_2 + \cdots + x_n}{n}$$

and then the upper and lower bounds of our interval are $\bar{x} + 1.96\sigma/\sqrt{n}$ and $\bar{x} - 1.96\sigma/\sqrt{n}$. Then, as is illustrated in Figure 7-7, 95 percent of the samples specify an interval containing the true value of μ. The other 5 percent of the samples specify intervals that do not contain μ.

Intervals specified in this way are known as *confidence intervals*. In practice, a confidence interval is specified by the data from a *single* sample. Obviously, this interval either will contain the value for μ or will not. Because of the randomness

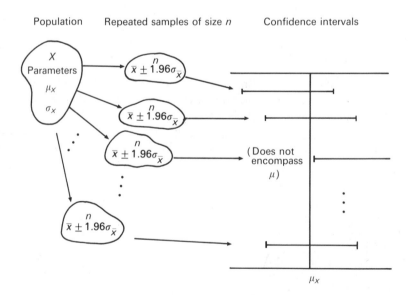

FIGURE 7-8. Interval estimates constructed from repeated samples of size n.

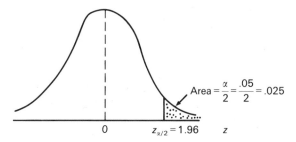

Sampling distribution of $\bar{X}$

FIGURE 7-9. Calculation of $z_{\alpha/2}$ for a 95 percent confidence interval.

inherent in sampling, we cannot be sure which of these two possibilities occurs. Since we do know, from the central limit theorem, that there is a .95 chance of selecting *a sample* that produces an interval containing μ, we are 95 percent confident in the interval obtained from the sample.

Confidence intervals can be constructed for any desired degree of confidence. Depending on the problem at hand, we may wish to be 90, 95, or 99 percent or even more confident in our result. Let $1 - \alpha$ be the desired degree of confidence specified as a proportion. For example, the 95 percent confidence interval would be .95, and the 99 percent confidence expressed as .99. Let us define $z_{\alpha/2}$ as the value of the standard normal table such that the area to the *right* of $z_{\alpha/2}$ is $\alpha/2$. For a 95 percent confidence interval we wish $\alpha/2 = .05/2 = .025$ of the area of the curve to the right of $z_{\alpha/2}$ (see Figure 7-9). From the standard normal table, $z_{.025} = 1.96$. Different confidence intervals imply different values for $z_{\alpha/2}$ and thus for the upper and lower bounds of the confidence interval. Then the general confidence interval for the mean is $\bar{x} \pm z_{\alpha/2}(\sigma/\sqrt{n})$. A simple example will explain this procedure.

EXAMPLE 7-1. A random sample of $n = 25$ shoppers at a supermarket reveals that patrons travel, on the average, 16 min from their homes to the supermarket. Furthermore, we are told that the population standard deviation for this variable is $\sigma = 4$, and these travel times are normally distributed. The problem is to construct the 99 percent confidence interval for the average travel time to this supermarket. The desired interval for .99 confidence is

$$\bar{x} - z_{\alpha/2}\frac{\sigma}{\sqrt{n}} \leq \mu \leq \bar{x} + z_{\alpha/2}\frac{\sigma}{\sqrt{n}}$$

which, by substitution of $\bar{x} = 16$, $\sigma = 4$, $n = 25$, and $\alpha = .01$, can be rewritten as

$$16 - z_{.005}\frac{4}{\sqrt{25}} \leq \mu \leq 16 + z_{.005}\frac{4}{\sqrt{25}}$$

From the normal table, $z_{.005} = 2.58$, so that the 99 percent confidence interval is given by

$$16 \pm 2.58\frac{4}{\sqrt{25}} = 16 \pm 2.064 = [13.936, 18.064]$$

TABLE 7-2
Confidence Intervals for μ for Varying Degrees of Confidence

Confidence level	α	$\dfrac{\alpha}{2}$	$z_{\alpha/2}$	Lower bound $\bar{x} - z_{\alpha/2}\dfrac{\sigma}{\sqrt{n}}$	Upper bound $\bar{x} + z_{\alpha/2}\dfrac{\sigma}{\sqrt{n}}$	Values of bound for travel time problem		Interval width
						Lower	Upper	
.90	.10	.05	1.65	$\bar{x} - 1.65\dfrac{\sigma}{\sqrt{n}}$	$\bar{x} + 1.65\dfrac{\sigma}{\sqrt{n}}$	14.688	17.312	2.624
.95	.05	.025	1.96	$\bar{x} - 1.96\dfrac{\sigma}{\sqrt{n}}$	$\bar{x} + 1.96\dfrac{\sigma}{\sqrt{n}}$	14.432	17.568	3.136
.99	.01	.005	2.58	$\bar{x} - 2.58\dfrac{\sigma}{\sqrt{n}}$	$\bar{x} + 2.58\dfrac{\sigma}{\sqrt{n}}$	13.936	18.064	4.128

TABLE 7-3
Confidence Intervals for μ with Varying Sample Sizes

Sample size n	Confidence level α	$\dfrac{\alpha}{2}$	Lower bound $\bar{x} - z_{\alpha/2}\dfrac{\sigma}{\sqrt{n}}$	Upper bound $\bar{x} + z_{\alpha/2}\dfrac{\sigma}{\sqrt{n}}$	Interval width
25	.05	.025	14.432	17.568	3.136
100	.05	.025	15.216	16.784	1.568
400	.05	.025	15.608	16.392	0.784

The value of μ may or may not be in this range. Our hope is that the random sample used to construct the interval is one of the 99 percent of all possible samples of size $n = 25$ that do contain the true value for μ.

The width of a confidence interval is controlled by two principal factors: the sample size n and the desired degree of confidence $z_{\alpha/2}$. Let us examine the impact of each of these factors on the intervals of travel time. First, note the changes in the width of the interval for varying degrees of confidence given in Table 7-2. As our confidence level increases, α decreases, $z_{\alpha/2}$ increases, and the width of the interval increases. This is expected. Note what happens as the level of confidence approaches 100 percent. The confidence interval approaches $[-\infty, +\infty]$, the entire real number line. This is because we can never be 100 percent sure that an interval will contain μ unless the interval contains the entire range of values of the population variable X. In practice, of course, only the range of permissible values for X would have to be included.

Table 7-3 summarizes the impact of sample size on interval width. Interval widths are calculated for sample sizes of 25, 100, and 400. As the sample size increases, the width of the confidence interval decreases. Notice, however, that the gains in accuracy decrease as sample size increases. When it is taken to the limit (that is, as sample size approaches infinity), the width of the interval approaches zero. Since the center of the interval is the sample mean, which approaches μ as n approaches infinity, the confidence interval for μ converges on the value of μ with larger sample sizes. This has great intuitive appeal. As our sample includes more and more members of some population, we are obviously increasingly knowledgeable about the values of population parameters. If we have included the entire population, there should be no doubt about the value of the population parameter; it can simply be calculated from known information.

There are thus two ways of decreasing the width of confidence intervals, that is, of making more accurate predictions. First, we can decrease the level of confidence. Second, and a usually more worthwhile alternative, we can increase the sample size. The formula for the confidence interval of the population mean μ can now be summarized.

DEFINITION: INTERVAL ESTIMATE OF POPULATION MEAN μ

Let $x_1, x_2, \ldots, x_n$ be a sample of size n randomly drawn from a normally distributed random variable X. The confidence interval with confidence level $1 - \alpha$ is

$$\bar{x} \pm z_{\alpha/2} \frac{\sigma}{\sqrt{n}} \qquad (7\text{-}3)$$

where $\bar{x}$ is the sample mean, σ is the population standard deviation, and $z_{\alpha/2}$ is the value of the standard normal distribution for which the area to the right of $z_{\alpha/2}$ is $\alpha/2$.

Interval Estimation for μ with X Not Normally Distributed

If the population random variable X is not normally distributed, then we cannot appeal to the strict form of the central limit theorem to justify the normality of the sampling distribution of $\bar{X}$. All we can say is that the sampling distribution is *approximately* normal. It follows that the confidence interval given in Equation (7-3) can also be considered *approximately* correct. As a rule of thumb, the approximation can be considered sufficiently close when sample size n exceeds 30.

Interval Estimation for μ with σ Unknown or $n < 30$

An extremely common problem in developing a confidence interval for μ is that the population standard deviation of the random variable X, or σ, is unknown. In any case, if it were to be known, so would the value of μ, since its calculation depends on μ. If the population standard deviation is not known, then formula (7-3) cannot be used. When σ is unknown, so, too, is the standard deviation of the sampling distribution of $\bar{X}$, or $\sigma_{\bar{X}} = \sigma/\sqrt{n}$. There is a simple way around this problem. First, the values of the sample $x_1, x_2, \ldots, x_n$ are used to estimate σ by

$$s = \sqrt{\frac{\sum_{i=1}^{n} (x_i - \bar{x})^2}{n - 1}} \qquad (7\text{-}4)$$

Recall that this is the most commonly used estimate of σ identified in Section 7.2. With s, we can now estimate $\sigma_{\bar{X}} = \sigma/\sqrt{n}$ by $s/\sqrt{n}$. Unfortunately, when this is done, we can no longer use the normal distribution to calculate the confidence interval.

Once the substitution of s for σ has been made, the distribution to be used in confidence limit calculations is the *t distribution*. This is also sometimes called the *Student's t distribution* because the statistician who originally uncovered it reported it in a statistical jounal, using the pseudonym *Student*. The t distribution is almost normal; it is symmetric but has greater variability than the normal. Since we are replacing the population parameter σ by an estimate s, more uncertainty is introduced into the sampling distribution, and we should expect greater variability in the distribution.

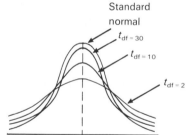

FIGURE 7-10. Normal distribution and the t distribution for 2, 10, and 30 df.

Now, the greater the sample size upon which our estimate of σ is based, the better the estimate s. Therefore, the bigger the sample size, the more the t distribution resembles the standard normal. The form of the t distribution thus depends on sample size n. Because we utilize $n - 1$ in the denominator for s in (7-4), the distribution of t is tabulated according to its *degrees of freedom* (df) rather than the sample size n. The degrees of freedom are always equal to the sample size minus 1, or $n - 1$. Figure 7-10 compares the standard normal with the t distributions for degrees of freedom 2, 10, and 30. At df $= 30$, note how similar the standard normal and t distributions are.

One way of quantifying the convergence of the standard normal and the t distributions with increasing degrees of freedom is to examine the value of t corresponding to a certain tail area and compare it to the standard normal. For example, we know that a value of $z = 1.96$ for the standard normal leaves .025 of the area under the curve in the tail. Now, if we compare this to the values of t for several members of the family of t distributions, we find

Degrees of freedom (df)	Value of t
1	12.706
2	4.303
3	3.182
10	2.228
20	2.086
30	2.042
40	2.021
60	2.000
120	1.980
$+\infty$	1.960

As the sample size and degrees of freedom increase, the value of t corresponding to an upper-tail probability of .025 converges to 1.96. Moreover, by the time df $= 30$ there is a close correspondence between z and t.

The use of the t distribution in constructing confidence intervals for μ with σ unknown is summarized as follows:

DEFINITION: CONFIDENCE INTERVAL ESTIMATION FOR μ WITH UNKNOWN POPULATION STANDARD DEVIATION σ

Let $x_1, x_2, \ldots, x_n$ be a sample of size n drawn from a population with an unknown population standard deviation σ. The confidence interval for μ at level of confidence $1 - \alpha$ is

$$\bar{x} \pm t_{\alpha/2, \, df = n - 1} \frac{s}{\sqrt{n}} \tag{7-5}$$

where $\bar{x}$ is the sample mean, s is the sample standard deviation estimated from (7-4), and $t_{\alpha/2, \, df = n - 1}$ is the value of the t distribution with degrees of freedom equal to $n - 1$, such that the area to the right of t is $\alpha/2$. It is assumed the population random variable X is normally distributed.

The confidence interval specified by (7-5) differs from (7-4) in two significant ways. First, because σ is unknown, it is replaced by the estimate s. Second, $z_{\alpha/2}$ is replaced by $t_{\alpha/2, \, df = n - 1}$. Whenever we use a value for t in our confidence intervals, it is always necessary to specify which t distribution is to be used by indicating the degrees of freedom. The notation $t_{\alpha/2, \, df = n - 1}$ (or more easily $t_{\alpha/2, \, n - 1}$) refers to the value of t with an upper-tail probability of $\alpha/2$ and with $n - 1$ degrees of freedom. Henceforth, the more compact notation is used.

As a first example of the use of confidence interval formula (7-5), let us return to the travel time example. The sample size is $n = 25$, and the mean is $\bar{x} = 16$. Now, let us change the situation slightly by assuming that the population standard deviation σ is unknown but has been *estimated* as $s = 4$. The confidence interval formula from the t distribution in (7-5) is

$$\bar{x} \pm t_{\alpha/2, \, n - 1} \frac{s}{\sqrt{n}}$$

which for $1 - \alpha = .95$, or 95 percent, confidence is equal to

$$16 \pm 2.064 \left(\frac{4}{\sqrt{25}} \right)$$

$$16 \pm 1.651$$

$$[14.349, \ 17.651]$$

The value of $t = 2.064$ is found in the t table for $t_{.025, 24}$ by using the column .025 and the row (for degrees of freedom) corresponding to 24. This interval is slightly wider than the interval found when σ was assumed known, $[14.432, 17.568]$. Since we are in a situation of greater uncertainty with σ unknown, this is expected. Because the t and the z values give very similar confidence intervals in large samples, the normal distribution with z is often used, even in cases where σ is unknown. For small samples, t should always be used.

Interval Estimation for π

For binomial random variables, the sample proportion of successes can be estimated by using

$$P = \frac{X}{n} \tag{7-6}$$

where X represents the number of successes in n trials. To define the confidence interval for π, it is necessary to specify the sampling distribution of the statistic P. Since P is an average of sorts, an appeal can be made to the central limit theorem. The statistic P is normally distributed for large sample sizes. To determine the parameters of this sampling distribution μ_P and σ_P, using a knowledge of the parameters of the binomial random variable X, we proceed as follows: First, note the mean of X, $E(X) = n\pi$ [see formula (5-11)], and therefore

$$\mu_P = E(P) = E\left(\frac{X}{n}\right) = \frac{1}{n}E(X)$$

$$\mu_P = \frac{1}{n}(n\pi) = \pi \tag{7-7}$$

As expected, the distribution of P is centered on π, the population proportion. This simply proves a point made in Section 7.2: P is an unbiased estimator of π. The standard deviation of the sampling distribution of P can be found in a similar way by using the formula for the variance (V) of a binomial random variable:

$$\sigma_P^2 = V\left(\frac{X}{n}\right) = V\left(\frac{1}{n} \cdot X\right)$$

$$\sigma_P^2 = \frac{1}{n^2}V(X) = \frac{n\pi(1 - \pi)}{n^2}$$

$$\sigma_P^2 = \frac{\pi(1 - \pi)}{n} \tag{7-8}$$

and thus

$$\sigma_P = \sqrt{\frac{\pi(1 - \pi)}{n}} \tag{7-9}$$

The derivation of σ_P^2 is based on the substitution of $V(X) = n\pi(1 - \pi)$ [see formula (5-12)] for a binomial random variable X. The principal difficulty here is that the standard deviation of P, or σ_P, itself requires a knowledge of π for estimation. This problem is solved by substituting our sample proportion p for π. The confidence interval for π can now be summarized.

DEFINITION: CONFIDENCE INTERVAL FOR π

Let X be a binomial random variable representing the number of successes in n trials where the probability of success in each trial is π. The confidence interval for π at a level of confidence $1 - \alpha$ is

$$p - z_{\alpha/2}\sqrt{\frac{p(1-p)}{n}} \leq \pi \leq p + z_{\alpha/2}\sqrt{\frac{p(1-p)}{n}}$$

or

$$p \pm z_{\alpha/2}\sqrt{\frac{p(1-p)}{n}} \qquad (7\text{-}10)$$

where p is the sample proportion of successes in n trials and $z_{\alpha/2}$ is the value of the standard normal distribution with an upper-tail probability of $\alpha/2$.

The confidence interval for π in (7-10) should be used only for large samples, say $n \geq 100$. For smaller sample sizes, the exact sampling distribution of the statistic P is the discrete binomial distribution. The binomial mass function (5-11) can be used to generate the exact sampling distribution, or else tables can be consulted.

EXAMPLE 7-2. An urban geographer interested in the proportion of residents in a city who moved within the last year conducts a survey. Summary statistics reveal 27 of the 270 people surveyed changed residences. What are the 95 percent limits on the true proportion of people who moved in the city?

Since the sample size of $n = 270 > 100$, we can utilize formula (7-10) with the estimate $p = \frac{27}{270} = .10$. The limits are therefore

$$.10 \pm 1.96\sqrt{\frac{(.10)(.90)}{270}}$$

$$.10 \pm .036$$

$$[.064, .136]$$

EXAMPLE 7-3. Prior to a referendum on a bylaw requiring mandatory recycling in a city, a random sample of $n = 100$ voters is taken and their voting intentions are obtained. Of those surveyed, 46 favor the bylaw, and 54 are opposed. What are the 90 percent confidence limits on the population proportion in favor of this bylaw?

Since $n \geq 100$, formula (7-10) can be used with $p = .46$, yielding

$$.46 \pm 1.65\sqrt{\frac{(.46)(.54)}{100}}$$

$$.46 \pm .082$$

$$[.378, .542]$$

Despite the fact that only 46 percent favor the bylaw, there is still the possibility that the true proportion in favor is a majority. This is because the interval includes $\pi = .5$.

Confidence Bounds and One-Sided Confidence Intervals

In certain instances, it is not an interval for a population parameter that is of interest, but either a lower or upper *bound* on the parameter. As an example, we may wish to say, "We are 95 percent confident that μ *does not exceed* some predetermined value." This is an example of an *upper confidence bound* on μ. Similarly, a *lower confidence bound* is set when we determine a value for a population parameter that μ *does exceed*, at a given level of confidence. We may be interested in specifying a lower confidence bound to ensure that some minimum standard is met. For example, the federal government may wish to specify gas mileage for newly produced vehicles in such a way that they are guaranteed to provide 40 mi/gal at, say, a confidence level of 99 percent. When the standard involved is a maximum, it may be appropriate to set an upper bound. In a related example, the federal government may wish to force car manufacturers to produce vehicles in such a way that the average damage (in dollars) to a vehicle involved in 5-mi/h front-end collision is no more than $300.

Upper and lower confidence bounds are determined in a way similar to a confidence interval. The major difference is that we are interested in only *one* end of the interval, not both. The formulas for the confidence interval are changed by altering the values of z or t used to specify the level of confidence. Since we are interested in only one end of the interval, it makes no sense to split the α into two $\alpha/2$, one for each end. Instead, the error α is placed at one end of the interval, depending on whether we are interested in a maximum or minimum bound. The formulas for the lower and upper confidence bounds for the parameters μ and π are given in Table 7-4. When we are interested in a lower confidence bound, α is placed on the left end of the interval. For an upper bound, it is placed on the right end of the confidence interval. This difference is graphically illustrated in Figure 7-11. In (*a*), the $100(1-\alpha)$ percent confidence limits are formed by splitting the α error and placing $\alpha/2$ on each side of the sampling distribution. In (*b*), the $100(1-\alpha)$ percent lower confidence bound

TABLE 7-4
Confidence Bounds for μ and π

Parameter	Lower $100(1-\alpha)$ percent confidence bound	Upper $100(1-\alpha)$ percent confidence bound	Notes
μ	$\bar{x} - z_\alpha \dfrac{\sigma}{\sqrt{n}}$	$\bar{x} + z_\alpha \dfrac{\sigma}{\sqrt{n}}$	1. σ known 2. X normally distributed or $n \geq 30$
μ	$\bar{x} - t_{\alpha, n-1} \dfrac{s}{\sqrt{n}}$	$\bar{x} + t_{\alpha, n-1} \dfrac{s}{\sqrt{n}}$	1. σ unknown 2. X normally distributed or $n \geq 30$
π	$p - z_\alpha \sqrt{\dfrac{p(1-p)}{n}}$	$p + z_\alpha \sqrt{\dfrac{p(1-p)}{n}}$	n is large

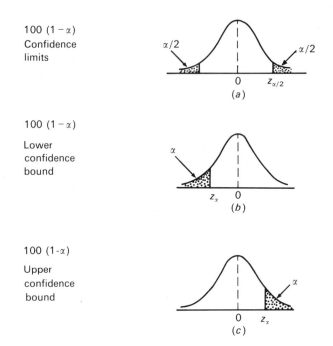

FIGURE 7-11. Placement of α error in confidence limits and confidence bounds.

places the entire α error in the lower, or left, tail. In (c), the upper confidence bound has the α error in the right, or upper, tail of the sampling distribution.

EXAMPLE 7-4. A candidate interested in reelection has decided that she will seek reelection if she can be 95 percent sure she will gain a majority, that is, at least 50 percent of the votes. She undertakes a random sample of 200 voters and finds 108 in favor of her, but 92 likely to vote for her opponent. Should she seek reelection?

From the survey, we have $n = 200$, $\alpha = .05$, and $p = \frac{108}{200} = .54$. Since n is large, the lower confidence bound specified in Table 7-5 can be used:

$$p - z_\alpha \sqrt{\frac{p(1 - p)}{n}}$$

$$.54 - 1.65 \sqrt{\frac{(.46)(.54)}{200}} = .54 - .058 = .482$$

Since she cannot guarantee to win a majority of voters with 95 percent certainty, she chooses to withdraw. Few politicians would use such high standards to make such a decision!

7.4. Sample Size Determination

One common question asked of statisticians is, "How large a sample should I take in my research?" This is an extremely complicated question involving many different issues. However, in certain restricted conditions, we can use our knowledge of sampling distribution and confidence intervals to determine appropriate sampling sizes. To determine the sample size necessary in a study, we must know three things:

1. Which population parameter is under study? Is it a mean μ, proportion π, or some other parameter?
2. How close do we wish the estimate made from our sample to be to the actual (or true) value of the population parameter?
3. How certain (or, in statistical jargon, confident) do we wish to be that our estimate is within the tolerance specified in item 2?

The answer to item 1 tells us which sampling distribution governs the statistic of interest. In this chapter we limit ourselves to the population mean μ and proportion π.

The procedure is illustrated within the context of sample size determination for estimating μ. Suppose we wish to take a sample which is capable of yielding a point estimate of μ that is no more than E units away from the actual value of μ. Short for *error*, E is a statement of the level of precision we desire. If E is small, we are saying we can tolerate only a small error in the estimate; if it is large, our required precision is less. Let us also assume that we wish to be 95 percent confident. In other words, a 5 percent chance of drawing a sample of size n with an estimate $\bar{x}$ for μ that is *more than E units away from μ* is acceptable.

To see how we can determine the sample size n meeting these requirements, let us examine the sampling distribution of $\bar{X}$ and the 95 percent confidence interval illustrated in Figure 7-12. We know that the interval $\mu \pm 1.96\sigma/\sqrt{n}$ contains exactly 95 percent of the values of the statistic $\bar{X}$. On this basis, the 95 percent confidence

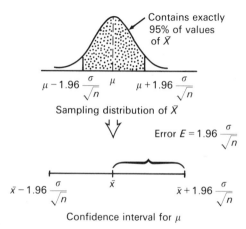

FIGURE 7-12. Relationship between sampling distribution confidence limit and error.

interval for μ with an estimate $\bar{x}$ is $\bar{x} \pm 1.96\sigma/\sqrt{n}$. If we wish to be no more than E units from μ with the estimate $\bar{x}$, then we can set

$$E = 1.96 \frac{\sigma}{\sqrt{n}}$$

which is one-half the interval width. In general, for a level of confidence $1 - \alpha$, we can set

$$E = z_{\alpha/2} \frac{\sigma}{\sqrt{n}} \tag{7-11}$$

we now must solve (7-11) for n. With a few straightforward algebraic manipulations, we find

$$n = \left(\frac{z_{\alpha/2}\sigma}{E}\right)^2 \tag{7-12}$$

This formula indicates that the required sample size depends on the desired level of confidence (the higher the confidence, the larger $z_{\alpha/2}$, and hence the larger n), the desired precision (the higher the precision, the smaller E, and hence the larger n), and the variability of the underlying population (the greater the variability in the population, the larger σ, hence the larger n). All these relations have great intuitive appeal.

The single difficulty with Equation (7-12) is that seldom will σ be known beforehand, except where a population census has been completed. If this were the case, μ would be known and not require estimation. Usually, then, we must substitute an estimate of σ to determine n. The obvious candidate is s. If a rough estimate for s is not available from a recent survey, then a small pilot study can be undertaken to generate s. This is a quite common procedure in much social science research. If no estimate of σ exists, then the following rough rule of thumb can be used: The *range* of many random variables is approximately 4σ. So, if we estimate the range of values for random variable X to be 100, then $\sigma = 25$. Usually it is unnecessary to resort to this crude procedure.

EXAMPLE 7-5. A soil scientist is interested in determining the number of soil samples in a specific soil horizon necessary to estimate the extractable P_2O_5 (measured in milligrams per 100 g). Previous research indicates the standard deviation of P_2O_5 is about .7. The desired precision is $E = .2$, and a confidence level for the study is set at $\alpha = .05$, or 95 percent.
Using formula (7-12) gives

$$n = \left[\frac{1.96(.7)}{.2}\right]^2 = \left(\frac{1.372}{.2}\right)^2 = 47.06$$

So 47 samples should be taken. Note that the entire procedure is based on the assumption that the random variable X is normally distributed, or that the required sample size is large enough ($n > 30$) and the central limit theorem applies.

An exactly analogous argument can be used to develop the following sample size formula for the population proportion π:

$$n = \left[\frac{z_{\alpha/2}\sqrt{p(1-p)}}{E} \right]^2 \tag{7-13}$$

This formula utilizes the estimate p for π in the confidence interval formula. Evidently some knowledge of the likely value of π, say from a previous study, can be used to make the necessary sample size decision. If no estimate of p is available, then an approximate sample size (in fact, a useful lower bound on n) can be found by assuming $p = .5$. With $p = .5$ the value of $\sqrt{p(1-p)}$ is maximized:

p	.2	.3	.4	.5
$\sqrt{p(1-p)}$	.4	.46	.49	.5

Therefore so is the value of n maximized.

EXAMPLE 7-6. An urban geographer is interested in determining the sample size necessary to estimate the proportion of households in a large metropolitan area who moved in the last year. Past experience indicates that approximately 20 percent of all residents in North American cities move each year. How many households should be sampled? Assume a confidence level of 95 percent and that the geographer will tolerate an error of $\pm .03$.

Using Equation (7-13) with an estimate of $p = .2$ yields

$$n = \left[\frac{1.96\sqrt{.2(.8)}}{.03} \right]^2 \simeq 682.95$$

or approximately 683 households. It might also be interesting to determine an upper bound on the sample size by assuming $p = .5$:

$$n = \left[\frac{1.96\sqrt{(.5)(.5)}}{.03} \right]^2 = 1067.11$$

Under these conditions, a very large sample of 1067 is required. If the researcher feels the percentage of people in the city who moved is likely to be larger than the 20 percent characteristic of North American cities, it would be wise to enlarge the sample.

7.5. Summary

This chapter has developed the ideas in statistical inference used to estimate the value of some population parameter. We distinguished between two types of statistical

TABLE 7-5
Summary of Point Estimators and Confidence Intervals for π and μ

Population parameter	Point estimator	Formula for confidence interval	Appropriate conditions
μ	$\bar{X}$	$\bar{x} \pm z_{\alpha/2}\dfrac{\sigma}{\sqrt{n}}$	Exact for any sample size when σ known and X normally distributed. Approximate when X is not normally distributed, but $n \geq 30$
μ	$\bar{X}$	$\bar{x} \pm t_{\alpha/2,\,n-1}\dfrac{s}{\sqrt{n}}$	Exact when σ unknown and X is normally distributed. Approximate when σ is unknown, X is not normal, but $n > 30$
π	P	$p \pm z_{\alpha/2}\sqrt{\dfrac{p(1-p)}{n}}$	Approximate when $n \geq 100$

estimation: point estimation and interval estimation. As is suggested by its name, a point estimate is a single value or point used to predict some population parameter, and it is based on some function of the sample values $x_1, x_2, \ldots, x_n$. To measure how good an estimate is, it is common to construct a confidence interval, or an interval between two points on the real number line that encompasses the value of the population parameter, with a prescribed level of confidence. The more commonly used confidence levels are 90, 95, and 99 percent, but the technique is general enough to produce intervals for *any* desired confidence level.

The formulas for point and interval estimates for μ and π are summarized in Table 7-5. The best point estimators have two important properties: They are unbiased and have minimum variances. These qualities are based on characteristics of the sampling distributions of the sample statistic under consideration. Confidence bounds are essentially one-sided confidence intervals. They are used to determine lower or upper bounds on the value of some population parameter, at a given level of confidence.

Finally, a knowledge of confidence intervals can be used to help solve one of the more perplexing problems in statistical inference, the determination of an appropriate sample size. The number of units necessary for a sample in a given situation depends on three factors: the population parameter involved (μ or π or σ^2), the error that the researcher can tolerate (how close to the parameter the estimate must be), and the level of confidence. As intuition suggestions, both increasing confidence and increasing accuracy imply larger sample sizes.

REFERENCES

R. S. Pindyck and D. L. Rubinfeld, *Econometric Models and Economic Forecasts*, 2d ed. (New York: McGraw-Hill, 1981).

FURTHER READING

Most introductory textbooks on statistical methods for geographers also contain the material covered in this chapter. Usually, these methods are discussed in the context of representative applications in various fields. The problems at the end of this chapter also indicate several possible applications of these techniques in geographical research.

PROBLEMS

1. Explain the meaning of the following concepts:
   a. Point estimation   f. Upper confidence bound
   b. Interval estimation   g. Lower confidence bound
   c. Unbiased estimator   h. Statistical estimator
   d. Efficient estimator   i. Statistical estimate
   e. Confidence interval

2. Discuss the two principal types of statistical estimation.

3. When should the t distribution be used in place of the z distribution in the construction of confidence intervals for the population mean?

4. The yields in metric tonnes per hectare of potatoes in a randomly selected sample of 10 farms in a small region are 32.1, 34.4, 34.9, 30.6, 38.4, 29.4, 28.9, 32.6, 32.9, 44.9. Assuming that these yields are normally distributed, determine a 99 percent confidence interval on the population mean yield.

5. Past experience shows that the standard deviation of the distances traveled by consumers to patronize Chinese restaurants is 2 km. How large a sample is needed to estimate the population mean distance traveled within 0.5 km? The probability of being correct should be set at .95.

6. The proportion of automobile commuters in a given neighborhood of a large city is unknown. A random sample of 50 households is taken, and 38 automobile commuters are found. Determine the 95 percent confidence interval of the proportion of commuters by automobile in the neighborhood.

7. A historical geographer is interested in the average number of children of households in a certain city in 1800. Rather than spend the time analyzing each entry in the city directory, she decides to sample randomly from the directory and estimate the size from this sample.

In a sample of 56 households, she finds the average number of children to be 4.46 with a standard deviation of 2.06.

a. Find the 95 percent confidence interval, using the t distribution.
b. Find the 95 percent confidence interval, using the z distribution.
c. Which interval is more appropriate? Why?

8. A geographer is asked to determine the sample size necessary to estimate the proportion of residents of a city who are in favor of declaring the city a nuclear-free zone. He is told the estimate must not differ from the true proportion by more than .05 with a 95 percent confidence coefficient. How large a sample should be taken? At 99 percent confidence?

9. Suppose we double the size of sample taken when trying to estimate the confidence interval for a mean. What will the effect be on the width of the confidence interval, assuming all other parameters (α level and standard deviation) are held constant?

10. A geography department is interested in purchasing a set of thermometers for use in its physical geography laboratories. Since it is important that these instruments be very accurate, the department places a condition on the supplier: The delivered thermometers must produce no more than a $0.25°C$ average error in measurement. To test whether this condition is met, the departmental technician compares the reading from a sample of 25 new thermometers to a superior recording thermometer and notes the error. The average error turns out to be $0.20°C$ with a standard deviation of $0.15°C$. Should the department accept delivery of the new thermometers? (*Hint*: Use a confidence bound.)

8

Parametric Statistical Inference:
Hypothesis Testing

In Chapter 7, the form of statistical inference known as *estimation* was introduced, and the techniques used for point and confidence interval estimation were developed for a few representative statistics. In this chapter, the form of statistical inference known as *hypothesis testing* is introduced. The methods used in hypothesis testing are closely related to confidence interval estimation. They are simply two ways of looking at the same problem, and both are based on the same theory. In Section 8.1, a highly structured method for testing hypotheses known as the *classical test of hypothesis* is outlined. A more recent and increasingly widely used variant of this procedure known as the *PROB-VALUE*, or *p-VALUE*, method is presented in Section 8.2. Hypothesis tests for population parameters μ and π are detailed in Sections 8.3 and 8.4. The link between hypothesis testing and confidence intervals is explained in Section 8.5.

In many practical problems, the principal concern is not to test a hypothesis about the value of some parameter for a *single* population. Rather, we may want to compare *two* populations with regard to some quantitative characteristic or parameter. For example, suppose we have measures of carbon dioxide concentration from different residential neighborhoods in some city. We wish to compare the two samples. The statistical question being asked is whether these two samples come from a single population or from two different populations. Or, in simpler terms, do these two neighborhoods have similar pollution levels, or are they significantly different? So-called two-sample tests of hypotheses for means and proportions are explained in Chapter 9. These tests also utilize the basic testing method developed in Sections 8.1 and 8.2.

8.1. Key Steps in the Classical Test of Hypothesis

A general conceptual framework for hypothesis testing is illustrated in Figure 8-1. We are interested in making inferences about the value of a population parameter θ. We hypothesize a value for this unknown population parameter, say $\theta = \theta_0$. A random sample of a given size n is collected having values $x_1, x_2, \ldots, x_n$. From these sample data a point estimator $\hat{\theta}$ is calculated. On the basis of $\hat{\theta}$, we evaluate our hypothesis by determining whether the sample value $\hat{\theta}$ does or does not support the

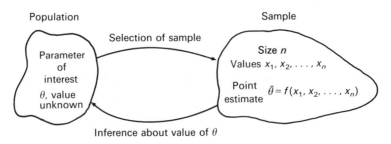

FIGURE 8-1. Sampling and statistical hypothesis testing.

contention that $\theta = \theta_0$. If $\hat{\theta}$ is close to θ_0, we might be led to believe the hypothesis is true. If $\hat{\theta}$ is very much different from θ_0, it is less likely that the hypothesis $\theta = \theta_0$ can be sustained. Recall that $\hat{\theta}$ might differ from $\theta = \theta_0$ for two reasons. First, the hypothesis $\theta = \theta_0$ may really be untrue. Second, $\hat{\theta}$ might differ from θ_0 just because of the effects of random sampling. A simple example can be used to illustrate the framework in Figure 8-1.

The average residential mobility in a North American city can be measured by the probability that a typical household in the city changes residence in a given year. Most studies have found that about 20 percent of the population of the city moves in any one year, or $\pi = .2$. On this basis, we would expect a typical *neighborhood* to have this overall rate of residential mobility. Let us now consider a specific neighborhood in a specific city. How could we determine whether this neighborhood has an average mobility rate? We first *hypothesize* a value of $\pi = \pi_0 = .20$. From a random sample of households in the neighborhood, we then determine the proportion who have moved into the neighborhood in the past year. Since this is a sample proportion, we call this value p, a specific value of the *estimator P*. Now, on the basis of p, we must decide whether this is a typical neighborhood (in terms of residential mobility). If p were close to our hypothesized value of .2, we would probably decide that the hypothesis was true and conclude it was a typical neighborhood. If, however, the sample proportion were $p = .5$, we would be less likely to believe the hypothesis

TABLE 8-1
The Six Steps of Classical Test of Hypothesis

Step 1. Formulation of hypotheses.

Step 2. Specification of sample statistic and its sampling distribution.

Step 3. Selection of a level of significance.

Step 4. Construction of a decision rule.

Step 5. Compute value of test statistic.

Step 6. Decision.

$\pi = .2$, we must always consider the probability this is an incorrect judgment. Why? Our conclusions must always recognize that our judgments are based on *random* samples.

The purpose of this section is to outline a rigorous procedure for testing any hypothesis of this type. A classical test of hypothesis follows the six steps defined in Table 8-1. We now consider each of these six steps, in turn, and then illustrate the procedure by using a straightforward example.

Formulation of Hypotheses

Statements of the hypothesis to be evaluated in a classical test *always* can be expressed in one of these three ways:

$$A: \; H_0: \; \theta = \theta_0 \qquad B: \; H_0: \; \theta \geq \theta_0 \qquad C: \; H_0: \; \theta \leq \theta_0$$

$$H_A: \; \theta \neq \theta_0 \qquad \quad H_A: \; \theta < \theta_0 \qquad \quad H_A: \; \theta > \theta_0$$

The symbol H stands for hypothesis. There are two parts to any hypothesis, labeled H_0 and H_A. In H_0, a statement is made about the value of some population parameter θ. The value θ_0 is the hypothesized, or conjectured, value for θ. In the first form, the statement $H_0: \; \theta = \theta_0$ means we are interested in deciding whether the value of the population parameter θ is equal to θ_0. In H_0, we always assert that θ does equal θ_0. This is called the *null hypothesis*. We either reject this null hypothesis or accept it, depending on the information we collect from the sample. If H_0 is not true, something else must be. As an alternative to H_0, we offer the statement $H_A: \; \theta \neq \theta_0$. Obviously, if we reject the statement $\theta = \theta_0$, then the only remaining possibility is $\theta \neq \theta_0$. Thus H_A is called the *alternate hypothesis*.

The null and alternate hypotheses for the example of residential mobility are

$$H_0: \; \pi = .2 \qquad (\pi_0 = .2)$$

$$H_A: \; \pi \neq .2 \tag{8-1}$$

This short form is an especially useful way of summarizing the hypotheses in a classical test. The population parameter of interest is clearly identified.

There are two other possible formats for H_0 and H_A in hypothesis sets *B* and *C*. Sometimes we are interested in asserting not an *exact* value for the population parameter θ, but a *range* of values. In format *B*, for example, we wish to test whether our sample is consistent with the statement $\theta \geq \theta_0$. That is, can we say the true population parameter is *at least* as large as θ_0? The alternative is that θ is definitely *less than* θ_0, as is expressed in H_A. In format *C*, we are interested in whether our sample supports the statement $\theta \leq \theta_0$ (that is, the true value of the population parameter is *no larger than* θ_0). In this case, the alternate hypothesis must be that θ is *larger than* θ_0.

Which of these three forms should we choose? The correct format depends on the question we wish to answer. To see this, let us return to our example of neighborhood residential mobility. If we express our hypotheses concerning residential mobility in the form of (8-1), we are simply interested in determining whether the neighborhood being studied is typical. Were we to express the hypotheses as in format B

$$H_0: \pi \geq .2 \qquad H_A: \pi < .2 \tag{8-2}$$

we would be testing a hypothesis about whether the neighborhood has a lower than average residential mobility. If we can reject H_0, then our alternate hypothesis tells us that the neighborhood is probably much more stable than the average. If the neighborhood under study is an older suburb where there is predominantly single-family housing, this may be a hypothesis well worth evaluating. Notice that by rejecting the null hypothesis H_0, we are in a position to assert neighborhood stability.

By placing our hypothesis in format C

$$H_0: \pi \leq .2 \qquad H_A: \pi > .2 \tag{8-3}$$

we could test the hypothesis of whether the neighborhood has a higher turnover than average. If we were studying the residential neighborhood around a major university, a high rate of mobility might be expected. In all three formats A, B, and C, what we are really interested in saying is stated in the alternate hypothesis H_A, which we prove by evaluating H_0 and then rejecting it, if possible. This apparently peculiar way of operating is actually the only way of proceeding. This becomes clear when we proceed to the second step in the hypothesis testing procedure.

Format A of the hypotheses in a statistical test is sometimes referred to as a *two-sided*, or *nondirectional*, *test*. This is because the alternate hypothesis does not say on which *side*, or *direction*, the true population parameter θ lies in relation to the hypothesized value θ_0. Formats B and C are usually called *directional hypotheses* because the alternate hypothesis H_A does specify on which side of θ_0 the true parameter lies. Sometimes format B is termed a *lower-tail test* and format C an *upper-tail test*, on the basis of the direction specified in H_A. The reasons for these names and the choice of formats among A, B, and C are explained further in the examples in this chapter.

Whenever sample data are used to choose between H_0 and H_A, there are risks of making the incorrect decision owing to sampling. These errors are known as *inferential errors*, because they arise when we make an incorrect inference (or

TABLE 8-2
Possible Decisions and Outcomes in a Classical Test of Hypothesis

	True state of nature	
Decision or conclusion reached	H_0 is true, H_A is false	H_0 is false, H_A is true
Reject H_0	Type I error	Correct decision
Do not reject H_A	Correct decision	Type II error

TABLE 8-3
Probabilities of Making a Correct or Incorrect Decision in a Test of Hypothesis

	True state of nature	
Decision reached	H_0 is true	H_0 is false
Reject H_0	α	$1 - \beta$
Do not reject H_0	$1 - \alpha$	β
Total probability	1	1

judgment, or conclusion) from the sample about the value of the population parameter. There are two possible types of error:

DEFINITION: TYPE I, OR α, ERROR
A Type I error in a statistical test of hypothesis is committed when a decision is made to reject a null hypothesis which is actually true. The probability of committing a Type I error is α (alpha).

DEFINITION: TYPE II, OR β, ERROR
A Type II error in a statistical test of hypothesis is committed when a decision is made to accept a null hypothesis which is false. That is, H_A is the correct conclusion. The probability of making a Type II error is denoted β (beta).

The way in which each type of error can arise in statistical hypothesis testing is illustrated in Table 8-2. Note that there are four possible outcomes resulting from any decision about the null hypothesis. In two cases, the decision may be the correct one. This occurs whenever we rightfully reject a false H_0 or accept a true H_0. In the other two cases we commit *either* a Type I or a Type II error. If we reject H_0 when it is true, we commit a Type I error; if we fail to reject H_0 when it is false, we commit a Type II error.

In adopting a method for choosing between H_0 and H_A, it is important to evaluate the risks of making both types of error. The chances of committing Type I and Type II errors are denoted α and β, respectively; that is, $\alpha = P(\text{Type I error})$ and $\beta = P(\text{Type II error})$. The relationship of α and β to the various choices about H_0 is indicated in Table 8-3. Note that we *must* select one of the two alternatives—either row 1 or row 2. Because our two choices are *complementary* and the total probability of each column must sum to 1, we can assign probabilities to the boxes as shown. Because we are going to base our decision about whether to reject H_0 on the basis of sample data, some risk must be accepted. There is no way to be completely certain of our conclusion short of taking a complete census of the population. The alternative approach is to *control* the error. The way in which this is done is explained in step 3 of the classical test of hypothesis.

Selection of Sample Statistic and Its Sampling Distribution

The second step of the classical procedure is to select the appropriate *sample statistic* upon which our decision to reject or not to reject the hypothesis is to be made. From Chapters 6 and 7 we know that the appropriate statistic to use is the unbiased, minimum variance estimator of the population parameter under study. When used in this way, this sample statistic is referred to as the *test statistic*. These test statistics are summarized in Table 7-1. It should make sense that tests concerning μ are based on the sample value of the estimator $\bar{X}$, and, similarly, tests of π are based on the estimator P and its sampling distribution, and so on.

Returning to the example of residential mobility, we conclude that the decision on the *proportion* of households in the neighborhood changing residence in the last year should be based on the sample proportion statistic P. What is the sampling distribution of P? The *exact* sampling distribution of P is given by the binomial distribution since P is a Bernoulli variable. For "small" samples we can use tables of the binomial in our hypothesis test. When the sample size is large, we can appeal to the central limit theorem for the approximate sampling distribution of P. If $n \geq 100$, the sampling distribution of P is approximately normal with mean $\mu_P = \pi$ and $\sigma_P = \sqrt{\pi(1-\pi)/n}$. Recall in Chapter 5 that we found the normal provided a good approximation to the binomial for large sample sizes. In this case, then, the sampling distribution of P is approximately normal with a mean of .2 and a standard deviation of $\sigma_P = \sqrt{.2(.8)/100} = .04$.

Selection of a Level of Significance

In this third step of the classical hypothesis testing procedure, we specify the chances we are willing to accept of making an inferential error. Of the two types of possible error, a Type I error is generally considered to be more serious and therefore should be controlled. That is, failing to reject a true hypothesis is considered a more significant error than rejecting a valid one. For this reason, the usual procedure is to select a very low value for α, the probability of committing a Type I error. Usually, a value of α is chosen from .10, .05, or .01. The actual value of α chosen for the test is known as its *significance level*.

DEFINITION: LEVEL OF SIGNIFICANCE
The level of significance of a classical test of hypothesis is the value chosen for α, the probability of making a Type I error.

Whenever we report a decision about the null hypothesis, we *always* state the level of significance of our result. We say, for example, that a null hypothesis is rejected at the .05 level or that it is *statistically significant* at the .05 level. This means that the probability we are making an α error is extremely low. It is very unlikely to have arisen due to chance, that is, random sampling.

The usual analogy in explaining the concept of a level of significance is based on the nature of justice in a court of law. In a court, a defendant is innocent until proved guilty beyond a reasonable doubt. The burden is placed entirely on the prosecution to prove any assertion. Why? Because the interests of justice are best served if we free a guilty person as opposed to imprisoning an innocent one. This is tantamount to saying that a Type I error is more serious than a Type II error. The reasonable doubt in a statistical test is the α level. If we are going to reject H_0, we must be very sure we are correct. This is why the usual levels of significance are so low. Thus, when we report a result as statistically significant, we are actually saying that it is *probably* correct. Using the word *probably* reflects our knowledge that we may be incorrect because of pure chance, owing to random sampling.

The level of significance chosen for a particular problem depends entirely on the nature of the problem and the uses to which the results will be put. We are likely to be very demanding if we are testing toxicity levels of drugs since the consequences of permitting a drug on the market having extremely serious side effects must clearly pass an exacting requirement. An α level of .0001 may be appropriate. For most social science purposes, it is common to evaluate hypotheses at .10, .05, or .01. The consequences of an error, for example, in the study of neighborhood mobility, cannot be considered to be very serious. For the example of residential mobility, a level of significance of $\alpha = .05$ is selected.

Construction of a Decision Rule

To this point, we can summarize the test of hypothesis as follows:

Step 1. H_0: $\pi = .20$ and H_A: $\pi \neq .20$.
Step 2. P is chosen as the sample or test statistic.
Step 3. $\alpha = .05$.

To construct a decision rule for H_0, it is now necessary to specify which values of the test statistic will lead us to conclude H_0 and which H_A. First, we assume H_0: $\pi = .20$ is true and generate the sampling distribution of P, as in Figure 8-2. Note that it is

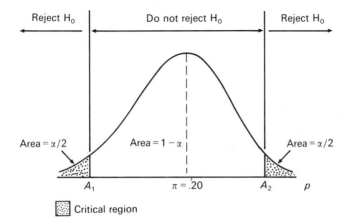

FIGURE 8-2. Sampling distribution of P, centered on hypothesized value $\pi = .20$.

centered over a value of $P = .20 = \pi$. The actual value of p generated in our sample will be close to .20 if H_0 is true and far away from .20 if H_0 is not true. For a given level of significance, we can divide the sampling distribution into two parts—those outcomes leading to an acceptance of H_0 and those leading to a rejection of H_0. Since we have specified a two-tailed test, we reject H_0 only if the sample outcome is in the most extreme parts of the sampling distribution. The most extreme outcomes occur in either tail of the distribution, so we place $\alpha/2$ of our possible outcomes in each tail. If the value of p is in the remaining $1 - \alpha$ percent of outcomes, we accept our null hypothesis. For this reason, the values of p which lead us to reject H_0 are known as the *critical region* of the test. When we construct a decision rule for our hypothesis, we are specifying the values for A_1 and A_2 in Figure 8-2. These are sometimes termed the *action limits*, or *critical values*, for the hypothesis.

It is easiest to see how this decision rule can be constructed by considering the example of residential mobility. Suppose we randomly sampled $n = 100$ households in the neighborhood and calculated the proportion of people who moved in the last year p. What values of p will lead us to conclude the neighborhood is more stable than the average (i.e., what is A_1?), and what values of p will lead us to conclude the neighborhood has a higher rate of residential mobility than normal (i.e., what is A_2?)? Let us solve this problem by generating the sampling distribution of P for $n = 100$, $\pi = .20$, and $\alpha = .05$. The appropriate sampling distribution is illustrated in Figure 8-3. First, note that the sampling distribution in this figure is drawn to be normal (or approximately so) since we know from step 2 that this is the appropriate sampling distribution for the test statistic P. Also we have placed $\alpha = .025$ of the outcomes for P in either tail of the distribution.

To determine the values of the critical limits A_1 and A_2, we use our knowledge of the normal distribution. First, we know this sampling distribution has a mean centered on $\pi = .20$. The standard deviation of this sampling distribution is

$$\sigma_P = \sqrt{\frac{\pi(1 - \pi)}{n}} = \sqrt{\frac{.2(.8)}{100}} = .04$$

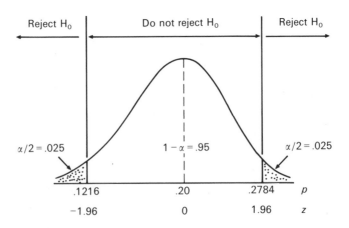

FIGURE 8-3. Determining critical region, limits for residential mobility example, $\alpha = .05$.

The critical limits on the sampling distribution are determined by the conditions

$$P(P < A_1) = \frac{\alpha}{2} \quad \text{and} \quad P(P > A_2) = \frac{\alpha}{2} \tag{8-4}$$

where $P(P < A_1) = \alpha/2$ means we must choose the value of A_1 such that the probability of the sample statistic P being less than A_1 is $\alpha/2$. Because the sampling distribution of P is normal, A_1 and A_2 are symmetric about $\pi = .2$.

To find A_1, we use the standardizing formula for the normal distribution:

$$z_{A_1} = \frac{A_1 - \pi}{\sigma_P} \tag{8-5}$$

And from the significance level α, we can say $z_{A_1} = -z(\alpha/2)$, the value of z which has $\alpha/2$ in the lower tail of the distribution. Similarly,

$$z_{A_2} = z_{\alpha/2} = \frac{A_2 - \pi}{\sigma_P} \tag{8-6}$$

can be used to determine z_{A_2}. Substituting the known values of π, α, and σ_P yields

$$-z_{.025} = \frac{A_1 - .2}{.04} \quad \text{and} \quad z_{.025} = \frac{A_2 - .2}{.04}$$

We find $z_{.025}$ by looking in the body of the normal table to find the z score which has .025 of the area under the curve in the tail. If there is .025 in the tail, there must be $.5000 - .0250 = .4750$ between the mean and $z_{.025}$. This corresponds to a z score of 1.96. Substituting $z_{.025} = 1.96$ into our equations, we find the critical limits for our decision rule:

$$-1.96 = \frac{A_1 - .2}{.04} \qquad 1.96 = \frac{A_2 - .2}{.04}$$

$$-.0784 = A_1 - .20 \qquad .0784 = A_2 - .2$$

$$.1216 = A_1 \qquad .2784 = A_2$$

The decision rule can now be summarized as follows:

Reject H_0: $\pi = .20$ if $p < .1216$ or $p > .2784$. Otherwise, do not reject H_0.

This is a very compact way of expressing what action we will take depending on the value of p generated in the sample.

There is an equivalent, widely used way of expressing the decision rule. Instead of making the decision rule based on the value of the sample statistic p, we express it in terms of z.

Reject H_0: $\pi = .20$ if $z < -1.96$ or $z > 1.96$. Otherwise, do not reject H_0.

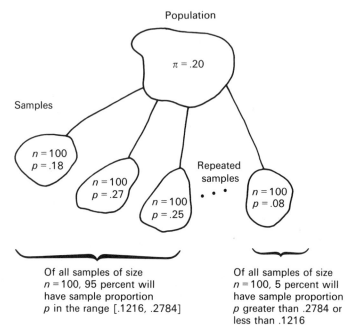

Of all samples of size
n = 100, 95 percent will
have sample proportion
p in the range [.1216, .2784]

Of all samples of size
n = 100, 5 percent will
have sample proportion
p greater than .2784 or
less than .1216

FIGURE 8-4. Interpreting the nature of the sampling distribution, residential mobility example.

In this case, we determine the value of z for the decision rule on the basis of the standardizing formula

$$Z = \frac{p - \pi}{\sigma_P} \qquad (8\text{-}7)$$

What we are doing in this simple transformation is taking the observed value of sample statistic p and locating it within the standard normal distribution. To develop the critical limits A_1 and A_2 in terms of p, we simply reworked (8-7) and solved for p.

Let us now explain the reasoning behind the use of these decision rules in greater detail. If the null hypothesis $\pi = .20$ were true, we would expect the value of statistic P to be fairly close to .20. For $\alpha = .05$, we would expect the value of P to be in the range [.1216, .2784] with probability $1 - .05 = .95$. If p were less than .1216, it would be very likely that the true value of π was less than .20; if p were greater than .2784, it would be very likely that π was greater than .20. This decision rule is thus based on the nature of the sampling distribution of the test statistic P if $\pi = .20$ is correct. This is expressed in a slightly different way in Figure 8-4.

Compute the Value of the Test Statistic

At this point, we collect our sample and calculate the *value* of the sample statistic P. Suppose that 26 of 100 randomly selected households in the neighborhood have moved in the past year. The value $p = \frac{26}{100} = .26$ is the value of the test statistic. We can use this value directly with the decision rule constructed in the previous step. Alternatively, if we wish to express or result in terms of z, we simply calculate

$$z = \frac{.26 - .20}{.04} = \frac{.06}{.04} = 1.50$$

and use this value in conjunction with the decision rule based on z.

Decision

The last step is to make the actual evaluation of the null hypothesis based on the sample. This is simply a matter of examining the calculated value of the test statistic in relation to the decision rule. Since $p = .26$ is between .1216 and .2784, it *does not* lie in the critical region. We conclude H_0 is true and cannot be rejected. If we were to reject H_0 on the basis of this sample, the probability of making a Type I error would exceed .05. Since only a significance level of $\alpha = .05$ is acceptable to us, we must not reject H_0. Even if the computed value of p is $p = .2783$, we must still accept H_0. In fact, this rigidity in our decision rule is one problem with the classical test of hypothesis. We can only reject or accept H_0. But surely a sample with $p = .26$ and a sample with $p = .20$ give us varying degrees of belief in H_0. In Section 8.2 we introduce the notion of PROB-VALUE method which gets around this problem. Instead of simply rejecting or accepting H_0, we determine our degree of belief in H_0, that is, we compute the value of α implicit in our sample.

What if $p = .31$ is the calculated sample value of p? In this instance, the decision is different. We reject H_0 since $p = .31 > .2784$. Rejecting H_0 in this case leads to two possibilities. Either we made a correct decision and $\pi \neq .20$, or we obtained a sample (probability $< .05$) that did come from a population with $\pi = .20$. In the latter case, we made a wrong decision and committed a Type I error. However, we should feel relatively secure, knowing the probability of such an occurrence is necessarily less than .05.

To evaluate the hypothesis in terms of a decision rule based on z, we note $z = 1.5$ calculated in the previous step is in the range $[-1.96, 1.96]$, and thus we must not reject H_0. Of course, any decision based on z is equivalent to one made by using p, and vice versa. Some texts in statistics always work with z, others with the value of sample statistic p, and still others both. The advantage of using p (or $\bar{x}$ or s, as the case may be) is that the decision being made is on the basis of the units in which the variable of interest is measured. In the current example, we could say that we will reject the hypothesis that the neighborhood has a typical level of residential mobility if the proportion of people who moved recently is less than .1216 or greater than .2784. To say the same thing in terms of some standardized normal variable removes the decision from the context of the problem at hand. Nevertheless, this is the more common approach.

The complete text of hypothesis for the example is summarized in Table 8-4. It provides a convenient format for classical tests for beginning students in inferential statistics. It is used throughout the remainder of this text. In subsequent examples in this chapter, many of the variations in the classical test are introduced. The

TABLE 8-4
Summary of Test of Hypothesis for Residential Mobility Example

Step 1. H_0: $\pi = .20$ and H_A: $\pi \neq .20$.

Step 2. P is chosen as sample statistic.

Step 3. $\alpha = .05$.

Step 4. Reject H_0 if $p < .1216$ or $p > .2784$; otherwise, do not reject H_0. (See Figure 8-3.)

Step 5. From random sample, $p = .26$

Step 6. Since $.1216 \leq p = .26 \leq .2784$, we do not reject H_0.

evaluation of a directional hypothesis, so-called lower- or upper-tail test, is explained in these examples.

8.2. PROB-VALUE Method of Hypothesis Testing

In Section 8.1 we outlined a strict procedure for hypothesis testing known as the classical method. This procedure is useful as a means of clarifying theoretical issues of extreme importance such as the distinction between a Type I and a Type II error. However, it is now rarely used by applied statisticians who increasingly opt for an alternative known as the PROB-VALUE, or p-VALUE, method. In this text we term the method *PROB-VALUE testing* rather than p-VALUE since the letter p is used in so many other contexts in statistics. What is wrong with hypothesis testing done in the classical way? First, it requires us to choose a value for α, the level of significance for the test. There are few instances where this value can be rationally set. The claim of critics of statistics that "You can prove anything with statistics" is hard to defend when we can easily think of cases where we would reject H_0 at $\alpha = .10$ but not at $\alpha = .05$. That it is necessary to arbitrarily select a level of significance leads to a second problem: The decision made at the final step of the classical test can only be to reject H_0 or not to reject it. The use of PROB-VALUE method is best explained by an example. For purposes of continuity, we reconsider the hypothesis test concerning residential mobility developed in Section 8.1.

PROB-VALUE Method for a Two-Sided Test

Figure 8-3 illustrates the decision rule developed for the hypothesis set

$$H_0: \pi = .20 \qquad H_A: \pi \neq .20$$

for the residential mobility problem. Selecting a significance level of $\alpha = .05$ led to this decision rule:

1. Reject H_0 if $p < .1216$ or $p > .2784$ (or $z < -1.96$ or $z > 1.96$).
2. Do not reject H_0 if $.1216 \leq p \leq .2784$ (or $-1.96 \leq z \leq 1.96$).

Here p is the value of the sample proportion statistic P, and z is the value of the standardized variable

$$Z = \frac{P - \pi_0}{\sqrt{\pi_0(1 - \pi_0)/n}}$$

Since our sample proportion is .26, we do not reject H_0. Now, the principal difficulty with the classical method is that it provides no information as to *exactly* how likely it is that the null hypothesis is correct. *Any* value of p in the critical region leads us to reject H_0. But surely a sample proportion $p = .8$ tells us that H_0 is less likely to be true than a sample proportion $p = .28$ (just on the edge of the critical region). It is this problem which the PROB-VALUE method is designed to solve.

Now, the actual sample proportion $p = .26$ leads us not to reject H_0. But since it is close to the critical limit of $p = .2784$, we might wonder, "If we were to reject H_0 on the basis of $p = .26$, how much faith would we have in this decision?" To see this, look at Figure 8-5. On this diagram we have established the critical region at the value of $p = .26$ on the upper side of $\pi_0 = .20$ and at $p = .14$ $(.20 - .06)$ on the lower side. Together these two values define another sort of critical region. If the area in this region is calculated, it is possible to determine the *probability* that a sample value *as extreme as* $p = .26$ would be obtained from a population with $\pi = .20$. This area can be easily determined from a knowledge of the normal distribution. First, convert the sample value of $p = .26$ to a standardized value of z, using

$$z = \frac{p - \pi_0}{\sqrt{\pi_0(1 - \pi_0)/n}} = \frac{.26 - .20}{\sqrt{.20(1 - .20)/100}} = 1.5$$

Next, determine the area to the right of $z = 1.5$. Using the normal table gives us, $.5000 - .4332 = .0668$. Since the distribution is symmetric, the area in both tails is $2(.0668) = .1336$. This is the PROB-VALUE attached to H_0 for this problem. The probability of .0668 is doubled because we are evaluating a two-sided alternative. Recall that in a classical test of hypothesis in a two-sided test the significance level of

PROB-VALUE = shaded area = 2(.0668) = .1336

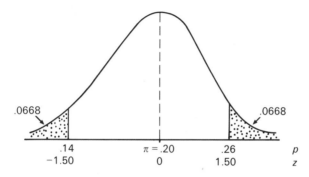

FIGURE 8-5. Determining the PROB-VALUE in two-sided hypothesis test, residential mobility example.

α was halved and $\alpha/2$ was placed in the two rejection regions of the tails of the sampling distribution.

The PROB-VALUE of .1336 can be interpreted in the following way: *it is the lowest value at which we could set the significance of the test and still be able to reject* H_0. If we reject H_0 on the basis of $p = .26$, we are saying that the probability of committing a Type I error of $\alpha = .1336$ is acceptable to us. In short, the PROB-VALUE indicates the *degree of belief in the null hypothesis*. As PROB-VALUE approaches 1, the belief in H_0 increases; as PROB-VALUE approaches 0, the belief in H_0 is less and less. Calculating a PROB-VALUE in a hypothesis testing problem is clearly a superior alternative to the classical test. For every possible value of the sample statistic, it is possible to find a PROB-VALUE. The researcher can then assess what action to take based on the value obtained in the problem at hand. This avoids the thorny problem of apparent dismissal of an alternate hypothesis simply because the sample statistic lay just outside the critical region.

However, setting a significance level and calculating a PROB-VALUE are not the only facts used in evaluating a null hypothesis. Suppose, for example, that we test the fairness of a coin by tossing it 10 times and recording the number of heads. If 8 heads appear, we reject the hypothesis that the coin is fair at $\alpha = .05$. There are at least two reasons why we might want to question this conclusion. First, the sample size is extremely small and hence possibly unreliable. Conclusions drawn from larger samples are much more dependable, and we should discount this conclusion appropriately. If, however, 800 heads are counted in 1000 tosses, we can certainly feel justified in rejecting H_0. Second, common sense has a definite role to play. The chances of a coin randomly chosen being unfair should certainly be considered remote. The fact that 8 heads appear might be attributed to chance. We know it is possible that a fair coin could yield 8 heads in 10 tosses; maybe this was just one of those times! The data are only one factor to be used in a judgment about a hypothesis. If the data, for example, indicated that an estimate of residential mobility in a transition neighborhood was very, very low, say 3 percent, this might say more about our sample than about the neighborhood. It is probably telling us we have a peculiar, unrepresentative sample. Certainly we should check this out before accepting a hypothesis at great odds with our expectations. Skepticism has a role to play in statistics.

Formally, PROB-VALUE can be defined in the following way:

DEFINITION: PROB-VALUE

The PROB-VALUE associated with a null hypothesis is equal to the probability that the value of the sample statistic is as extreme as the value observed, if the null hypothesis is true. It can be loosely interpreted as degree of belief in the null hypothesis.

Virtually all statistical packages now report PROB-VALUE. These values report what the sample data tell us about the credibility of a null hypothesis, rather than forcing a decision about H_0 on the basis of a possibly arbitrarily defined standard or level of significance. Determining a PROB-VALUE does not preclude

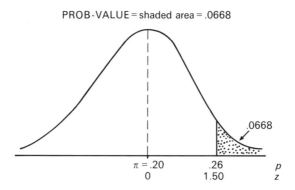

PROB-VALUE = shaded area = .0668

.0668

$\pi = .20$ .26 p
0   1.50 z

FIGURE 8-6. Determining the PROB-VALUE in one-sided hypothesis test, residential mobility example.

performing a classical test. Rejecting null hypotheses with **PROB-VALUES** less than .05 is equivalent to a classical test of hypothesis with $\alpha = .05$.

PROB-VALUE Method for a One-Sided Test

Determining the **PROB-VALUE** in a one-sided test of hypothesis, either lower or upper tail, differs in only one respect from the method discussed above. Suppose the hypothesis being evaluated in the residential mobility example is

$$H_0: \pi \le .20 \qquad H_A: \pi > .20$$

So we are examining whether the neighborhood has a higher mobility rate than average. The **PROB-VALUE** associated with the sample value $p = .26$ is .0668, as illustrated in Figure 8-6. There is no need to double the probability in the tail in a one-sided test *since the critical region is always placed in one tail or the other*, depending on whether it is a lower- or an upper-tail test. The fact that the **PROB-VALUE** is lower in this version of the hypothesis may reflect some prior belief we have about the true value of π in this neighborhood. We might have formulated an upper-tail test simply because the neighborhood is in transition. A less extreme result is needed to support this hypothesis. A value of $p = .26$ indicates the neighborhood is experiencing higher mobility than average. A **PROB-VALUE** of .0668 (as opposed to .1336) better reflects our evaluation of H_0.

8.3. Hypothesis Tests concerning the Population Mean μ

In the remainder of this chapter, a series of hypothesis tests are described for the population parameters μ and π. Examples of all tests illustrate typical situations where these tests may be used. In these examples, an effort is made to clarify many of the theoretical issues discussed in Sections 8.1 and 8.2. Thus, examples vary between lower- and upper-tail directional (or one-sided) hypotheses and two-sided hypotheses. At times, the decision rules are based on the value of the sample statistics such as $\bar{x}$ and p; in other cases the decision rules are expressed as standardized values of t or z. Finally, some questions are worded as classical tests of hypotheses, others in terms of PROB-VALUE. This variety of problem formats reinforces the notions and issues in hypothesis testing developed to this point.

The first test concerns hypotheses related to the population mean μ. The sample mean statistic $\bar{X}$ is the appropriate test statistic in this case, and its sampling distribution is specified in the central limit theorem. Just as we distinguished between two cases for specifying a confidence interval for μ, hypothesis tests of μ must be based on the knowledge of the population standard deviation σ:

Case 1. Population standard deviation σ is known.
Case 2. Population standard deviation σ is unknown.

Case 1: Population Standard Deviation σ Known

In Chapter 6 we showed that *if a population random variable X is normally distributed,* then the sample mean statistic is also normally distributed with mean $E(\bar{X}) = \mu_{\bar{X}} = \mu$ and standard deviation $\sigma_{\bar{X}} = \sigma/\sqrt{n}$. Tests of hypothesis concerning μ can therefore be evaluated by using the standard normal statistic

$$Z = \frac{\bar{X} - \mu}{\sigma_{\bar{X}}} = \frac{\bar{X} - \mu}{\sigma/\sqrt{n}} \tag{8-8}$$

This standard normal statistic finds extensive application in hypothesis testing. Under certain conditions, it can be used to generate decision rules for tests of sample proportions from Equation (8-7). In fact, for many different sample statistics θ, the general form of the standard normal statistic

$$Z = \frac{\hat{\theta} - \theta}{\sigma_\theta} \tag{8-9}$$

can be used to construct confidence intervals or test hypotheses. In (8-9), θ is the hypothesized value of the population parameter and the mean of the sampling distribution of the sample statistic $\hat{\theta}$. The denominator σ_θ is the standard deviation of the sampling distribution of $\hat{\theta}$, sometimes called the *standard error* of the sample statistic. The purpose of the standardized statistic Z is to locate the value of the sample statistic within this sampling distribution.

The value of Z obtained from Equation (8-9) for a particular sample tells us how many standard deviations (or standard errors) from the mean of the sampling distribution the sample lies. The higher the absolute value of z, the less likely the occurrence of the sample and the less likely that H_0 can be accepted. A decision rule based on $\alpha = .05$ for *any* sample statistic that is normally distributed will *always* be as follows for a two-sided test:

1. Reject H_0: $z < -1.96$ or $z > 1.96$.
2. Do not reject H_0: $-1.96 \leq z \leq 1.96$.

This is so because the standard score $z = 1.96$ has a tail probability of .025 and hence .05 in the two tails of the sampling distribution. However, we continue to formulate decision rules for hypotheses in terms of both the sample statistic itself ($\bar{x}$ or p) and standardized values of t or z.

Let us return to hypothesis testing for μ by considering the following example.

EXAMPLE 8-1. The average household size in a certain city is 3.2 persons with a standard deviation of $\sigma = 1.6$. A firm interested in estimating weekly household expenditures on food takes a random sample of $n = 100$ households. To check whether the sample reflects the average household size in the city, the firm calculates the average household size of the sample to be 3.6 persons. Assuming that the random variable called household size is normally distributed, test the hypothesis that the firm's sample is representative of the average household size in the city. What PROB-VALUE can be attached to this hypothesis?

Solution. The values of $\mu = 3.2$ and $\sigma = 1.6$ are given for a normally distributed random variable. The sample of $n = 100$ households yields $\bar{x} = 3.6$. The null and alternate hypotheses are

$$H_0: \mu \neq 3.2 \qquad H_A: \mu \neq 3.2$$

A two-sided test is specified since the firm would probably be concerned if they oversampled large families (hence $\bar{x} > 3.2$) or oversampled smaller households (hence $\bar{x} < 3.2$). The sampling distribution of $\bar{X}$ is illustrated in Figure 8-7, and the action limits for $\bar{x}$ are

$$-1.96 = \frac{A_1 - 3.2}{1.6/\sqrt{100}} \qquad 1.96 = \frac{A_2 - 3.2}{1.6/\sqrt{100}}$$

$$A_1 = 2.89 \qquad\qquad A_2 = 3.51$$

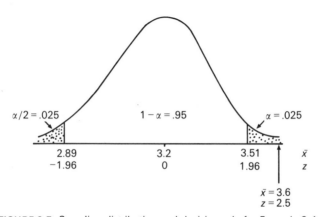

FIGURE 8-7. Sampling distribution and decision rule for Example 8-1.

This leads to these decision rules:

1. Reject H_0: $\bar{x} < 2.89$ or $\bar{x} > 3.51$ (or $z < -1.96$ or $z > 1.96$).
2. Do not reject H_0: $2.89 \leq \bar{x} \leq 3.51$ (or $-1.96 \leq z \leq 1.96$).

The value of the test statistic is $\bar{x} = 3.60$ and is indicated by an arrow within the sampling distribution of Figure 8-7. Since it lies in the critical region, we can reject H_0; the firm appears to have a sample of households with family sizes not representative of the metropolitan area.

To determine the PROB-VALUE for H_0, the sample value of $\bar{x} = 3.6$ is first standardized:

$$z = \frac{3.6 - 3.2}{1.6/\sqrt{100}} = 2.5$$

From the normal table, a value of $z = 2.5$ leaves $.5000 - .4938 = .0062$ in one tail of the distribution. Because we have a two-tail test, this value is doubled, so that the PROB-VALUE for H_0 is $2(.0062) = .0124$. We conclude it is highly unlikely that this is a representative sample.

A final note concerning this example. Since household size is a discrete variable, it is exceedingly unlikely that the variable is normally distributed. However, with the large sample size, it is still likely that $\bar{X}$ is normally distributed because of the implications of the central limit theorem. Our results can be interpreted as being approximately correct. Whenever the underlying distribution of the random variable is unknown or is not normal, tests of hypotheses or PROB-VALUE calculations should be considered only approximate. The approximation becomes increasingly accurate as the sample size increases.

Case 2: Population Standard Deviation σ Unknown

If σ is unknown, it must be estimated from the sample by using the estimator S. The standardized random variable

$$t = \frac{\bar{X} - \mu}{S/\sqrt{n}} \tag{8-10}$$

is t-distributed with $n - 1$ degrees of freedom. This result was first presented in Chapter 7. For a normally distributed random variable X, Equation (8-8) can be used to test hypotheses and determine PROB-VALUES. When X is not normally distributed, the use of (8-10) can still be considered to give approximate results.

EXAMPLE 8-2. A commuter train advertisement asserts that the time on the train between a certain station and the downtown terminal is 23 min. A random sample of $n = 30$ trains in a single month yields an average time of 26 min with a standard deviation of 6 min. Test the assertion that the trains are really slower than advertised, using $\alpha = .05$. Determine the PROB-VALUE.

Solution. Viewing this as a hypothesis testing problem, we can put

$$H_0: \mu \leq 23 \qquad H_A: \mu > 23$$

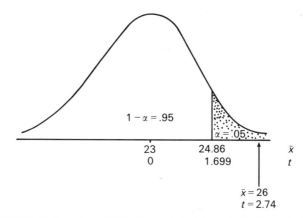

$1 - \alpha = .95$

$\alpha = .05$

23   24.86 $\bar{x}$
0   1.699 t

$\bar{x} = 26$
$t = 2.74$

FIGURE 8-8. Sampling distribution and decision rule for Example 8-2.

So by rejecting H_0 we can make the assertion we would like: Average times are longer than advertised. Based on a sample of size $n = 30$, the value of $\bar{x} = 26$, and the estimated $s = 6$, we utilized the standardized t statistic in this test. The sampling distribution of t with df $= 29$ and $\alpha = .05$ is illustrated in Figure 8-8. Note that the region of rejection is placed in the upper tail. The critical value of t from the t table is 1.699. The sample mean of $\bar{x} = 26$ leads to a test statistic value of

$$t = \frac{26 - 23}{6/\sqrt{30}} = \frac{3}{1.095} = 2.74$$

which is clearly in the critical region. We therefore reject the null hypothesis and conclude trains are slower than advertised.

    Calculating the PROB-VALUE by using the t table is not so straightforward as calculating it from the normal table. Looking across the row corresponding to df $= n - 1 = 29$, we see the following standardized t values and tail probabilities:

df	.10	.05	.025	.01	.005
			Tail probability		
29	1.311	1.699	2.042	2.457	2.750

The computed value of $t = 2.74$ lies between .01 and .005, so the PROB-VALUE for this test can be reported as $.01 <$ PROB-VALUE $< .005$. It is possible to interpolate between these values, if necessary. Most statistical packages on the computer give exact PROB-VALUES.

    The tests of hypothesis concerning the population parameter μ are formally summarized in Table 8-5. The table distinguishes between two cases depending on whether σ is known beforehand. Both cases require X to be normally distributed random variable, but are approximate if X is not normally distributed. Usually, a sample size of at least 30 is required. If the distribution of X is highly skewed, multimodal, or nearly rectangular, larger samples are usually required.

TABLE 8-5
Single-Sample Test of a Mean

Specification
A single random sample of size n is drawn from a population, and a value of $\bar{x}$ is calculated. If variable X is normally distributed, then the following test gives exact **PROB-VALUES** and hypothesis test evaluations. If X is not normal, the test gives approximately correct results if $n \geq 30$. The greater the departure from normality, the larger the sample size required to utilize this test. Two cases are distinguished.

Case 1. σ is known.
Case 2. σ is unknown, and a value for S is estimated from the sample.

Hypotheses
H_0: $\mu = \mu_0$; the population mean has some given value μ_0.
H_A: (a) $\mu \neq \mu_0$ (two-tailed).
  (b) $\mu < \mu_0$ (lower tail), implies H_0: $\mu \geq \mu_0$.
  (c) $\mu > \mu_0$ (upper tail), implies H_0: $\mu \leq \mu_0$.

Test statistic

$$\begin{array}{cc}
\textit{Case 1} & \textit{Case 2} \\[4pt]
Z = \dfrac{\bar{X} - \mu_0}{\sigma/\sqrt{n}} & t = \dfrac{\bar{X} - \mu_0}{S/\sqrt{n}} \quad df = n - 1
\end{array}$$

These equations give exact results when X is normal.

Decision rule
For case 1, reject H_0 if $z < -z_{\alpha/2}$ or if $z > z_{\alpha/2}$; otherwise, do not reject H_0 for a two-tailed test. For a lower-tail test, reject H_0 if $z < -z_\alpha$; otherwise, do not reject H_0. For an upper-tail test, reject H_0 if $z > z_\alpha$; otherwise, do not reject H_0.

For case 2, simply substitute t and z. The degrees of freedom are always $n - 1$.

Case 1 Sampling Distributions

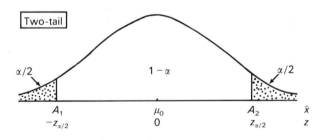

where $A_1 = \mu_0 - z_{\alpha/2}\dfrac{\sigma}{\sqrt{n}}$

$A_2 = \mu_0 + z_{\alpha/2}\dfrac{\sigma}{\sqrt{n}}$

TABLE 8-5 (continued)

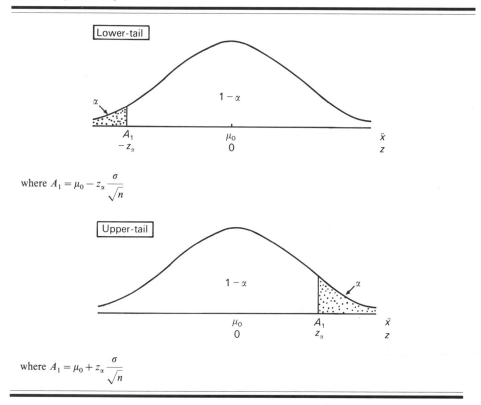

where $A_1 = \mu_0 - z_\alpha \dfrac{\sigma}{\sqrt{n}}$

where $A_1 = \mu_0 + z_\alpha \dfrac{\sigma}{\sqrt{n}}$

8.4. Hypothesis Tests concerning the Population Proportion π

This test was used as the example in the explanation of the classical test of hypothesis in Section 8.1. The test is formally specified in Table 8-6. Let us consider another example to illustrate an upper-tail version of this problem.

EXAMPLE 8-3. Household surveys of residents of a suburban neighborhood have *consistently* found the proportion of residents in the neighborhood taking public transit to be .16. The city council, at the urging of neighborhood residents, increased the bus service to this neighborhood. However, the council vowed to discontinue the service if there was not an increase in patronage. Three months after the introduction of the new service, the council hired a consultant to undertake a survey of $n = 200$ residents to determine their current modes of transportation. Forty-two residents indicated they took public transit. Should the council continue the service?

Solution. Since the proportion of residents taking public transit has consistently been .16 and the council has specified that the proportion must *increase*, we can specify the two hypotheses as follows:

$$H_0: \pi \leq .16 \qquad H_A: \pi > .16$$

TABLE 8-6
Single-Sample Test of a Proportion

Specification
A single random sample of size n is drawn from a population, and the value p of the sample proportion statistic P is calculated. When n is larger (that is, greater than 100), the normal approximation to the binomial applies and this test can be used. For small samples, a proportion is tested by using the binomial distribution.

Hypotheses
H_0: $\pi = \pi_0$, the population proportion has some given value π_0.
H_A: (a) $\pi \neq \pi_0$ (two-tailed).
(b) $\pi < \pi_0$ (lower-tail), implies H_0: $\pi \geq \pi_0$.
(c) $\pi > \pi_0$ (upper tail), implies H_0: $\pi \leq \pi_0$.

A common form of this test is the upper-tail test with $\pi_0 = .50$. This can be used to test whether a certain population proportion is a majority.

Test statistic

$$Z = \frac{P - \pi_0}{\sqrt{\pi_0(1 - \pi_0)/n}}$$

Some statisticians use P to estimate π_0 in the standard error of this equation.

Decision rule
Reject H_0 if $z < -z_{\alpha/2}$ or if $z > z_{\alpha/2}$; otherwise, do not reject H_0 for a two-tailed test. For a lower-tail test, reject H_0 if $z < -z_\alpha$; otherwise, accept H_0. For an upper-tail test, reject H_0 if $z > z_\alpha$; otherwise, accept H_0.

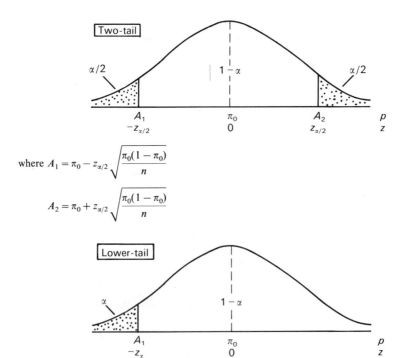

where $A_1 = \pi_0 - z_{\alpha/2}\sqrt{\dfrac{\pi_0(1 - \pi_0)}{n}}$

$A_2 = \pi_0 + z_{\alpha/2}\sqrt{\dfrac{\pi_0(1 - \pi_0)}{n}}$

TABLE 8-6 (continued)

where $A_1 = \pi_0 - z_\alpha \sqrt{\dfrac{\pi_0(1 - \pi_0)}{n}}$

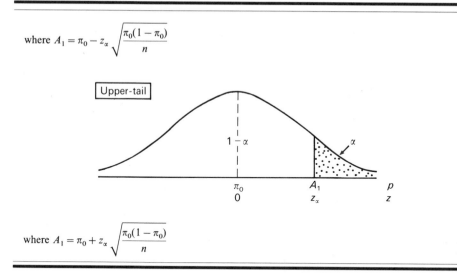

where $A_1 = \pi_0 + z_\alpha \sqrt{\dfrac{\pi_0(1 - \pi_0)}{n}}$

This is an upper-tail test. The sampling distribution of P is shown in Figure 8-9. With $\pi = .16$, the standard deviation of the sampling distribution for a sample size of $n = 200$ is $\sqrt{.16(.84)/200} = .026$. For a significance level of $\alpha = .05$, this leads to this decision rule

Reject H_0: $p > .203$; otherwise, do not reject H_0.

since the critical or action limit is found by using

$$1.645 = \frac{A_1 - .16}{.026}$$

$$.203 = A_1$$

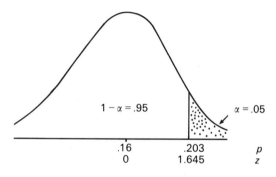

FIGURE 8-9. Sampling distribution and decision rule for Example 8-3.

Note that the region for rejection is placed entirely in the upper tail of the distribution. The value of the sample statistic is $p = \frac{42}{200} = .21$. Using this decision rule, a reputable city council would certainly continue bus service. A skeptical council would undoubtedly authorize a new survey!

8.5. Relationship between Hypothesis Testing and Confidence Interval Estimation

Hypothesis testing and confidence interval estimation are closely related methods of statistical inference. The *significance* level of a test of hypothesis is closely related to the *confidence* in a confidence interval. The fact that we denote the level of significance as α and the confidence level as $1 - \alpha$ [or, more usually, $100(1 - \alpha)$, to express it as a percentage] is no coincidence. The relationship between these two procedures for statistical inference is best explained by a simple example. For this purpose, let us reconsider Example 8-1, which deals with household sizes. In this example, a test of the representativeness of a sample with $\bar{x} = 3.6$ is made against a known (or hypothesized) population mean of $\mu_0 = 3.2$ with $\sigma = 1.6$. The sample size is $n = 100$.

When it is viewed as a two-tail test of hypothesis of

$$H_0: \mu = 3.2 \qquad H_A: \mu \neq 3.2$$

the appropriate sampling distribution is the normal, with test statistic

$$z = \frac{\bar{x} - \mu_0}{\sigma/\sqrt{n}} = \frac{3.6 - 3.2}{1.6/\sqrt{100}} = 2.5$$

Based on this value for z, we can show that the PROB-VALUE for the two-tail test is .0124.

Now, for comparative purposes, let us construct both the 95 and the 99 percent confidence intervals on μ, using the sample $\bar{x} = 3.6$ and $\sigma = 1.6$:

95 percent interval: $\qquad \bar{x} \pm z_{.025} \dfrac{\sigma}{\sqrt{n}} = 3.6 \pm 1.96 \dfrac{1.6}{\sqrt{100}} = [3.28, 3.91]$

99 percent interval: $\qquad \bar{x} \pm z_{.005} \dfrac{\sigma}{\sqrt{n}} = 3.6 \pm 2.58 \dfrac{1.6}{\sqrt{100}} = [3.19, 4.01]$

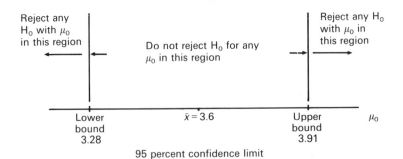

FIGURE 8-10. Relationship between confidence interval estimation and hypothesis testing for Example 8-1.

TABLE 8-7
Confidence Interval Estimation and Hypothesis Testing Relations for μ and π

Form of null hypothesis	Confidence interval estimation	Decision concerning μ_0 or π_0
$H_0: \mu = \mu_0$ $\pi = \pi_0$	$100(1 - \alpha)$ percent confidence interval	Reject any H_0 for which μ_0 or π_0 does *not* lies in the interval.
$H_0: \mu \geq \mu_0$ $\pi \geq \pi_0$	$100(1 - \alpha)$ percent *upper* confidence limit	Reject any H_0 for which μ_0 or π_0 exceeds upper confidence limit.
$H_0: \mu \leq \mu_0$ $\pi \leq \pi_0$	$100(1 - \alpha)$ percent *lower* confidence limit	Reject any H_0 for which μ_0 or π_0 is less than lower confidence limit.

Notice that the 95 percent confidence interval does not contain the hypothesized value of $\mu_0 = 3.2$, but the 99 percent interval does. Since the PROB-VALUE was found to be .0124, this is the expected result. This interpretation can now be extended as follows. If the level of confidence is *greater* than $100(1 - \text{PROB-VALUE})$, in this case $100(1 - .0124) = 98.76$, then the confidence interval will contain $\mu_0 = 3.2$; and if the level of confidence is *less* than 98.76, it will not. In the first case, we cannot reject H_0 since $\mu_0 = 3.2$ is contained in the interval; in the latter case we reject H_0 since 3.2 is not in the interval. This rule is illustrated in Figure 8-10. This relationship is generalized and summarized in Table 8-7 to include all three forms of the null hypothesis for the population parameters μ and π.

8.6. Summary

In the form of statistical inference known as hypothesis testing, a value of a population parameter is assumed or hypothesized, and the sample information is used to see whether this hypothesis is tenable. One of three forms of a null hypothesis is chosen, depending on the nature of the problem under study. A six-step procedure known as the classical test of hypothesis has been outlined and applied to several single-sample tests for $\bar{x}$ and p. The selection of a level of significance depends on the risk involved in committing a Type I error. The probability of committing a Type II error, β, can also be determined for a statistical test. One of the few controls over these types of error is the sample size used in the test. Both error risks can be reduced by the use of a larger sample.

A variant of the classical test known as PROB-VALUE has been explained as well as the interrelationships among the classical test, PROB-VALUE method, and confidence interval estimation. By using one of these procedures, it is possible to evaluate hypotheses and determine their statistical significance. Any evaluation must include a statement of this significance level, although it can be expressed as a PROB-VALUE or made implicit in a confidence interval. However, one must always be careful to differentiate between statistical significance and practical significance. Just because some result is statistically significant does not necessarily give it any practical significance. A great deal of criticism is often leveled at statisticians because a few poor practitioners confuse these two issues.

Proving the statistical significance of a meaningless hypothesis is of no value. Only a thorough understanding of the problem under study can lead to the specification of intelligent hypotheses worthy of statistical evaluation. For example, it would be foolish to take a sample to test whether the proportion of people who moved in a neighborhood within the last year exceeded 80 percent, or $\pi = .80$. Before we select the sample, we must be sure that such a rate of mobility is impossible. It makes little sense to test whether the mobility exceeds this unrealistically high level. Other examples of statistical and practical significance are presented in Chapter 9.

FURTHER READING

The material covered in this chapter is contained in most elementary textbooks of statistical methods. Additional examples of situations in which hypothesis testing can be undertaken are described in the texts listed at the end of Chapter 2 and in the problems.

PROBLEMS

1. Explain the meaning of the following terms or concepts:
   a. Classical test of hypothesis   f. Test statistic
   b. Null hypothesis   g. Critical, or rejection, region
   c. Alternate hypothesis   h. Action limits (or critical values)
   d. Directional hypothesis   i. Decision rule
   e. Significance level   j. PROB-VALUE method

2. Differentiate between (a) a one-sided (or one-tail) and a two-sided test, (b) a Type I and a Type II error, and (c) statistical and practical significance.

3. List the steps in the classical test of hypothesis.

4. When should the t distribution rather than the z distribution be used in a hypothesis test of population mean?

In each of the following questions, follow the steps in the classical test of hypothesis outlined in this chapter. Where necessary, modify this procedure to incorporate the requirements of the PROB-VALUE method.

5. Department of Agriculture and Livestock Development researchers in Kenya estimate that yields in a certain district should approach the following amounts, in metric tons per

hectare: groundnuts, .50; cassava, 3.70; beans, .30. A survey undertaken by a district agricultural officer on farm holdings headed by women reveals the following results:

	$\bar{x}$	s
Groundnuts	0.40	.03
Cassava	2.10	.83
Beans	0.40	.30

There are 100 farm holdings in the sample.
a. Test the hypothesis that these farm holdings are producing at target levels of this district for each crop.
b. Test the hypothesis that the farm holdings are producing at lower than target levels.
c. Why would you want to be extremely careful in interpreting the results of this study, even if great care were taken in selecting a truly random sample?

6. A stream has been monitored weekly for a number of years, and the total dissolved solids in the stream averages 40 parts per million and is constant throughout the year. Following a recent change in land use in the drainage basis of the stream, a fluvial geomorphologist finds that the mean parts per million of dissolved solids in a 25-week sample to be 52 with a standard deviation of 32. Has there been a change in the average level of dissolved solids in this stream?

7. Mean annual water consumption per household in a certain city is 6800 L. The variance is 1,440,000. A random sample of 40 households in one neighborhood reveals a mean of 8000.
a. Test the hypothesis that these households have (i) different and (ii) higher consumption levels than the average city household.
b. Comment on the practical significance of this result.
c. What is the PROB-VALUE for each of the hypotheses in part (a)?

8. An exhaustive survey of all users of a wilderness park taken in 1960 revealed that the average number of persons per party was 2.6. In a random sample of 25 parties in 1985, the average was 3.2 persons with a standard deviation of 1.08.
a. Test the hypothesis that the number of persons per party has changed in the last 25 yr.
b. Determine a PROB-VALUE result.

9. A manufacturer of barometers believes that the production process yields 2.2 percent defectives. In a random sample of 250 barometers, 7 are defective.
a. To test whether the manufacturer's belief is confirmed by this sample, calculate the PROB-VALUE for a test of hypothesis of H_0: $\pi = .022$.
b. At which of the following significance levels will the null hypothesis be rejected? (i) 20, (ii) .10, (iii) .05, (iv) .01, (v) .001

10. The average score of geography majors on a standard graduate school entrance examination is hypothesized to be 600; that is, H_0: $\mu = 600$. A random sample of $n = 75$ students is selected. The decision rule used to reject H_0 is

   1. Reject H_0: $\bar{x} < 575$ or $\bar{x} > 625$
   2. Accept H_0: $575 \leq \bar{x} \leq 625$

Assume that the standard deviation is 100 and $\bar{X}$ is normally distributed. Find α for this test.

11. A shipper declares that the probability that one of the shipments is delayed is .05. A customer notes that in her first $n = 200$ shipments, 12 have been delayed.
   a. Test the assertion of the shipper that $\pi = .05$ at $\alpha = .05$.
   b. Calculate the **PROB-VALUE**.

9

Parametric Statistical Inference: Two-Sample Tests

Hypothesis testing procedures are not limited to situations in which a single sample is drawn and a statement about some quantitative characteristic of the population evaluated. In a large number of practical inference problems, the goal is to compare two populations with regard to some quantitative characteristic such as the mean. For example, we may wish to compare the average distances traveled to purchase clothing by males and by females. To do this, we would take a random sample of male shoppers and an independent random sample of female shoppers, and then we would undertake a test of hypothesis, using the sample means. The procedures used in hypothesis testing can easily be extended to handle this sort of inferential problem. Because we are testing the value of some characteristic in two samples, these problems are known as *two-sample tests*. They could equally well be labeled *two-population tests* since the fundamental question being asked is, Do these two samples come from the same population or from two different populations? For the shopping behavior example, the question we are trying to answer is, Do women and men travel, on the average, the same distance to purchase this item (are they from the same population?), or do they travel different distances (are they from two populations?)?

In Section 9.1 of this chapter we consider the two-sample difference-of-means test, and in Section 9.2 we introduce the difference-of-means test for paired samples. The two-sample difference-of-proportions test is detailed in Section 9.3. Before we summarize the chapter, hypothesis tests for the population variance are explained for both one- and two-sample situations.

9.1. Two-Sample Difference-of-Means Test for $\mu_1 - \mu_2$

For two-sample tests, it is necessary to use additional subscripts to distinguish between the two samples and populations. Thus, $\bar{X}_1$ and $\bar{X}_2$ are the random variables corresponding to the sample mean statistics for sample 1 and sample 2, respectively, and $\bar{x}_1$ and $\bar{x}_2$ are the values of these random variables calculated from the two samples. Similarly, μ_1 and μ_2 are the population mean values of population 1 and population 2, respectively. There are n_1 units sampled from the first population and n_2 units drawn from the second population. To complete the notation, we use a double subscript to label the sample values x_{ij}. The first subscript, i, is the population from which the sample value has been drawn ($i = 1, 2$), and subscript j designates the jth sample unit drawn from the ith population. The necessary data for a two-sample difference-of-means test are of the following form:

Sample from population 1	Sample from population 2
x_{11}	x_{21}
x_{12}	x_{22}
.	.
.	.
.	.
x_{1n_1}	x_{2n_2}

From these we calculate the values of sample means statistics:

$$\bar{x}_1 = \frac{\sum_{j=1}^{n_1} x_{ij}}{n_1} \qquad \bar{x}_2 = \frac{\sum_{j=1}^{n_2} x_{ij}}{n_2}$$

The tests described in this section are based on the assumption that the population random variables X_1 and X_2 are normally distributed. If this cannot be assumed, then the test will still be approximately correct as long as n_1 and n_2 are large, say $n_1, n_2 \geq 30$.

The basic inferential framework for comparing two populations means is illustrated in Figure 9-1. There are two populations with means μ_1 and μ_2 and variances σ_1^2 and σ_2^2, respectively. From each population a sample has been drawn, and the values of the sample mean and sample variance are calculated. For sample 1 these values are $\bar{x}_1$ and s_1^2, and for sample 2 the values are $\bar{x}_2$ and s_2^2. The most general form of the hypothesis in the two-tail case can be stated as

$$H_0: \mu_1 - \mu_2 = D_0 \qquad H_A: \mu_1 - \mu_2 \neq D_0 \qquad (9\text{-}1)$$

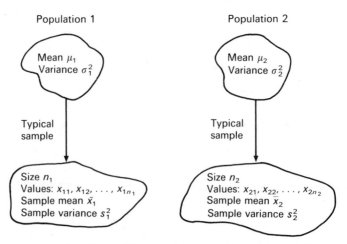

FIGURE 9-1. Basic inferential framework for comparing two population means.

That is, the hypothesis is couched in terms of the difference between the two population means, $\mu_1 - \mu_2$. In the null hypothesis, we may assume a value for this difference D_0. For example, we could test the hypothesis that the difference in the mean annual rainfall at two climatic stations is 10 cm. If we define the value of D_0 to be zero, as is most commonly done, then the null hypothesis can be rewritten as H_0: $\mu_1 - \mu_2 = 0$, or H_0: $\mu_1 = \mu_2$. Expressed in this manner, the null hypothesis asserts that there is no difference between the means, implying the two samples are not from two populations but from one population. In the more general case where D_0 takes on some nonzero value, we are testing whether the two population means differ by a given amount.

There are also directional forms of both hypothesis:

Lower tail: $\qquad\qquad\qquad$ H_0: $\mu_1 - \mu_2 \geq 0$ $\qquad$ H_A: $\mu_1 - \mu_2 < 0$

Upper tail: $\qquad\qquad\qquad$ H_0: $\mu_1 - \mu_2 \leq 0$ $\qquad$ H_A: $\mu_1 - \mu_2 > 0$

These forms can be used whenever we wish to assert which population mean has the higher value. Notice that we can rewrite H_0: $\mu_1 - \mu_2 \geq 0$ as H_0: $\mu_1 \geq \mu_2$ for the lower-tail test. If the two samples were mean annual rainfall at two stations, for example, a directional hypothesis might be useful to test an assertion that annual rainfall at station 1 exceeds that of station 2.

We have already shown that the best estimator of a population mean μ is the statistic or estimator $\bar{X}$. It should not be surprising that the best point estimator of the difference $\mu_1 - \mu_2$ is $\bar{X}_1 - \bar{X}_2$. The estimator has unbiased minimum variance properties. Thus, if it is found from the samples of annual rainfall totals that the difference between the average rainfall at two climatic stations is 100 cm/yr, this is the best estimate that we use to estimate the difference in the population means $\mu_1 - \mu_2$. To construct confidence intervals or to test hypotheses concerning $\mu_1 - \mu_2$, it is necessary to distinguish among three cases.

Case 1. σ_1^2 and σ_2^2 are known.
Case 2. σ_1^2 and σ_2^2 are unknown but equal.
Case 3. σ_1^2 and σ_2^2 are unknown but unequal.

Case 1 can be differentiated from cases 2 and 3 on the same basis as was necessary for confidence intervals and hypothesis tests for the population mean μ discussed in Chapters 7 and 8. When σ^2 is known, the confidence intervals and tests are based on the standard normal variable Z; but when it is unknown, they are based on the t distribution. This distinction is also followed in two-sample tests. The difference between the second and third cases is explained in the discussion below.

Case 1: σ_1^2 and σ_2^2 Known

When the values of the population variances are known, the following result can be used as the basis for generating the standard error of the sampling distribution of the difference $\bar{X}_1 - \bar{X}_2$.

DEFINITION: VARIANCE OF THE DIFFERENCE BETWEEN TWO RANDOM VARIABLES

Let X_1 and X_2 be two independent random variables with variances $\sigma^2_{X_1}$ and $\sigma^2_{X_2}$, respectively. Then the variance of the difference between X_1 and X_2, $V(X_1 - X_2)$, is the sum of the variances of X_1 and X_2:

$$V(X_1 - X_2) = \sigma^2_{X_1} + \sigma^2_{X_2} \qquad (9\text{-}2)$$

If we apply this result to the difference between the two population mean estimators $\bar{X}_1$ and $\bar{X}_2$, then

$$\sigma^2_{\bar{X}_1 - \bar{X}_2} = V(\bar{X}_1 - \bar{X}_2) = \frac{\sigma^2_1}{n_1} + \frac{\sigma^2_2}{n_2} \qquad (9\text{-}3)$$

and

$$\sigma_{\bar{X}_1 - \bar{X}_2} = \sqrt{\frac{\sigma^2_1}{n_1} + \frac{\sigma^2_2}{n_2}} \qquad (9\text{-}4)$$

Recall that we found the variance of $\bar{X}$ to be σ^2/n in Chapter 7.

Having derived the results for the mean and standard error of the estimator of $\mu_1 - \mu_2$, we are now in a position to specify both the confidence interval and hypothesis testing formulas. Following the conventions established in Chapter 7, we formalize the confidence interval estimate of $\mu_1 - \mu_2$ as follows:

DEFINITION: CONFIDENCE INTERVAL ESTIMATE FOR $\mu_1 - \mu_2$, WITH σ^2_1 AND σ^2_2 KNOWN

Let X_1 and X_2 be two population random variables. The $100(1 - \alpha)$ percent confidence interval estimate for the difference between the population means $\mu_1 - \mu_2$ is

$$\bar{x}_1 - \bar{x}_2 \pm z_{\alpha/2} \sqrt{\frac{\sigma^2_1}{n_1} + \frac{\sigma^2_2}{n_2}} \qquad (9\text{-}5)$$

where $\bar{x}_i$, σ^2_i, and n_i are the sample mean, population variance, and sample size associated with the ith population, respectively. The value $z_{\alpha/2}$ is the point on the standard normal such that $P(Z \geq z_{\alpha/2}) = \alpha/2$. This is an exact interval where X_1 and X_2 are normal variables and approximate for nonnormal X_1 and X_2 if $n_1, n_2 \geq 30$.

The hypothesis test is formally detailed in Table 9-1. Let us illustrate the use of both the confidence interval and hypothesis test in an example.

EXAMPLE 9-1. A geographer interested in comparing the shopping patterns of women and men interviewed 200 randomly selected persons ($n_1 = n_2 = 100$) and determined the distances traveled by each respondent to the store at which the last major clothing purchase (defined as any purchase greater than $25) was made. The

TABLE 9-1
Two-Sample Difference-of-Means Test, Case 1: σ_1^2 and σ_2^2 Known

---

Specification

There are two independent random samples of size n_1 and n_2 drawn from two samples for which the two values of the sample mean statistic $\bar{X}$ are determined. Denote these two values as $\bar{x}_1$ and $\bar{x}_2$. The variables X_1 and X_2 are normally distributed. The test is approximate for nonnormal X if $n_1, n_2 \geq 30$.

Hypotheses

H_0: $\mu_1 - \mu_2 = D_0$, the population means differ by D_0.
H_A: (a) $\mu_1 - \mu_2 \neq D_0$ (two-tailed).
    (b) $\mu_1 - \mu_2 < D_0$ (lower tail), implies H_0: $\mu_1 - \mu_2 \geq D_0$.
    (c) $\mu_1 - \mu_2 > D_0$ (upper tail), implies H_0: $\mu_1 - \mu_2 \leq D_0$.

Usually, D_0 is set equal to zero, and the test is used to examine the hypothesis that $\mu_1 - \mu_2 = 0$, or equivalently $\mu_1 = \mu_2$.

Test statistic

$$Z = \frac{\bar{X}_1 - \bar{X}_2 - D_0}{\sqrt{\sigma_1^2/n_1 + \sigma_2^2/n_2}}$$

Decision rule

Reject H_0 if $z < -z_{\alpha/2}$ or if $z > z_{\alpha/2}$; otherwise, accept H_0 for a two-tailed test. For a lower-tail test, reject H_0 if $z < -z_\alpha$; otherwise, accept H_0. For an upper-tail test, reject H_0 if $z > z_\alpha$; otherwise, accept H_0.

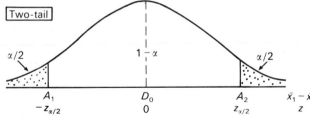

where $A_1 = D_0 - z_{\alpha/2}\sqrt{\dfrac{\sigma_1^2}{n_1} + \dfrac{\sigma_2^2}{n_2}}$

$A_2 = D_0 + z_{\alpha/2}\sqrt{\dfrac{\sigma_1^2}{n_1} + \dfrac{\sigma_2^2}{n_2}}$

where $A_1 = D_0 - z_\alpha\sqrt{\dfrac{\sigma_1^2}{n_1} + \dfrac{\sigma_2^2}{n_2}}$

(continued)

TABLE 9-1 (continued)

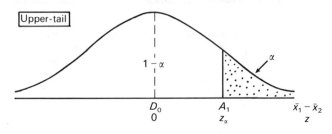

where $A_1 = D_0 + z_\alpha \sqrt{\dfrac{\sigma_1^2}{n_1} + \dfrac{\sigma_2^2}{n_2}}$

TABLE 9-2
Test of Hypothesis for Example 9-1

Hypotheses
H_0: $\mu_1 - \mu_2 \geq -2$, the mean shopping trip length of men is not more than 2 mi shorter than that of women.
H_A: $\mu_1 - \mu_2 < -2$

Test statistic and test to be employed
This is a case 1 difference-of-means test, so the standard Z statistic is applied.

Level of significance
Set $\alpha = .10$.

Decision rule

If $z < -1.645$, reject H_0;
otherwise, accept H_0.

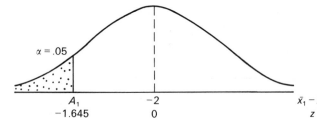

$A_1 = -2 - \dfrac{1.645}{.556} = -4.91$

Computation of test statistic

$z = \dfrac{6.2 - 14.8 + 2}{\sqrt{\frac{16}{100} + \frac{16}{100}}} = \dfrac{-6.6}{.57} = -11.58$

Decision
Since $z = -11.58 < -1.645$, reject H_0.

average distance traveled by men was 6.2 mi and by women 14.8 mi. Assuming that the population variances of both populations are equal and that $\sigma_1^2 = \sigma_2^2 = 16$, (a) determine the 95 percent confidence limits on the mean difference traveled by women and men on these shopping trips and (b) test the hypothesis that the mean difference in trip lengths exceeds 2 mi and women make longer trips.

Solution. (a) We are given $n_1 = n_2 = 100$, $\bar{x}_1 = 6.2$, $\bar{x}_2 = 14.8$, and $\sigma_1^2 = \sigma_2^2 = 16$. If the trip lengths are normally distributed, then Equation (9-5) can be used to generate an exact confidence interval. If they are not normal, then (9-5) still provides an approximate interval. Substituting the given information, with $z_{.025} = 1.96$, we see the interval is

$$6.2 - 14.8 \pm 1.96 \sqrt{\tfrac{16}{100} + \tfrac{16}{100}}$$

$$-8.6 \pm 1.11$$

$$[-9.71, -7.49]$$

Thus women's shopping trips average 7.49 to 9.71 mi more than men's shopping trips. Note that the sign of the difference is merely determined by the larger of the two means.

(b) In this test we are surely interested in a lower-tail test, since we wish to know whether women make longer trips. We would expect $\bar{x}_1 - \bar{x}_2$ to be negative and in the lower tail of the distribution. Moreover, we set $D_0 = -2$ since we wish to show that this difference exceeds 2 mi. The complete test for this hypothesis is presented in Table 9-2.

Case 2: σ_1^2 and σ_1^1 Unknown but Equal

In this second case, the two population variances are unknown, but they are known to be equal, so $\sigma_1^2 = \sigma_2^2 = \sigma$. The standard error, the sampling distribution of $\mu_1 - \mu_2$, can be simplified to

$$\sqrt{\frac{\sigma_1^2}{n_1} + \frac{\sigma_2^2}{n_2}} = \sqrt{\sigma^2\left(\frac{1}{n_1} + \frac{1}{n_2}\right)} = \sigma\sqrt{\frac{1}{n_1} + \frac{1}{n_2}} \qquad (9\text{-}6)$$

It is now necessary to estimate σ from the sample data. Since there is one common variance, it seems reasonable that the information in both samples variances s_1^2 and s_2^2 should be combined to make a single estimate of σ^2. This *pooled variance* is given by

$$s_p^2 = \frac{(n_1 - 1)s_1^2 + (n_2 - 1)s_2^2}{(n_1 - 1) + (n_2 - 1)} \qquad (9\text{-}7)$$

and is weighted average of the two sample variances s_1^2 and s_2^2. It follows that

$$\sigma_{\bar{X}_1 - \bar{X}_2} = s_p \sqrt{\frac{1}{n_1} + \frac{1}{n_2}} \qquad (9\text{-}8)$$

for this case.

Unfortunately, the test statistic is no longer normally distributed and takes the form of a t statistic

$$t = \frac{\bar{X}_1 - \bar{X}_2 - D_0}{S_p\sqrt{1/n_1 + 1/n_2}} \tag{9-9}$$

where S_p^2 is the sample pooled standard deviation statistic. The degrees of freedom are $n_1 + n_2 - 2$, the divisor in the standard deviation statistic. Similar changes are also made for the confidence interval formula (9-5).

Case 3: σ_1^2 and σ_2^2 Unknown but Unequal

This is the most difficult of the three cases. Although the details need not concern us here, a t statistic of the form

$$t = \frac{\bar{X}_1 - \bar{X}_2 - D_0}{\sqrt{S_1^2/n_1 + S_2^2/n_2}} \tag{9-10}$$

with degrees of freedom given by

$$df = \frac{(s_1^2/n_1 + s_2^2/n_2)^2}{(s_1^2/n_1)^2/(n_1 - 1) + (s_2^2/n_2)^2/(n^2 - 1)} \tag{9-11}$$

gives approximately correct results. The three cases for the difference-of-means test are summarized in Table 9-3.

TABLE 9-3
Summary of the Sample Difference-of-Means Tests

Case	Conditions	Test statistic value
1	σ_1^2 and σ_2^2 known	$z = \dfrac{\bar{x}_1 - \bar{x}_2 - D_0}{\sqrt{\sigma_1^2/n_1 + \sigma_2^2/n_2}}$
2	σ_1^2 and σ_2^2 unknown but equal	$t = \dfrac{\bar{x}_1 - \bar{x}_2 - D_0}{s_p\sqrt{1/n_1 + 1/n_2}}$
		$df = n_1 + n_2 - 2$
		s_p^2 given by (9-7)
3	σ_1^2 and σ_2^2 unknown but unequal	$t = \dfrac{\bar{x}_1 - \bar{x}_2 - D_0}{\sqrt{s_1^2/n_1 + s_2^2/n_2}}$
		df given by (9-11)

9.2. The *t* Test for Paired Samples

All three cases of the difference-of-means test presented to this point assume that the random samples collected from the two populations are independently drawn. There are many instances, particularly in experimental situations, where the observations are not independent but are paired.

EXAMPLE 9-2. A cartographer tests the time taken by introductory students to perform a given set of tasks involving the extraction of information from a set of maps. At the end of the course, the same students are given an identical test and the times recorded. Have the students learned how to use maps more effectively? In particular, do they now take *less* time to extract information from maps? It is clear that the two sets of sample times are not independent. In fact, the values of time in one sample are paired with another value of time in the second sample. Whenever measurements are taken from the same individuals, the samples are seldom independent.

The analysis of such paired observation data is done differently than if the two samples were independent. Let us denote the pairs of values in the experiment as (x_{i_j}, x_{2_j}), where x_{1_j} is the time taken by student j in the first test and x_{2_j} the time taken by student j in the second test. The matched-pairs t test for this case examines

$$d_j = x_{1_j} - x_{2_j} \qquad j = 1, 2, \ldots, n$$

the differences in time taken by each of the n students. These values may be positive, zero, or negative, depending on whether the student has taken less time, the same time, or more time than in the previous test, respectively. If this cartographer is at all worth her salt, she would hope to find that, on the average, the d_j's are positive and test times are decreased. This is tantamount to making an hypothesis concerning the value of

$$\bar{d} = \frac{\sum_{j=1}^{n} d_j}{n}$$

Making the assumption that the sample of differences $d_j, j = 1, 2, \ldots, n$, is a random sample from a normal population with mean $\mu_1 - \mu_2 = 0$, we see the test statistic

$$t = \frac{\bar{d} - 0}{S_d / \sqrt{n}} \qquad (9\text{-}12)$$

is a value from the t distribution with $n - 1$ degrees of freedom. The standard deviation of the differences is estimated by

$$S_d = \sqrt{\frac{\sum_{j=1}^{n} (d_j - \bar{d})}{n - 1}}$$

as usual. The formal specification of the t test for paired samples is given in Table 9-4. The test is generalized to allow hypotheses concerning any value for the mean difference in the populations $\mu_d = \mu_1 - \mu_2$ to be made. For example, we could test whether the time taken by these cartography students has declined by more than, say, 5 min.

TABLE 9-4
The t Test for Paired Samples

Specification
There are n paired values (x_{i_j}, x_{2_j}). The test examines the differences $d_j = x_{i_j} - x_{2_j}$. The sample mean difference $\bar{d}$ is used in this test as well as to construct confidence intervals.

Hypotheses
H_0: $\mu_1 - \mu_2 = \mu_d$, the difference in means is equal to some hypothesized difference μ_d.
H_A: (a) $\mu_1 - \mu_2 \neq \mu_d$ (two-tailed).
     (b) $\mu_1 - \mu_2 < \mu_d$ (lower tail), implies H_0: $\mu_1 - \mu_2 \geq \mu_d$.
     (c) $\mu_1 - \mu_2 > \mu_d$ (upper tail), implies H_0: $\mu_1 - \mu_2 \leq \mu_d$.

Normally, μ_d is taken to be zero, and the test is used to examine the significance of the difference $\bar{d}$. In this instance, a lower-tail test examines the hypothesis that the mean has declined, and an upper-tail test examines the hypothesis that the mean has increased.

Test statistic

$$t = \frac{\bar{d} - \mu_d}{S_d/\sqrt{n}}$$

with $n - 1$ degrees of freedom.

Decision rule
Reject H_0 if $t < -t_{\alpha/2}$ or if $t > t_{\alpha/2}$; otherwise, do not reject H_0 for a two-tailed test. For a lower-tail test, reject H_0 if $t < -t_\alpha$; otherwise, do not reject H_0. For an upper-tail test, reject H_0 if $t > t_\alpha$; otherwise, do not reject H_0.

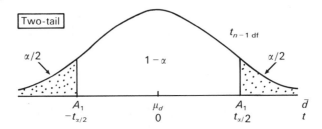

where $A_1 = \mu_d - t_{\alpha/2} \dfrac{S_d}{\sqrt{n}}$

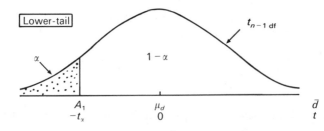

where $A_1 = \mu_d - t_\alpha \dfrac{S_d}{\sqrt{n}}$

TABLE 9-4 (continued)

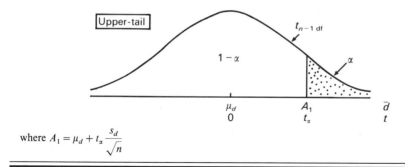

where $A_1 = \mu_d + t_\alpha \dfrac{s_d}{\sqrt{n}}$

Suppose the times, in minutes, taken by the students in the two tests are as follows:

Student	First time	Second time	Difference
1	16	15	1
2	23	21	2
3	17	16	1
4	14	15	−1
5	16	15	1
6	21	19	2
7	19	18	1
8	24	10	14
9	26	15	11
10	19	20	−1

First, we compute $\bar{d} = \frac{31}{10} = 3.10$ and $s_d = 5.11$. This leads to a value of the test statistic

$$t = \frac{3.10 - 0}{5.11/\sqrt{10}} = \frac{3.10}{1.62} = 1.91$$

The sampling distribution and decision rule for a directional hypothesis are illustrated in Figure 9-2. Based on an α level of .05, the critical value for $t = 1.833$ with 9 degrees of freedom, the null hypothesis is rejected, and we conclude that test times have decreased. It may seem peculiar that this is an upper-tail test since we hypothesize a *decrease* in the test times. However, since the test times in the first test will be longer than those in the second test, the difference is positive and an upper-tail test is called for.

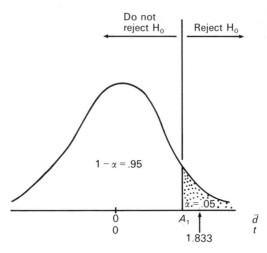

FIGURE 9-2. Sampling distribution and decision rule for Example 9-2.

9.3. Two-Sample Difference-of-Proportions Test for $\pi_1 - \pi_2$

The final test in this chapter is used to compare two population proportions π_1 and π_2. Just as the two-sample difference-of-means test examines the difference between two population means $\mu_1 - \mu_2$, this test is concerned with the difference $\pi_1 - \pi_2$. Inferences concerning the population parameters π_1 and π_2 are based on the sample proportions p_1 and p_2, respectively, since we know from Chapter 7 that P is the best point estimator of the population parameter π. Under the null hypothesis, it is assumed that the difference between π_1 and π_2 is D_0, that is, $H_0: \pi_1 - \pi_2 = D_0$, where D_0 is some prespecified value. In the simplest of all cases, D_0 is set equal to zero, and the null hypothesis can be expressed as $H_0: \pi_1 - \pi_2 = 0$ (or $\pi_1 = \pi_2$). In this case, we are saying that our two samples are drawn from a single population with population parameter π. Our alternate hypothesis can take one of three forms. As in the differences-of-means test, we can assert

$$H_A: \pi_1 - \pi_2 \neq 0 \quad \text{(two-tail)}$$
$$H_A: \pi_1 - \pi_2 < 0 \quad \text{(lower tail)}$$
$$H_A: \pi_1 - \pi_2 > 0 \quad \text{(upper tail)}$$

Examples of a two-tailed test and a directional hypothesis are given later in this section.

The central limit theorem can again be used to specify the nature of the sampling distribution for the difference of proportions. The best point estimator of $\pi_1 - \pi_2$ is the difference of sample proportion statistics $P_1 - P_2$. Whenever the two sample sizes n_1 and n_2 are both greater than 100, the test statistic

$$Z = \frac{P_1 - P_2 - D_0}{\sqrt{P_1(1 - P_1)/n_1 + P_2(1 - P_2)/n_2}} \tag{9-13}$$

is a standardized normal random variable. The complete test is summarized in Table 9-5.

TABLE 9-5
Two-Sample Difference-of-Proportions Test

Specification
There are two independent random samples of size n_1 and n_2 from which two values of the sample proportion statistic P are determined. Denote these two values as p_1 and p_2. The test can be used when $n_1, n_2 \geq 100$.

Hypotheses
H_0: $\pi_1 - \pi_2 = D_0$, the true population proportions differ by the amount D_0 (two-tailed).
H_A: (a) $\pi_1 - \pi_2 \neq D_0$ (two-tailed)
    (b) $\pi_1 - \pi_2 < D_0$ (lower tail), implies H_0: $\pi_1 - \pi_2 \geq D_0$.
    (c) $\pi_1 - \pi_2 > D_0$ (upper tail), implies H_0: $\pi_1 - \pi_2 \leq D_0$.

Normally, D_0 is taken to be zero, and the test is used to examine the hypothesis that $\pi_1 - \pi_2 = 0$, or equivalently $\pi_1 = \pi_2$.

Test statistic

$$Z = \frac{P_1 - P_2 - D_0}{\sqrt{P_1(1 - P_1)/n_1 + P_2(1 - P_2)/n_2}}$$

Decision rule
Reject H_0 if $z < -z_{\alpha/2}$ or if $z > z_{\alpha/2}$; otherwise, accept H_0 for a two-tailed test. For a lower-tail test, reject H_0 if $z < -z_\alpha$; otherwise, accept H_0. For an upper-tail test, reject H_0 if $z > z_\alpha$; otherwise, accept H_0.

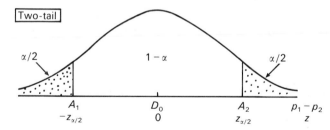

where $A_1 = D_0 - z_{\alpha/2} \sqrt{\dfrac{p_1(1 - p_1)}{n_1} + \dfrac{p_2(1 - p_2)}{n_2}}$

$A_2 = D_0 + z_{\alpha/2} \sqrt{\dfrac{p_1(1 - p_1)}{n_1} + \dfrac{p_2(1 - p_2)}{n_2}}$

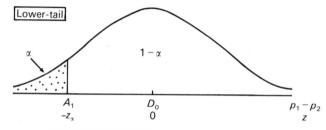

where $A_1 = D_0 - z_\alpha \sqrt{\dfrac{p_1(1 - p_1)}{n_1} + \dfrac{p_2(1 - p_2)}{n_2}}$

(continued)

TABLE 9-5 (continued)

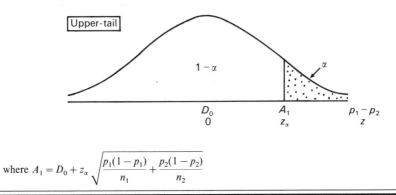

where $A_1 = D_0 + z_\alpha \sqrt{\dfrac{p_1(1-p_1)}{n_1} + \dfrac{p_2(1-p_2)}{n_2}}$

EXAMPLE 9-3. A study is conducted to determine whether an equal proportion of two groups of farmers in a particular agricultural region adopted a new innovation in fertilizer for corn. The first group of farmers is composed of Mennonites, and the second group consists of all other farmers in the region. A random sample of 100 Mennonite farmers produced a sample proportion of .68 who said they utilized the fertilizer in the current growing season. A random sample of 144 from the second group yielded a sample proportion of .58. Test the hypothesis that the two population proportions π_1 and π_2 are equal at the $\alpha = .05$ level of significance.

Since the two sample sizes are greater than 100, the two-sample difference-of-proportions test can be applied. We are given $p_1 = .68$ (and therefore $1 - p_1 = .32$), $p_2 = .58$ (and $1 - p_2 = .42$), $n_1 = 100$, and $n_2 = 144$. The test statistic value is

$$z = \frac{.68 - .58 - 0}{\sqrt{.68(.32)/100 + .58(.42)/100}}$$

$$= \frac{.10}{\sqrt{.0022 + .0024}} = \frac{.10}{.068} = 1.47$$

Since $z = 1.47$ falls in the acceptance region $-1.96 \le z \le 1.96$, we conclude that Mennonite farmers are as likely to adopt the fertilizer innovation as non-Mennonite farmers.

EXAMPLE 9-4. A study is to conducted to compare the proportions of residents in the inner city and in suburban areas who favor amalgamation of the region into a single metropolitan area. It is expected that a lower proportion of suburban households will favor the proposal for amalgamation. A random telephone survey of 900 inner-city residents indicates 79 percent favor amalgamation, whereas only 32 percent of the 680 suburban residents favor the proposal. Test the hypothesis that the proportion of inner-city residents favoring the proposal *is greater* that the proportion of suburbanites who do.

TABLE 9-6
Test of Hypothesis for Example 9-4

Hypotheses
H_0: $\pi_1 \leq \pi_2$ (or $\pi_1 - \pi_2 \leq 0$, or $D_0 \leq 0$), the proportion of residents favoring amalgamation in the inner city is not greater than the proportion in suburban areas.
H_A: $\pi_1 > \pi_2$, it is greater.

Test statistic and test to be employed
This is a difference-of-proportions test; since n_1, $n_2 \geq 100$, the standard Z statistic given by (9-13) can be applied.

Level of significance
Set $\alpha = .05$.

Decision rule
If $z > 1.96$, reject H_0;
if $z \leq 1.96$, accept H_0

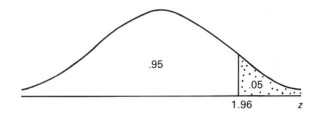

.95   .05   1.96 z

Computation of test statistic

$$z = \frac{.79 - .32 - 0}{\sqrt{(.79)(.21)/900 + (.68)(.32)/680}}$$

$$= \frac{.47}{\sqrt{.0002 + .0003}} = \frac{.47}{.002} = 21.36$$

Decision
Since $z > 1.96$, we reject H_0 and conclude that a larger proportion of inner-city residents than suburban residents favor amalgamation.

This is clearly a directional hypothesis, with the critical region in the upper tail. We expect a positive difference in $\pi_1 - \pi_2$. A formal test of hypothesis for this problem is summarized in Table 9-6. If the PROB-VALUE method is employed, we can only say that PROB-VALUE < .00004 since the largest value of z in the normal table of the Appendix has the value of $z = 4.0$, leaving a tail probability of $.50000 - .49996 = .00004$.

9.4. Inferences concerning Population Variances

To this point, we have not emphasized statistical inference problems related to the population variance σ^2, except to note that the statistic S^2 is its best estimator. In this section, details of the sampling distribution of the statistics S^2 are presented. Also an appropriate confidence interval is derived, and the one-sample test of hypothesis of S^2 against some hypothesized value σ_0^2 is detailed. Finally, inferences concerning two population variances σ_1^2 and σ_2^2 are shown to focus on the ratio of the variances σ_1^2/σ_2^2 rather than their difference $\sigma_1^2 - \sigma_2^2$.

Sampling Distribution of S^2

Statistical theory provides an important theorem concerning the sampling distribution of S^2.

If a random sample of size n is selected from a normal population with variance σ^2, then the standardized variable

$$\frac{(n-1)S^2}{\sigma^2}$$

can be shown to be distributed according to the χ^2 distribution with $v = n - 1$ degrees of freedom, where S^2 is the sample variance.

The χ^2, or chi-square, distribution is a continuous probability distribution with one parameter, v, which is called *degrees of freedom*. The degrees of freedom can be any positive integer, and there is one distribution for each value of v. The random variable associated with the χ^2 distribution is denoted $\chi^2(v)$ and can take on any value $0 \le \chi^2(v) \le \infty$. Figure 9-3 illustrates the χ^2 probability density function for three different values of v. Just as with the t distribution, there is a family of χ^2 distributions, and the exact shape of each individual distribution depends on its degrees of freedom.

Two characteristic features of the χ^2 distribution are illustrated in Figure 9-3. First, it is unimodal. Second, it is positively skewed, although less so as the number of degrees of freedom increases. The mean or expected value is $E[\chi^2(v)] = v = n - 1$, and it is located to the right of the mode since the distribution is skewed. As the number of degrees of freedom increases, the shape of the χ^2 distribution approaches that of the normal distribution.

The χ^2 table in the Appendix contains *percentiles* of selected χ^2 distributions

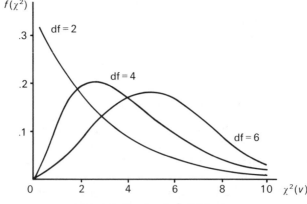

FIGURE 9-3. Family of χ^2 distributions.

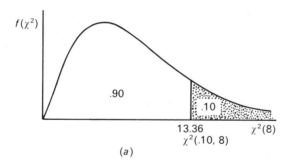

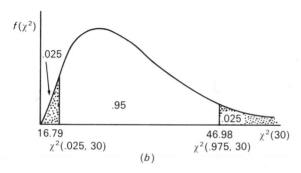

FIGURE 9-4. Graph for Example 9-5.

for degrees of freedom from 1 to 100. Let $\chi^2(a, v) = \chi^2(a, n - 1)$ denote the $100 \cdot a$ percentile of a χ^2 random variable with $v = n - 1$ degrees of freedom. That is,

$$P[\chi^2(v) \le \chi^2(a, v)] = a$$

It is important to distinguish $\chi^2(v)$, the chi-square random variable, from a percentile or point on the distribution. Two examples illustrate this relationship.

EXAMPLE 9-5. Find $\chi^2(0.90, 8)$. The situation is depicted in Figure 9-4(a). The 90th percentile of the χ^2 distribution with 8 degrees of freedom, from the table in the Appendix, is 13.36. Therefore, we can say that the probability that $\chi^2(8)$ is less than 13.36 is .90.

EXAMPLE 9-6. Find $\chi^2(.025, 30)$ and $\chi^2(.975, 30)$. This situation is depicted in Figure 9-4(b). These two values of χ^2 will isolate 2.5 percent of the values of $\chi^2(30)$ in each tail. From the table, we find $\chi^2(.025, 30) = 16.79$ and $\chi^2(.975, 30) = 46.98$. That is, 2.5 percent of the values of $\chi^2(30)$ are less than 16.79, and 2.5 percent are greater than 46.98.

Confidence Interval Estimate for σ^2

Using our knowledge of the sampling distribution of S^2 and the χ^2 distribution, it is now possible to specify the confidence interval for σ^2:

DEFINITION: CONFIDENCE INTERVAL ESTIMATE OF POPULATION VARIANCE σ^2
The interval estimate of the population variance σ^2 with $1 - \alpha$ confidence is given by

$$\frac{(n-1)s^2}{\chi^2(1-\alpha/2, n-1)} \leq \sigma^2 \leq \frac{(n-1)s^2}{\chi^2(\alpha/2, n-1)} \qquad (9\text{-}14)$$

where s^2 is the sample variance calculated from n randomly drawn values in a normally distributed population and where $\chi^2(1-\alpha/2, n-1)$ and $\chi^2(\alpha/2, n-1)$ are the values of $\chi^2(n-1)$ which locate one-half of α in each tail of the χ^2 distribution.

It is quite complicated to specify the *exact* confidence interval for the population standard deviation, but an approximate interval can be obtained by taking the square roots of the upper and lower bounds of the population variance estimate.

EXAMPLE 9-7.   A thermometer is calibrated to estimate temperatures in such a way that the standard deviation of the errors is no more than 0.1°C. If the thermometer is properly calibrated, then the variance of the errors between the actual and recorded temperatures must be less than or equal to $[(0.1)^2]°C^2$. A sample of 31 readings is taken, and the sample variance s^2 is found to be .017. Find a 95 percent confidence interval for the population variance σ^2.

Using the confidence interval formula (9-14) with $\alpha = .05$ and thus $\alpha/2 = .025$, we must find the values on the χ^2 distribution equal to $\chi^2(.025, 30)$ and $\chi^2(.975, 30)$. From the table these values are found to be 16.79 and 46.98, respectively. The appropriate confidence interval is therefore

$$\frac{30(.017)}{46.98} \leq \sigma^2 \leq \frac{30(.017)}{16.79}$$

or $.0109 \leq \sigma^2 \leq .0304$. We are thus 95 percent confident that the true variance is within this interval. An *approximate* interval for the standard deviation is ($\sqrt{.0109}$, $\sqrt{.0304}$), or $.1044 \leq \sigma \leq .1743$. On this basis, we may wish to question the manufacturer's claim that the standard deviation of errors is no more than 0.1°C. However, we also note that the lower bound is only approximate and extremely close to the proposed standard. Further testing may be called for.

Single-Sample Tests concerning σ^2

Hypothesis tests concerning the population variance σ^2 are based on the χ^2 distribution, with the usual provision that the population be normally distributed (or at least approximately so when the sample size is greater than 100). Table 9-7 outlines the appropriate formulas for both one- and two-tail versions of this test.

TABLE 9-7
Hypothesis Tests concerning the Population Variance

Specification
A sample of size n is taken from a normal population, and a value for S^2 is calculated from the data.

Hypotheses

	Two-tailed	Lower tail	Upper tail
H_0:	$\sigma^2 = \sigma_0^2$	$\sigma^2 \le \sigma_0^2$	$\sigma^2 \ge \sigma_0^2$
H_A:	$\sigma^2 \ne \sigma_0^2$	$\sigma^2 > \sigma_0^2$	$\sigma^2 < \sigma_0^2$

In all cases, σ_0^2 is the hypothesized value of the population variance.

Test statistic
$$\chi^2 = \frac{(n-1)S^2}{\sigma_0^2}$$
with degrees of freedom $v = n - 1$.

Decision rule
Reject H_0 if $\chi^2 < \chi^2(\alpha/2, n-1)$ or $\chi^2 > \chi^2(1 - \alpha/2, n-1)$; otherwise, accept H_0 for a two-tailed test. For a lower-tail test, reject H_0 if $\chi^2 < \chi^2(\alpha, n-1)$; otherwise, accept H_0. For an upper-tail test, reject H_0 if $\chi^2 > \chi^2(1 - \alpha, n-1)$; otherwise, accept H_0.

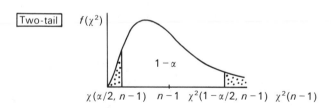

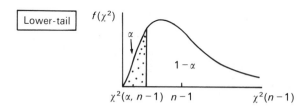

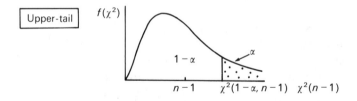

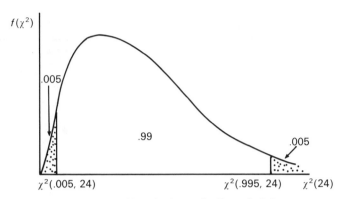

FIGURE 9-5. Hypothesis test for Example 9-8.

EXAMPLE 9-8. A fertilizer manufacturer claims that the variability in corn yield with the application of fertilizer GMC on a specific seed hybrid, as measured by yield variance, is equal to 16. The standard deviation of corn yield in a sample of 25 plots at an agricultural college is 4.47 bu/acre, and thus $s^2 = 19.9809$. Assume that the population distribution is normal, use a level of significance of .01, and test whether the sample data support the contention that the variance is 16.

This is a two-tailed version of the hypothesis test for σ^2 detailed in Table 9-7. We have $n = 25$, $\sigma_0^2 = 16$, and $s^2 = 19.9809$. The test statistic is therefore $\chi^2 = (25 - 1)(19.9809)/16 = 29.97135$. The two action limits for this hypothesis are, from the χ^2 table in the Appendix, $\chi^2(.005, 24) = 9.89$ and $\chi^2(.995, 24) = 45.56$. As we see in Figure 9-5, the test statistic lies in the region of acceptance, and so we must conclude that the manufacturer's claim is supported by these data.

Inferences concerning Two Population Variances σ_1^2 and σ_2^2

To make comparisons between two population variances, it is conventional to examine the ratio of the variances σ_1^2/σ_2^2, rather than the difference $\sigma_1^2 - \sigma_2^2$. For example, to test the hypothesis $H_0: \sigma_1^2 = \sigma_2^2$, the hypothesis is restated as $\sigma_1^2/\sigma_2^2 = 1$, rather than $\sigma_1^2 - \sigma_2^2 = 0$ or $\sigma_1^2 = \sigma_2^2$. Both the confidence interval estimator and hypothesis testing procedures are based on the ratio of the two variances. Under certain conditions, the statistic

$$F = \frac{s_1^2/\sigma_1^2}{s_2^2/\sigma_2^2} \tag{9-15}$$

is a random variable with an F distribution.

The F distribution is illustrated in Figure 9-6. It is a positively skewed (to the right) distribution and takes on values from 0 to $+\infty$. Note that all values in (9-15) must be positive. A pair a degrees of freedom are associated with F, one for the numerator and one for the denominator. These degrees of freedom are termed r_1 and r_2, respectively. There are thus a family of F distributions, one for each pair of values

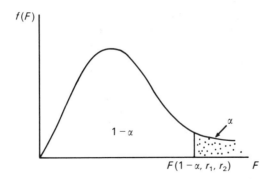

FIGURE 9-6. The F distribution.

of r_1 and r_2. The F distribution in the Appendix is the cumulative density function denoted by $F(1 - \alpha, r_1, r_2)$, that is,

$$P(F) \leq F(1 - \alpha, r_1, r_2) = 1 - \alpha$$

For example, if $r_1 = 10$ and $r_2 = 8$, then for $\alpha = .05$ we find $F(.95, 10, 8) = 3.35$. To find this value in the table, we proceed down the column of $r_1 = 10$ until we hit the row for $r_2 = 8$ degrees of freedom. There are several critical values for each pair of degrees of freedom: $1 - \alpha = .50, .90, .95, .975, .99, .995,$ and $.999$. Cumulative values on the left-hand side of the F distribution can be found from the relation

$$F(1 - \alpha, r_1, r_2) = \frac{1}{F(\alpha, r_2, r_1)}$$

where the degrees of freedom on the right-hand side are switched from (r_1, r_2) to (r_2, r_1). For example, if we wish to find $F(.05, 8, 10)$, we could use the relation

$$F(.05, 8, 10) = \frac{1}{F(.95, 10, 8)} = \frac{1}{3.35} = .2985$$

In general, then, $F(1 - \alpha, r_1, r_2) \neq F(1 - \alpha, r_2, r_1)$.

Now, the the null hypothesis H_0: $\sigma_1^2 = \sigma_2^2$ or $\sigma_1^2/\sigma_2^2 = 1$ is true, then the F statistic

$$F = \frac{s_1^2/\sigma_1^2}{s_2^2/\sigma_2^2}$$

reduces to

$$F = \frac{s_1^2}{s_2^2}$$

and all hypotheses concerning this ratio are based on this statistic. Significance tests are detailed in Table 9-8.

EXAMPLE 9-9. A study is undertaken to determine whether the *total fertility rate*, or the total number of children born to a woman over her lifetime, differs between two

TABLE 9-8
Hypothesis Test for Population Variance Ratio σ_1^2/σ_2^2

Specification
Independent random samples of size n_1 and n_2 are selected from a normal population. Estimates of the variance are calculated from each sample as s_1^2 and s_2^2.

Hypotheses

	Two-tailed	Lower tail	Upper tail
H_0:	$\dfrac{\sigma_1^2}{\sigma_2^2} = 1$	$\dfrac{\sigma_1^2}{\sigma_2^2} \geq 1$	$\dfrac{\sigma_1^2}{\sigma_2^2} \leq 1$
H_A:	$\dfrac{\sigma_1^2}{\sigma_2^2} \neq 1$	$\dfrac{\sigma_1^2}{\sigma_2^2} < 1$	$\dfrac{\sigma_1^2}{\sigma_2^2} > 1$

Test statistic

$$F = \frac{s_1^2}{s_2^2}$$

with degrees of freedom $(r_1, r_2) = (n_1 - 1, n_2 - 1)$.

Decision rule
Reject H_0 if $F < F(\alpha/2, r_1, r_2)$ or if $F > F(1 - \alpha/2, r_1, r_2)$; otherwise, accept H_0 for a two-tailed test. For a lower-tail test, reject H_0 only if $F < F(\alpha, r_1, r_2)$. For an upper-tail test, reject H_0 only if $F > F(1 - \alpha, r_1, r_2)$.

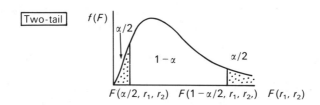

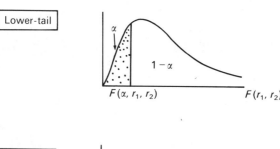

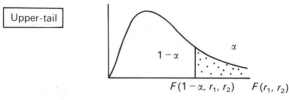

tribes located in different regions of Kenya. A sample is taken of 31 randomly selected women (all above child-rearing age) of the Samburu tribe, and the fertility rate is found to be $\bar{x}_1 = 3.84$, with a standard deviation of $s_1 = 1.02$. From a randomly selected sample of 25 Kisii tribe members, the fertility rate is estimated to be $\bar{x}_2 = 6.91$ with a standard deviation of 2.31. Test the hypothesis that fertility rates between the two tribes are equal at the $\alpha = .05$ level of significance. Assume that the variable called total fertility rate is normally distributed.

Since σ_1^2 and σ_2^2 are unknown, the first step in evaluating this difference-of-means test must be to determine whether the sample data tend to support the hypothesis $\sigma_1^2 = \sigma_2^2$. If this hypothesis can be accepted, then the difference-of-means test can be undertaken by using the procedures of case 2 outlined in Section 9.1. If not, then the methodology of case 3 must be applied. The null and alternate hypotheses are

$$H_0: \frac{\sigma_1^2}{\sigma_2^2} = 1$$

$$H_A: \frac{\sigma_1^2}{\sigma_2^2} \neq 1$$

and the appropriate test statistic is

$$F = \frac{s_1^2}{s_2^2}$$

Since this is a two-tailed test, the appropriate decision rule is to reject H_0: if $F < F(\alpha/2, n_1 - 1, n_2 - 1)$ or if $F > F(1 - \alpha/2, n_1 - 1, n_2 - 1)$, where $\alpha = .05$, $n_1 = 31$, and $n_2 = 24$. From the F table in the Appendix, these values are found to be $F(.025, 30, 24) = 1/F(.975, 24, 30) = 1/2.14 = .4673$ and $F(.975, 30, 24) = 2.21$. The situation is depicted in Figure 9-7.

The test statistic is $F = 1.02^2/2.31^2 = .1950$. Since F falls in the region where H_0 is rejected, we must conclude that $\sigma_1^2 \neq \sigma_2^2$ and apply the case 3 method for the

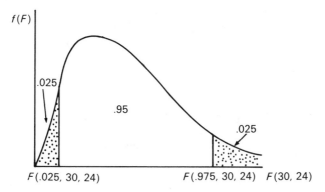

FIGURE 9-7. Sampling distribution for Example 9-9.

difference-of-means test. For the difference-of-means test, the test statistic is

$$t = \frac{3.84 - 6.91}{\sqrt{1.02^2/31 + 2.31^2/25}} = -6.177$$

which is clearly highly significant.

9.5. Summary

In this chapter, statistical hypothesis testing procedures have been developed for testing the difference between two population means $\mu_1 - \mu_2$, the difference between two population proportions $\pi_1 - \pi_2$, and the ratio of two population variances σ_1^2/σ_2^2. For the difference-of-means test, three cases are differentiated on the basis of the most appropriate assumption for the two population variances. The suitability of these assumptions can be examined by using the test for the equality of variances outlined in Section 9.4. In addition, two variants of the difference-of-means test are identified, depending on the nature of the underlying sample design. In some instances, the two samples may be *paired* or *matched*, and it is best to use a matched-pairs t test rather than the usual independent random samples difference-of-means test, for this case.

   In testing one- or two-sample hypotheses, it is important to distinguish between the practical and the statistical significance of the results. These two types of significance are not equivalent, nor do they always occur simultaneously. For example, suppose a random sample of police emergency response times was taken in 1980. Then another random sample of these times was taken in 1985. Using a two-sample difference-of-means test, one is able to show, in a upper-tail test, that response times have increased. Suppose the result is highly significant, say a PROB-VALUE of .0001. Is this of any concern? Whether this conclusion has any practical significance depends on the exact null hypothesis being examined. If, as usual, the null hypothesis is expressed as $\mu_1 - \mu_2 \leq 0$ or $\mu_1 \leq \mu_2$, the result may have no practical significance. This tells us nothing about *how much* response times have increased, only that they have. An increase of 10 s is of no practical concern, but an increase of 3 or 4 min may be. Proving the statistical significance of meaningless hypotheses is of no value. Only a thorough understanding of the problem under study can lead to the specification of intelligent hypotheses worthy of statistical evaluation.

PROBLEMS

1. Explain the meaning of (a) pooled variance and (b) paired samples.

2. Using the F table given in the Appendix, determine the following F statistics:

   a. $F(.95, 4, 3)$ d. $F(.95, 9, 1)$ g. $F(.01, 15, 15)$
   b. $F(.99, 15, 15)$ e. $F(.95, 4, 8)$ h. $F(.05, 30, 15)$
   c. $F(.95, 3, 4)$ f. $D(.975, 8, 10)$ i. $F(.95, 13, 11)$

3. Determine the χ^2 statistics from the χ^2 table in the Appendix for the following cases:

a. $\chi^2(.90, 5)$ e. $\chi^2(.95, 50)$ i. $\chi^2(.95, 5)$
b. $\chi^2(.10, 5)$ f. $\chi^2(.99, 23)$ j. $\chi^2(.05, 40)$
c. $\chi^2(.025, 25)$ g. $\chi^2(.90, 20)$ k. $\chi^2(.01, 4)$
d. $\chi^2(.975, 25)$ h. $\chi^2(.10, 20)$ l. $\chi^2(.975, 2)$

4. Fertilizer XLC is applied to eight identical plots of land planted with corn. The same corn seed is also applied to eight other identical plots of land for a control treatment. At the end of the growing season, the total yield for each plot of land is measured. The data are shown in the following table:

Plot No.	Treatment	
	Control	Fertilizer XLC
1	75	86
2	66	80
3	68	83
4	71	87
5	73	82
6	74	79
7	70	81
8	71	78

a. Use a t test to determine whether yields on XLC-treated plots differ from those of control plots.
b. Test the hypothesis that using fertilizer XLC improves corn yield.

5. Hourly traffic volumes are monitored over the evening rush hour on a sample of 10 residential streets. Shortly after the survey, a series of turning restrictions are put in place in an attempt to lower traffic volumes. Hourly traffic volumes are again monitored on the same sample of streets, on the same day of the week during identical weather conditions. The volumes are summarized as follows:

Street	Hourly traffic volume	
	Before	After
Fisher Street	650	600
Pine Street	240	300
Foul Bay Road	860	750
First Avenue	350	320
Magnolia Street	500	500
Renfrew Road	600	420
Hartley Boulevard	570	400
Fort Street	420	400
Pearson Street	480	450
Snedecor Road	530	560

a. Have the turning restrictions decreased traffic volumes?
b. Is this a fair test? Think of reasons why it might be considered to be invalid. (*Hint*: Might it violate any assumptions of statistical testing?)

6. The minimum daily temperature recorded at two locations in the first week of February is determined for a random sample of $n = 60$ different days over a 100-yr climate record. Based on these samples, the following statistics are computed:

	Sample size	Mean minimum temperature, °C	Standard deviation
Location A	$n = 60$	$\bar{X} = 7.3$	$s = 4.2$
Location B	$n = 60$	$\bar{X} = 6.8$	$s = 3.7$

a. Assuming that the population variances at the two locations are equal, find a 95 percent confidence interval on the mean temperature difference between the two locations, $\mu_A - \mu_B$. Is one location colder than the other, on average, during this period? Explain.
b. Is the assumption of equal variances reasonable? Test the equality of these two variances at $\alpha = .05$. Determine an appropriate PROB-VALUE for this test.

7. In several empirical studies of the patronage of consumers in rural areas, the assumption of perfect rationality is examined by comparing the actual distance traveled by a consumer to purchase a given good or service to the distance of the closest location offering that service to the consumer. Consider the actual and closest distances for the following set of rural residents:

	Distance, km	
Consumer	Actual	Closest
1	7	4
2	13	8
3	3	3
4	4	3
5	5	2
6	5	4
7	9	8
8	7	6
9	7	5
10	8	3
11	13	2
12	4	4

a. Test the hypothesis that consumers tend to patronize the closest outlet for this service.
b. Should it be a one- or two-tailed test? Does it matter?
c. Is this a reasonable way to evaluate this assumption of rationality in consumer behavior? Can you think of others?

10

Nonparametric Statistics

All the methods of statistical inference discussed so far have required specific assumptions about the nature of the population from which the samples were obtained. We select a population parameter of interest (μ or σ or π), collect a random sample, calculate a point estimator ($\bar{x}$ or s or p), construct a sampling distribution, and then employ hypothesis testing decision rules or confidence interval formulas. The most common assumption is that the population distribution function of the random variable is normal with mean μ and variance σ^2. Because these methods require a number of assumptions about one or more population parameters, they are termed *parametric statistical methods*. The tests described in this chapter are referred to as *nonparametric*, or *distribution-free*, *tests*. Compared with parametric tests, nonparametric tests require less stringent assumptions about the nature of the underlying population distribution function of the random variable being analyzed.

There are two situations in which nonparametric methods can be applied. When the level of measurement of the variable is nominal or ordinal, most parametric methods cannot be applied. It is impossible for a categorical or ordinal variable to be normally distributed. This is the first situation where nonparametric statistical methods prove useful. The second situation occurs when the nature of the underlying population distribution is unknown or left unspecified. Usually, this occurs when we wish to use a parametric test but we cannot meet its exacting requirements. For example, we might find that a variable has a highly skewed or even bimodal distribution and suspect the underlying population distribution to be nonnormal. In fact, one class of nonparametric tests, *goodness-of-fit tests*, can be used to test the assumption of normality. An observed frequency distribution can be compared to the distribution we might expect according to some theory or by some assumption. Quite frequently, one of the first steps in parametric testing is to use a goodness-of-fit test to see whether the assumption of normality is warranted. When the assumption of normality is rejected by a goodness-of-fit test, the alternative is to employ a nonparametric procedure.

In Section 10.1 comparisons are made between parametric and nonparametric procedures. The advantages and disadvantages of each type of method are described. Which type of test should be used in a specific case is shown to depend on a number of conditions. Sections 10.2 to 10.5 describe several nonparametric procedures useful

315

in four different situations. Section 10.2 discusses the nonparametric equivalents to several parametric tests presented in Chapter 9. Goodness-of-fit tests are presented in Section 10.3. In Section 10.4, tests dealing with contingency tables or cross-classified count data are described. A common assumption in all statistical methods is randomness. One specific test of this assumption is explained in Section 10.5. The chapter is summarized in Section 10.6.

10.1. Comparison of Parametric and Nonparametric Tests

Both parametric and nonparametric tests have points of superiority. The choice of statistical test to be undertaken depends on several factors. Some advantages and disadvantages are outlined in the following paragraphs.

Statistical Efficiency

When a data set satisfies all the assumptions of a given parametric test, the parametric test should be applied. Any nonparametric alternative is almost always significantly *less powerful* than the comparable parametric procedure. The probability of making a Type I error is usually appreciably lower with parametric tests. However, violations of a parametric test's assumptions can change the sampling distribution of the test statistic and thus the power of the test.

Robustness

A test is said to be *robust* against the violation of a certain assumption if the power of the test is not significantly affected by the violation. For example, the one-sample *t* test described in Chapter 9 is still powerful when the population is nonnormal. In fact, because of the central limit theorem, the larger the sample, the less critical the assumption of normality. The power of the test, in turn, depends on the extent to which the violation of the assumption alters the sampling distribution of the test statistic. Robustness is a matter of degree. Violations of some assumptions, such as normality, may be less critical than a violation of, say, the homogeneity of variance. When a specific assumption is violated, it is important to check, first, how large the violation is and, second, the sensitivity of the test to the departure. Where the violation is significant, it is often advantageous to revert to a nonparametric test which does not require that particular assumption. Or it may be possible to *transform* the variable so that the assumption is not violated. Applying a transformation can sometimes correct for nonnormality. An example of this procedure is illustrated in Section 10.3.

Scope of Application

Since they are generally based on fewer or less restrictive conditions than parametric tests, nonparametric tests can be legitimately applied to a much larger set of populations. Even when it is known that the populations are skewed or bimodal or rectangular, most nonparametric tests can still be applied. Also, statistics calculated from variables measured at the nominal or ordinal scale can be tested by using nonparametric methods. This is a distinct advantage in survey design. It can be difficult to extract interval-scale responses from mail questionnaires or personal interviews. Preference ratings or approve/disapprove questions typically yield ordinal and/or nominal responses. As an example, consider the problem of getting respondents in a mail questionnaire to accurately answer a question about their annual income. For reasons of confidentiality, many respondents are willing to answer a question concerning income only if it allows them to check a box corresponding to a range of income, such as "From $10,000 to $15,000."

Simplicity of Derivation

The derivation of the sampling distribution of the test statistics utilized in parametric tests is based on quite complicated theorems such as the central limit theorem. Usually, the form of the sampling distribution is a function of the form of the population probability distribution. Given the mathematical complexity of the normal distribution [recall the equation for the normal density function, (5-30)], many users of parametric statistical methods cannot comprehend the derivation of the key results upon which the tests are based. Some proponents of nonparametric statistics suggest that many who employ parametric statistics must therefore apply them in a "cookbook" manner.

In contrast, the sampling distributions of the test statistics used in nonparametric tests are derived from quite basic combinatorial arguments and are easily understood. Several examples which illustrate the simplicity of the derivations are given in this chapter. For this reason, nonparametric tests are an inviting alternative to many geographers and other social scientists. The claim of superiority based on simplicity must be tempered by the fact that nonparametric tests are also misused in some social science applications. At times, they are inappropriately applied or else interpreted incorrectly.

Ease of Application

In addition, the actual arithmetic involved in operationalizing nonparametric tests is less complicated. The most common operations are counting, ranking, addition, and subtraction. Less sophisticated users find these tests alluring, to say the least. The availability of these statistical tests in the standard software packages such as SAS encourages many researchers to report the results from these packages—even if they are unclear what the tests are actually saying about the variables being analyzed. Simplicity has proved to be a two-edged sword.

Sample Sizes

The size of the available sample is an important consideration in determining whether to use a parametric or nonparametric test. When sample size is extremely small, say

n is less than 10 or 15, violations of the assumptions of parametric tests are particularly hard to detect (see Section 10.3 for an example). Yet a violation of an assumption can have extremely deleterious effects under these conditions. The use of nonparametric tests in these situations is often advisable. In general, the larger the sample, the more advantageous a parametric test becomes. The implications of the central limit theorem support this assertion. Where sample size is large, the presence of nonnormality is often insignificant.

In sum, parametric tests have a significant advantage over nonparametric tests when their assumptions are met. Violations of these assumptions introduces inexactness into the test. The seriousness of the violation depends on the size of the sample, the specific assumption violated, and the robustness of the test to the particular violation.

10.2. One-, Two-, and k-Sample Tests

The first set of nonparametric procedures that we discuss are the one- and two-sample nonparametric equivalents to the t tests described in Chapter 9. They are used whenever an interval-scaled data set violates one of the assumptions of the appropriate parametric test or when the variables of the data set are at an ordinal level of measurement.

Sign Test for the Median

The simplest parametric test discussed in Chapter 9 compares the sample mean $\bar{x}$ and some hypothesized (or known) value of the population mean μ, specified in the null hypothesis. The *sign test* is used to test the value of the sample median against some hypothesized value of the population median η_I (Greek letter eta), specified in H_0. Let us denote the true population median as η. The sample median M is an unbiased consistent estimator of η, just as the sample mean $\bar{x}$ estimates μ. The null hypothesis in the sign test for the median is that the hypothesized value of the median η_I is equal to the true median η, that is, H_0: $\eta_I = \eta$. If H_0 is true, then in any sample of n observations we are just as likely to get an observation above the value η_I as below it. If an observation x_i lies above η_I, then the difference $x_i - \eta_I$ has a positive sign. If it lies below η_I, then the difference has a negative sign. Thus, the null hypothesis suggests we should get about an equal number of plus and minus signs. Put another way, the probability of getting a negative sign is $p = P(X < \eta_I) = .5$ In any random sample we will not get an equal number of plus and minus signs. Or, the probability of getting a negative sign is $p = P(X_i < \eta_I) \neq .5$. In any random sample we will not necessarily get an equal number of positive and negative signs. The question boils down to this: How few minus signs (or, equivalently, how few plus signs) must there be before we can reject the null hypothesis that η_I is the true median?

The test statistic is simply the number of observations in the sample with values below the value of the population median specified in the null hypothesis. Call this number B where B is a *binomial random variable* with $\pi = .5$. There are only two outcomes, and there are n observations or trials. Thus, the sign test turns out to be equivalent to a test of the population proportion of $\pi = .5$. The test is summarized in Table 10-1.

TABLE 10-1
Sign Test for the Median

Specification

The n observations are compared with the value of the median hypothesized in H_0. The number of observations below the hypothesized median is counted. Call this number B. Equivalently, estimate $p = B/n$.

Hypotheses

H_0: $\eta = \eta_I$, or equivalently $p = P(X_i < \eta_I) = .5$, the true median equals the hypothesized median.
H_A: (a) $\eta \neq \eta_I$, or equivalently $p = P(X_i < \eta_I) \neq .5$ (two-tail).
    (b) $\eta < \eta_I$, or equivalently $p = P(X_i < \eta_I) < .5$ (one-tail).
    (c) $\eta > \eta_I$, or equivalently $p = P(X_i < \eta_I) > .5$ (one-tail).

Test statistic

The proportion of the n observations below the median is p, and $B =$ the number of observations below the median.

Decision rule

For small samples, the binomial distribution can be used to test this hypothesis. The exact PROB-VALUE for B can be determined from binomial tables. For large samples, the normal approximation to the binomial is used, and a test for P is utilized.

Large-sample test

Reject H_0 if $p < .5 - z_{\alpha/2}\sqrt{(.5)(.5)/n}$ or $p > .5 + z_{\alpha/2}\sqrt{(.5)(.5)/n}$ for a two-tail test. For one-tail, reject H_0 if $p < .5 + z_\alpha \sqrt{(.5)(.5)/n}$ for $H_A(b)$, and reject H_0 if $p > .5 - z_\alpha \sqrt{(.5)(.5)/n}$ for $H_A(c)$.

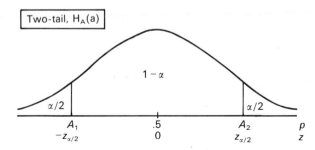

where $A_1 = .5 - z_{\alpha/2}\sqrt{\dfrac{(.5)(.5)}{n}}$ $A_2 = .5 - z_{\alpha/2}\sqrt{\dfrac{(.5)(.5)}{n}}$

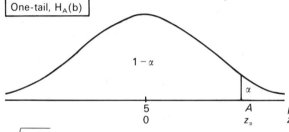

where $A = .5 + z_\alpha \sqrt{\dfrac{(.5)(.5)}{n}}$

(continued)

TABLE 10-1 (continued)

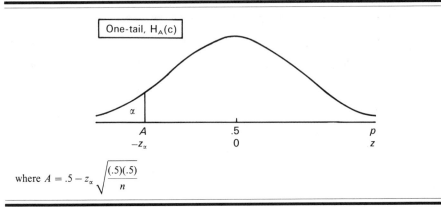

where $A = .5 - z_\alpha \sqrt{\dfrac{(.5)(.5)}{n}}$

EXAMPLE 10-1. The travel time of an individual's journey to work is randomly sampled on 20 different workdays over 3 months. Each day, the total door-to-door travel time is recorded, and the following values are obtained: 22, 23, 25, 27, 27, 28, 28, 29, 29, 31, 33, 37, 37, 38, 39, 42, 43, 43, 43, 58. For purposes of simplicity, the times have been sorted in ascending order. Do these data support the null hypothesis that the true median travel time is 30 min for this trip, at the .05 level of significance?

First, it is necessary to calculate the number of minus signs, or observations below the hypothesized median of $\eta_I = 30$. Since the observations are in order, a simple count reveals $B = 9$ observations below the hypothesized median of 30. Or, equivalently, the probability of a negative sign in this sample is $p = B/n = \frac{9}{20} = .45$. Since $np = 20(.45) = 9$ is greater than 5, the sampling distribution of P is closely approximated by the normal (see Table 8-6). Since the null hypothesis makes the assumption that $\pi = .5$, we estimate the standard deviation of the sampling distribution of p as $\sqrt{(.5)(.5)/20} = \sqrt{.0125} = .112$. The standard hypothesis testing procedure for a test of p versus π, as described in Chapter 9, is followed in Table 10-2. The null hypothesis cannot be rejected, and the decision rule leads to the conclusion that the median travel time is 30 min.

EXAMPLE 10-2. A recreation geographer randomly samples $n = 9$ visitors to a neighborhood park. Each visitor is asked to rate satisfaction with the park on a five-point scale with labels *completely dissatisfied, dissatisfied, neutral, satisfied,* and *completely satisfied.* The responses are then scored by assigning a value of 1 to *completely dissatisfied,* 2 to *dissatisfied,* and so on. The nine responses are 2, 3, 1, 2, 1, 5, 3, 1, and 2. Are these respondents satisfied with this neighborhood park?

A test of hypothesis for this question uses $H_0: \eta \le 3$ versus the alternate hypothesis $H_A: \eta > 3$. We are interested in whether we can say, at $\alpha = .05$, that park users *are* satisfied and therefore use a one-tail test. The sequence of signs is $-0---+0--$. There are two observations with a value of zero. By convention, these observations are omitted, and the sample size is reduced to $n = 7$ observations with $B = 6$ minus signs. The probability of obtaining a negative sign is

TABLE 10-2
Sign Test for Example 10-1

Hypotheses
H_0: $\eta_I = 30$, the median travel time is 30 min.
H_A: $\eta_I \neq 30$, the median travel time is not 30 min.

Test statistic and test to be employed
H_0 can be tested by using the sign test where the test statistic is the proportion of observations with a negative sign, p.

Level of significance
$\alpha = .05$.

Decision rule (two-tailed test)

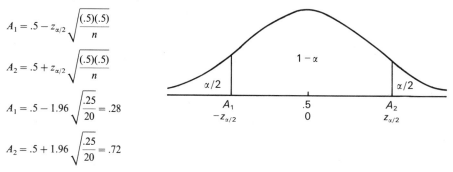

$$A_1 = .5 - z_{\alpha/2}\sqrt{\frac{(.5)(.5)}{n}}$$

$$A_2 = .5 + z_{\alpha/2}\sqrt{\frac{(.5)(.5)}{n}}$$

$$A_1 = .5 - 1.96\sqrt{\frac{.25}{20}} = .28$$

$$A_2 = .5 + 1.96\sqrt{\frac{.25}{20}} = .72$$

If $.28 \leq p \leq .72$, conclude H_0; if $p < .28$ or $p > .72$, conclude H_A.

Computation of test statistic
$p = \frac{9}{20} = .45$

Decision
Since $.28 \leq .45 \leq .72$, we accept the null hypothesis and conclude the median $\eta_I = 30$.

$p = \frac{6}{7} = .86$. Given the small sample size, we use the binomial tables directly. For $\pi = .5$ and $n = 7$ these probabilities are as follows:

x	0	1	2	3	4	5	6	7
$P(x)$	.0078	.0547	.1641	.2734	.2734	.1641	.0547	.0078

To construct a decision rule for $\alpha = .05$, we must include all outcomes in the lower tail whose combined probabilities are less than or equal to .05. Since outcomes 0 and 1 taken together, $.0078 + .0547 = .0625$, exceeds .05, only the outcome of 0 can be used to reject the null hypothesis. The decision rule is therefore: If $B = 0$, conclude H_A; if $1 \leq B < 7$, conclude H_0. Since $B = 6$, we must accept the null hypothesis. Notice that the selection of an α level is quite difficult in this case. Unlike the normal or the t distribution, the sampling distributions of nonparametric test statistics often take large discrete jumps, particularly where the sample size is small. The use of PROB-VALUE method is superior to classical hypothesis tests in these cases.

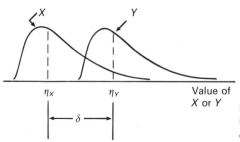

FIGURE 10-1. Two population distributions with identical shapes differing in the shift parameter δ.

Mann-Whitney Test

The Mann-Whitney test is useful for comparing two populations on the basis of independent random samples. By assuming only that the two population distributions have the same shape, the test is used to determine whether the two populations have the identical location or position. The situation is depicted in Figure 10-1. The two populations are labeled X and Y and have the same shape. The exact form of the distribution need not be specified. If the two populations are identical, then the shift parameter δ (Greek letter delta) between the two medians η_X and η_Y is zero. If $\delta > 0$, as in Figure 10-1, then the median (or location, or position) of population Y is to the right of population X. If $\delta < 0$, then the distribution of Y is to the left of X. This test is comparable to the two-independent-sample t test described in Chapter 9. It is possible to specify either a one- or a two-tail version of the hypotheses depending on whether the alternate hypothesis specifies the sign of δ.

Let $x_1, x_2, \ldots, x_{n_1}$ and $y_1, y_2, \ldots, y_{n_2}$ be two independent random samples of size n_1 and n_2, respectively. Combine the n_1 observations of X from the first sample with the n_2 observations of Y, and arrange them in ascending order. Then assign ranks to the observations in this combined sample. Rank 1 is assigned to the lowest-valued observation, and rank $n_1 + n_2$ is assigned to the highest-valued observation. If the variable is collected at the ordinal level, then it is important that the rankings be obtained over the combined set of observations. For example, we might have respondents rank a series of 10 neighborhoods in terms of residential desirability. After the responses are recorded, the 10 neighborhoods are divided into two categories, say inner-city and suburban, creating two different samples. It is now possible to test the preferences of the respondents to inner-city or suburban locations. This would be impossible if the rankings were obtained separately for each type of neighborhood.

The Mann-Whitney test is based on the idea that if the two independent samples are drawn from the same population, then the average ranks of the two samples should be almost equal. The test statistic is simply the sum of the ranks from the first sample of X. Denote this sum S. The smallest possible value of S is the sum of the first n_1 ranks. This occurs when all the observations in the first sample are smaller than all the observations in the second sample. So the minimum value is $S = 1 + 2 + \cdots + n_1 = n_1(n_1 + 1)/2$. The largest possible value of S is the sum of the last n_1 ranks, or $(n_2 + 1) + (n_2 + 2) + \cdots + (n_2 + n_1)$. Therefore the largest possible value of S is $n_1 n_2 + n_1(n_1 + 1)/2$. For example, suppose $n_1 = n_2 = 5$. The two extreme values of S occur in the following ways:

	Minimum S		Maximum S	
X	Y	X	Y
Sample 1	Sample 2	Sample 1	Sample 2
1			1
2			2
3			3
4			4
5			5
	6	6	
	7	7	
	8	8	
	9	9	
	10	10	
$\overline{15}$		$\overline{40}$	

The minimum value is $S = 1 + 2 + 3 + 4 + 5 = 5(5+1)/2 = 15$. The maximum value is $S = 6 + 7 + 8 + 9 + 10 = 5(5) + 5(5+1)/2 = 40$. If the calculated value of S is near the minimum value of S, then the distribution of X is to the left of Y and $\delta > 0$ (as in Figure 10-1). Conversely, if Y is to the right of X and $\delta < 0$, then the calculated value of S will be near the maximum value of S. If $\delta = 0$, as is hypothesized in H_0, then all possible orderings of the combined samples are equally likely and S will be between these two extremes.

To derive hypothesis testing rules or confidence intervals, it is necessary to know the sampling distribution of S when $\delta = 0$. For small samples, it is possible to work out the exact sampling distribution of S by determining the value of S for all possible orderings of the $n_1 + n_2$ observations. For example, if $n_1 = 2$ and $n_2 = 3$, the value of S for each possible permutation of the five values could be calculated in the following format:

	Observation ranks	
Sample 1	Sample 2	S
1, 2	3, 4, 5	3
1, 3	2, 4, 5	4
1, 4	2, 3, 5	5
2, 3	1, 4, 5	5
.	.	.
.	.	.
.	.	.
4, 5	1, 2, 3	9

TABLE 10-3
The Mann-Whitney Test

Specification
The test compares two independent random samples of size n_1 and n_2 drawn from the two populations of X and Y, respectively. The measurement scales of X and Y are at least ordinal. Ranks are assigned to the *combined* sample; the lowest-valued observation is assigned the rank 1 and the highest observation the rank $n_1 + n_2$.

Hypotheses
H_0: $\delta = 0$, the two random samples are drawn from the same population.
H_A: (a) $\delta \neq 0$, the two random samples are drawn from different populations.
   (b) $\delta > 0$, the sample Y lies to the right of X, or $\eta_Y < \eta_X$.
   (c) $\delta < 0$, the sample Y lies to the left of X, or $\eta_Y < \eta_X$.

Test statistic
The sum of the ranks from the sample X is S.

Decision rule
For small sample sizes with $n_1, n_2 < 10$, the exact **PROB-VALUE** can be determined by generating the sampling distribution of S. This is accomplished by considering all possible arrangements of the $n_1 + n_2$ observations. Tables of S are available in several advanced nonparametric textbooks. For large samples, the normal approximation is used.

Large-sample test
Reject H_0 if

$$S < \frac{n_1(n_1 + n_2 + 1)}{2} - z_{\alpha/2}\sqrt{\frac{n_1 n_2(n_1 + n_2 + 1)}{12}}$$

or if

$$S > \frac{n_1(n_1 + n_2 + 1)}{2} + z_{\alpha/2}\sqrt{\frac{n_1 n_2(n_1 + n_2 + 1)}{12}}$$

for a two-tailed test. For one-tail test, reject H_0 if

$$S < \frac{n_1(n_1 + n_2 + 1)}{2} - z_{\alpha}\sqrt{\frac{n_1 n_2(n_1 + n_2 + 1)}{12}}$$

for H_A(b), and reject H_0 if

$$S > \frac{n_1(n_1 + n_2 + 1)}{2} + z_{\alpha}\sqrt{\frac{n_1 n_2(n_1 + n_2 + 1)}{12}}$$

for H_A(c).

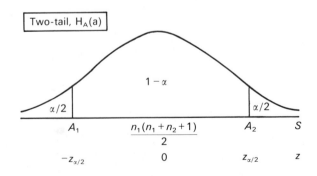

TABLE 10-3 (continued)

where $A_1 = \dfrac{n_1(n_1 + n_2 + 1)}{2} - z_{\alpha/2}\sqrt{\dfrac{n_1 n_2(n_1 + n_2 + 1)}{12}}$

$A_2 = \dfrac{n_1(n_1 + n_2 + 1)}{2} + z_{\alpha/2}\sqrt{\dfrac{n_1 n_2(n_1 + n_2 + 1)}{12}}$

One-tail, $H_A(b)$

where $A = \dfrac{n_1(n_1 + n_2 + 1)}{2} - z_{\alpha}\sqrt{\dfrac{n_1 n_2(n_1 + n_2 + 1)}{12}}$

One-tail, $H_A(c)$

where $A = \dfrac{n_1(n_1 + n_2 + 1)}{2} + z_{\alpha}\sqrt{\dfrac{n_1 n_2(n_1 + n_2 + 1)}{12}}$

Tables of this sampling distribution are available in specialized texts on nonparametric statistics listed at the end of this chapter. When both n_1 and n_2 exceed 10, the sampling distribution of S is approximately normal with mean

$$E(S) = \frac{n_1(n_1 + n_2 + 1)}{2} \tag{10-1}$$

$$\sigma^2(S) = \frac{n_1 n_2(n_1 + n_2 + 1)}{12} \tag{10-2}$$

The hypothesis that $\delta = 0$ can then be tested by using the conventional format described in Chapter 9. This procedure is summarized in Table 10-3. The use of the test is illustrated in Example 10-3.

TABLE 10-4
Calculations for Mann-Whitney Test for Example 10-3

Combined ordered observations		Combined ranks	
X	Y	X	Y
	9		1
10		2	
11		3	
12	12	4.5	4.5
	14		6
15		7	
16		8	
17		9	
18		10	
20		11	
	21		12
22	22	13.5	13.5
27		15	
	28		16
	29, 29		17.5, 17.5
	31		19
	58		20
$n_1 = 10$	$n_2 = 10$	$\sum = 83$	$\sum = 127$

$$S = 83$$

$$E(S) = \frac{10(10 + 10 + 1)}{2} = 105$$

$$\sigma^2(S) = \frac{10(10)(10 + 10 + 1)}{12} = 175$$

$$\sigma(S) = 13.23$$

EXAMPLE 10-3. Carbon monoxide concentrations are sampled at 20 intersections in the residential areas of a city. All intersections have approximately equal average daily traffic volumes. Ten of the intersections are controlled by yield signs, and 10 are controlled by stop signs. The concentrations of carbon monoxide (in parts per million) are as follows:

Intersections with yield signs	Intersections with stop signs
10	31
15	9
16	21
17	28
22	14
27	12
11	58
12	29
18	22
20	29

Because the stop signs cause more cars to travel at lower speeds, encourage periods of engine idling, and create more acceleration-deceleration phases, it is hypothesized that these intersections will have higher concentrations than those with yield signs. Is this hypothesis supported by the data?

The calculations required for the Mann-Whitney test of the carbon monoxide pollution levels are shown in Table 10-4. In the first two columns, the data are ordered within the combined set. In the last two columns, these observation values are replaced by the appropriate ranks. Notice the convention used when ties are encountered. Two observations have carbon monoxide concentrations of 12 parts per million. These two observations are ranked 4th and 5th. Each is assigned the average rank of $(4 + 5)/2 = 4.5$. The convention of assigning tied observations the average rank can be used for two, three, or more tied observations. For example, if three observations were tied for what would be the 5th, 6th, and 7th ranks, all three would be assigned a rank of $(5 + 6 + 7)/3 = 6$. However, the test assumes there are *no* ties in the data and is really only approximately correct when ties exist. The decision rule for a hypothesis test in a case where numerous ties exist requires modification. These modifications are discussed in the nonparametric statistics texts cited under Further Reading.

The sum of the ranks of X yields $S = 83$. Since n_1 and n_2 are greater than or

TABLE 10-5
Test of Hypothesis for Example 10-3

Hypotheses
H_0: $\delta = 0$, the two samples have the same carbon monoxide concentration.
H_A: $\delta < 0$, intersections with yield signs have lower concentrations of carbon monoxide.

Test statistic and test to be employed
H_0 can be tested by using the Mann-Whitney test. Since the stop sign observations have a much longer variation and a possible extreme value (58 ppm), it is preferred to the parametric alternative. The test statistic is S, the sum of ranks of yield sign observations.

Level of significance
$\alpha = .05$.

Decision rule

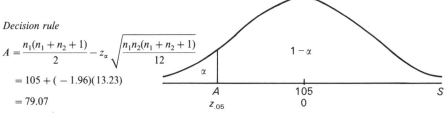

$$A = \frac{n_1(n_1 + n_2 + 1)}{2} - z_\alpha \sqrt{\frac{n_1 n_2(n_1 + n_2 + 1)}{12}}$$

$$= 105 + (-1.96)(13.23)$$

$$= 79.07$$

If $S < 79.07$, reject H_0; if $S \geq 79.07$, conclude H_0.

Computation of test statistic
From Table 10-4, $S = 83$.

Decision
Since $S = 83 > 79.07$, we accept the null hypothesis and conclude that intersections controlled by yield signs do not have lower carbon monoxide concentrations.

equal to 10, the normal approximation to the sampling distribution of S is utilized: $E(S) = 105$ and $\sigma(S) = 13.23$. The formal test of hypothesis is detailed in Table 10-5. Since $S = 83$ lies above the critical value of 79.07, the null hypothesis is accepted, and we conclude that concentration of carbon monoxide at intersections controlled by stop signs is about the same as that at intersections controlled by yield signs.

Wilcoxon Signed-Ranks Test

The *Wilcoxon signed-ranks test* is the nonparametric equivalent to the matched-pairs t test presented in Chapter 9. The sample data consist of n matched pairs (x_i, y_i) of an interval-scaled variable. Like the matched-pairs t test, this nonparametric test utilizes the differences $D_i = y_i - x_i$. These differences are related to the individual populations of X and Y in the manner illustrated in Figure 10-2. When the y_i's are larger than the x_i's, as in Figure 10-2, the differences tend to be positive and the median difference η_D is also positive. Or, if the y_i's are generally smaller than the x_i's, then the differences are mostly negative and $\eta_D < 0$. The median difference may be thought of as a parameter which measures how far apart the distributions of X and Y are located. When the median difference is zero, the two distributions are in the same location. The sign of the median indicates the relative location of population X relative to population Y. If the distribution of difference is *symmetric*, then $\eta_D = \mu_D$ and the Wilcoxon signed-ranks test is equivalent to the matched-pairs t test for the difference of population means.

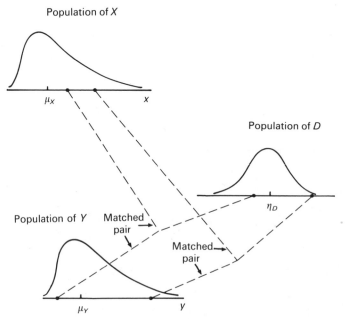

FIGURE 10-2. Wilcoxon signed-ranks test.

The calculation of the test statistic is quite simple. First, calculate the differences D_i, $i = 1, 2, \ldots, n$. Eliminate any observations with $D_i = 0$, and decrease the sample size appropriately. Consider now the absolute value of the differences $|D_i|$. These absolute values are ranked, with the lowest rank assigned to the smallest difference. Ties are handled in the manner described in the Mann-Whitney test. To each rank we also attach a sign of plus or minus according to whether the difference is positive or negative. The test statistic T is simply the sum of the signed ranks.

Under the null hypothesis of $\eta_D = 0$, we are just as likely to get a positive difference as a negative difference. Therefore, the rank associated with any absolute difference has an equal probability of being positive or negative. The expected value of T is therefore zero. The maximum value of T occurs when all the differences are positive. In this instance T is the sum of the ranks $1, 2, \ldots, n$. The sum of the first n integers is $n(n + 1)/2$, and this is the maximum value of T. The minimum value of T is $-n(n + 1)/2$ and occurs when all differences are negative. For a two-sided test, values of T close to zero lead us to conclude H_0: $\eta_0 = 0$. Values of T close to $n(n + 1)/2$ or $-n(n + 1)/2$ lead to the conclusion H_A: $\eta_D \neq 0$. A one-tail version of the test can also be employed.

The sampling distribution of T under H_0 has mean and variance

$$E(T) = 0 \tag{10-3}$$

$$\sigma^2(T) = \frac{n(n + 1)(2n + 1)}{6} \tag{10-4}$$

The sampling distribution depends only on the number of differences n. The exact sampling distribution of T can be easily generated for small sample sizes. For example, for $n = 3$ the sampling distribution is as follows:

Signed ranks			T
1	2	3	6
-1	2	3	4
1	-2	3	2
1	2	-3	0
-1	-2	3	0
-1	2	-3	-2
1	-2	-3	-4
-1	-2	-3	-6

Tables of T are available but are generally not needed unless n is very small. For $n \geq 10$, the sampling distribution of T is approximately normal with mean and variance given by Equations (10-3) and (10-4). The Wilcoxon signed-ranks test is completely specified in Table 10-6.

EXAMPLE 10-4. Ten water samples are taken at different sites on a river just downstream from the location of a proposed cannery. All samples are taken on the same day during summer drought conditions. Water quality measurements of the samples are recorded. The measure of water quality employed is the dissolved oxygen (DO)

TABLE 10-6
Wilcoxon Signed-Ranks Test

Specification
The n matched pairs (X_i, Y_i) are interval-scaled variables. Let $D_i = Y_i - X_i$. The absolute values of the differences D_i are then marked. A sign is attached to each rank, depending on whether it is positive or negative.

Hypotheses
H_0: $\eta_D = 0$, the population median difference equals zero.
H_A: (a) $\eta_D \neq 0$, the population median difference is not zero.
  (b) $\eta_D > 0$, there is a positive difference, and distribution Y lies to the right of X.
  (c) $\eta_D < 0$, there is a negative difference, and distribution X lies to the right of Y.

Test statistic
$$T = \min (T^+, T^-)$$

Decision rule
Tables of the exact sampling distribution of T are available in many statistics textbooks. However, these tables are rarely needed since the normal approximation is applicable for sample sizes as small as $n = 10$. Reject H_0 if

$$T < -z_{\alpha/2} \sqrt{\frac{n(n + 1)(2n + 1)}{6}}$$

or if

$$T > z_{\alpha/2} \sqrt{\frac{n(n + 1)(2n + 1)}{6}}$$

for a two-tail test. For H_A(b), reject H_0 if

$$T > z_{\alpha} \sqrt{\frac{n(n + 1)(2n + 1)}{6}}$$

and for H_A(c) reject H_0 if

$$T < -z_{\alpha} \sqrt{\frac{n(n + 1)(2n + 1)}{6}}$$

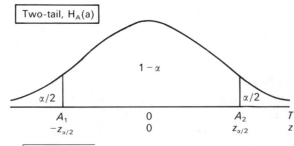

where $A_1 = -z_{\alpha/2} \sqrt{\dfrac{n(n + 1)(2n + 1)}{6}}$

$A_2 = z_{\alpha/2} \sqrt{\dfrac{n(n + 1)(2n + 1)}{6}}$

TABLE 10-6 (continued)

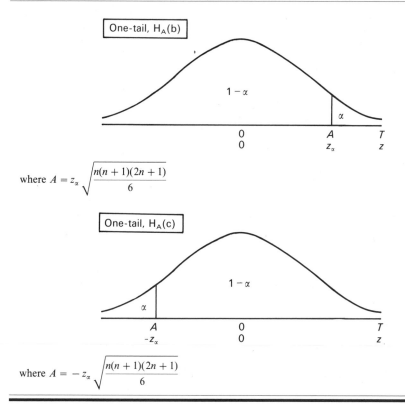

where $A = z_\alpha \sqrt{\dfrac{n(n+1)(2n+1)}{6}}$

where $A = -z_\alpha \sqrt{\dfrac{n(n+1)(2n+1)}{6}}$

TABLE 10-7
Dissolved Oxygen Levels at 10 Selected Sites,
Summer Drought Conditions

Site	Dissolved oxygen, mg/L	
	Before	After
1	5.0	6.0
2	7.2	5.5
3	4.5	3.0
4	7.0	5.0
5	7.0	6.2
6	2.4	2.5
7	3.0	0
8	4.0	3.1
9	2.8	2.2
10	4.2	3.0

TABLE 10-8
Calculations for Wilcoxon Signed-Ranks Test for Example 10-4

Site i	Before x_i	After y_i	Difference D_i	Rank	Signed rank
1	5.0	6.0	+1.0	5	+5
2	7.2	5.5	−1.7	8	−8
3	4.5	3.0	−1.5	7	−7
4	7.0	5.0	−2.0	9	−9
5	7.0	6.2	−0.8	3	−3
6	2.4	2.5	+0.1	1	+1
7	3.0	0	−3.0	10	−10
8	4.0	3.1	−0.9	4	−4
9	2.8	2.2	−0.6	2	−2
10	4.2	3.0	−1.2	6	−6
					$T = -43$

TABLE 10-9
Wilcoxon Signed-Ranks Test for Example 10-4

Hypotheses
H_0: $\eta_D = 0$, there is no change in the median DO levels.
H_A: $\eta_D \neq 0$, the median DO has changed.

Test statistic and test to be employed
Given the small sample size and the apparent nonnormality of the data, the Wilcoxon signed-ranks test is employed. Test statistic is T, the sum of the signed ranks.

Level of significance
$\alpha = .05$.

Decision rule

$$A_1 = 0 - z_{\alpha/2} \sqrt{\frac{n(n+1)(2n+1)}{6}}$$

$$A_2 = 0 + z_{\alpha/2} \sqrt{\frac{n(n+1)(2n+1)}{6}}$$

$$A_1 = -1.96 \sqrt{\frac{10(11)(21)}{6}} = -38.46$$

$$A_2 = +1.96 \sqrt{\frac{10(11)(21)}{6}} = +38.46$$

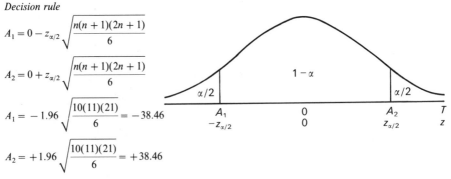

If $-38.46 \leq T \leq 38.46$, conclude H_0; if $T < -38.46$ or $T > +38.46$, conclude H_A.

Computation of test statistic
From Table 10-7, $T = -43$.

Decision
Since $T < -38.46$, reject H_0; conclude there has been a change in DO levels.

content in micrograms per liter. Label these 10 samples *before*. Shortly after the opening of the cannery and its waste treatment system, new DO measurements are taken at the same locations. Label these samples *after*. The water quality measurements are listed in Table 10-7. Has there been a change in DO levels?

The calculations for the Wilcoxon signed-ranks test of these data are illustrated in Table 10-8. The first three columns contain the (x_i, y_i) pairs and the differences D_i. The next column contains the ranks of the absolute differences, the lowest (rank 1) for observation 6 and the highest (rank 10) for observation 7. Eight of the differences have negative signs, and two have positive signs. The signed ranks are shown in the last column. The test statistic T is the sum of the signed ranks, or -43. It is near the lower extreme of T which ranges from -55 to $+55$ for sample size $n = 10$. Since there are 10 differences, the normal approximation can be used. A complete test of hypothesis is given in Table 10-9. The calculated test statistic of $T = -43$ leads to the rejection of H_0 at $\alpha = .05$. It is concluded that DO levels in the river decreased with the opening of the cannery. To calculate the two-sided PROB-VALUE for this problem, we convert T to Z, using $Z = T - 0/\sigma(T)$. Substituting $T = -43$ and $\sigma(T) = 19.62$ yields $Z = -2.19$. From the normal table we see that the two-sided PROB-VALUE equals $2(.0143) = .0286$.

The Wilcoxon signed-ranks test can also be used to test whether the median of a set of numbers equals some specified value, say $\eta = \eta_I$. If $x_i, i = 1, 2, \ldots, n$, represents the observations in a random sample from a population, then define the differences $D_i = x_i - \eta_I$. These differences are then used to test whether $\eta_D = 0$. If the hypothesis that $\eta_D = 0$ cannot be rejected, it follows that the median of the x_i's must be equal to η_I. Used in this way, the Wilcoxon signed-ranks test is an alternative to the sign test for the median.

Kruskal-Wallis Test

The *Kruskal-Wallis test* extends the Mann-Whitney test to the case when there are more than two populations. It is the nonparametric equivalent to one version of a powerful parametric test known as the *analysis of variance*, called ANOVA. The data for this test can be arranged to the following format:

Sample 1	Sample 2	. . .	Sample k
x_{11}	x_{21}	. . .	x_{k1}
x_{12}	x_{22}		x_{k2}
x_{1n_1}	x_{2n_2}	. . .	x_{kn_k}

where x_{ij} is the value of jth observation in the ith sample. In the k samples there are $n = n_1 + n_2 + \cdots + n_k = \Sigma_{i=1}^k n_i$ observations. The test utilizes the ranks of the individual observations within the *combined* set. If the measurement scale is ordinal, then the ranks must be obtained within the combined set, just as in the Mann-Whitney test.

The Kruskal-Wallis test is also very simple. It is based on a comparison of the sums of the ranks for each of the k independent samples. The first step is to rank the observations. Let $r(X_{ij})$ be the rank of the jth observation of the ith sample within the combined set of n observations. Ties are handled in the usual way. Define r_i as the sum of the rankings in the ith sample:

$$r_i = \sum_{j=1}^{n_i} r(x_{ij}) \tag{10-5}$$

Under the null hypothesis that the k samples come from the same population, the average ranks in each sample should be approximately equal. If the average ranks differ, then this supports the alternative hypothesis that *at least one* of the samples comes from a different population than the other samples. If only the locations of the population distributions differ, and not the shapes, then the Kruskal-Wallis test is equivalent to testing the equality of the k population means $\mu_1 = \mu_2 = \cdots = \mu_k$.

The test statistic for the Kruskal-Wallis test is

$$H = \frac{12}{n(n+1)} \sum_{i=1}^{k} \frac{r_i^2}{n_i} - 3(n+1) \tag{10-6}$$

if there are no ties in the k samples. If there are numerous ties in the data, then a correction factor must be used. The sampling distribution of H depends primarily on the sum of the ranks in each sample r_i. For small sample sizes, the sampling distribution of H can be generated by considering all distinguishable combinations of the n rankings for the k groups. For example, consider the smallest possible problem, a $k = 3$ sample problem with $n_1 = 2$, $n_2 = n_3 = 1$, and $n = 4$ observations. The possible combinations of rankings are shown in the following table:

	Sample		
1	2	3	H
1, 2	3	4	2.7
1, 2	4	3	2.7
3, 4	1	2	2.7
3, 4	2	1	2.7
2, 3	1	4	2.7
2, 3	4	1	2.7
1, 3	2	4	1.8
1, 3	4	2	1.8
2, 4	1	3	1.8
2, 4	3	1	1.8
1, 4	2	3	0.3
1, 4	3	2	0.3

Notice that arrangements of the rankings *within* a sample are ignored. For example, no distinction can be made between these two combinations

Sample 1	Sample 2	Sample 3	Sample 1	Sample 2	Sample 3
1, 2	3	4	2, 1	3	4

TABLE 10-10
Kruskal-Wallis Test

Specification
There are k independent random samples with n_i, $i = 1, 2, \ldots, k$, observations. In total there are
$n = \Sigma_{i=1}^{k} n_i = n_1 + n_2 + \cdots + n_k$ observations. The individual observations are ranked within the
complete data set of n observations. Calculate the sum of the ranks r_i for each of k samples.

Hypotheses
H_0: $\eta_1 = \eta_2 = \eta = \eta_k$, all samples are drawn from the same population. If the population distributions
have the same shape and variance, then this is equivalent to a test of the k
population means $\mu_1 = \mu_2 = \cdots = \mu_k$.
H_A: At least one median differs from the rest.

Test statistic

$$H = \frac{12}{n(n+1)} \sum_{i=1}^{k} \frac{r_i^2}{n_i} - 3(n+1)$$

Decision rule
Tables of the critical values of H for $k = 3$ samples where all $n_i \leq 5$ are available in most nonparametric textbooks. For all other cases the distribution of H is approximately χ^2 with $k - 1$ degrees of
freedom. Reject H_0 if $H > \chi^2(1 - \alpha, k - 1)$. This is a one-tail test.

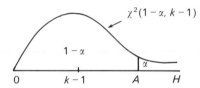

where $A = \chi^2(1 - \alpha, k - 1)$.

since in both cases the two lowest-ranked observations are in the first sample. The
value of H is equal in these cases. Under the null hypothesis that the three samples
come from the same population, each of these combinations of rankings is equally
likely. If the actual sample contained the rankings 1, 2; 3; 4 with a calculated value
of $H = 2.7$, then we see from the enumeration table that it has a $\frac{6}{12}$, or 50 percent,
chance of occurring. Thus, if we reject the null hypothesis of equality, we do so at an
$\alpha = .50$ level of significance.

However, it is not always necessary to consult detailed tables of H for a
Kruskal-Wallis test. Whenever $k > 3$ and/or $n_i > 5$ for at least one sample, the sampling distribution of H is χ^2 with $k - 1$ degrees of freedom. The Kruskal-Wallis test
is summarized in Table 10-10.

EXAMPLE 10-5. Suppose the sample of carbon monoxide concentrations utilized in
the Mann-Whitney test example is augmented by 10 more observations. These 10

observations are from intersections with no traffic controls. This augmented data set is now of the form of the Kruskal-Wallis input:

Intersections with yield signs	Intersections with stop signs	Intersections with no control
10	31	17
15	9	14
16	21	12
17	28	8
22	14	6
27	12	19
11	58	13
12	29	23
18	22	5
20	29	7

TABLE 10-11
Calculations for the Kruskal-Wallis Test for Example 10-5

Intersections with yield signs		Intersections with stop signs		Intersections with no controls	
Observation value	Rank	Observation value	Rank	Observation value	Rank
				5	1
				6	2
				7	3
				8	4
		9	5		
10	6				
11	7				
12	9	12	9	12	9
				13	11
		14	12.5	14	12.5
15	14				
16	15				
17	16.5			17	16.5
18	18				
				19	19
20	20				
		21	21		
22	22.5	22	22.5		
				23	24
27	25	28	26		
		29, 29	27.5, 27.5		
		31	29		
		58	30		
$r_1 = 153$		$r_2 = 210$		$r_3 = 102$	

$$H = \frac{12}{30(31)}\left(\frac{153^2}{10} + \frac{210^2}{10} + \frac{102^2}{10}\right) - 3(31) = 7.53$$

TABLE 10-12
Kruskal-Wallis Test for Example 10-5

Hypotheses
H_0: All three samples come from the same population. Carbon monoxide concentrations are the same at intersections with stop signs, yield signs, and no controls.
H_A: At least one of the samples differs from the rest.

Test statistic and test to be employed
The test statistic of the Kruskal-Wallis test H.

Level of significance
$\alpha = .05$.

Decision rule
$A = \chi^2(1 - \alpha, k - 1)$
$\quad = \chi^2(0.95, 2) = 5.99$
Reject H_0 if $H > 5.99$; if $0 \le H \le 5.99$, accept H_0.

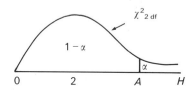

Computation of test statistic
From Table 10-11, $H = 7.53$.

Decision
Since $H = 7.53 > 5.99$, reject H_0 and conclude that carbon monoxide concentrations tend to vary with intersection controls.

Suppose that we have not yet analyzed any part of this data set and do not know the results of the Mann-Whitney test. In fact, the best procedure for analyzing k-sample data is to test the k-sample data for significance and then to perform two-sample tests on pairs of samples if it can be shown that more than one population is involved.

The necessary calculations are illustrated in Table 10-11. Notice how ties are handled. For the carbon monoxide value of 12 parts per million, there are three tied observations, one in each sample. The rank assigned to each is $(8 + 9 + 10)/3 = 9$. The sum of the ranks in the three samples of $r_1 = 153$, $r_2 = 210$, and $r_3 = 102$ is 465. For problems calculated by hand, a useful computational check is to compare the sum of the ranks of the k samples $\Sigma_{i=1}^{k} r_i$ to the sum of the first n ranks $n(n + 1)/2$. These two should be equal. In this example $\Sigma_{i=1}^{k} r_i = 465 = 30(31)/2$. The complete test of hypothesis is given in Table 10-12. The calculated value of $H = 37.53$ leads to a rejection of the null hypothesis at $\alpha = .05$. We now know that at least one of these samples differs from the other two. The obvious questions are: Does only one sample differ from the other two, and if so, which one? and Are all three samples from different populations? We already know that intersections with yield signs have lower concentrations of carbon monoxide than intersections with stop signs. To answer this question completely, however, Mann-Whitney tests between each pair of samples are required.

10.3. Goodness-of-Fit Tests

It is frequently necessary in empirical research to test the hypothesis that a random variable has a specified theoretical probability distribution. This hypothesis may be of interest for two reasons: to confirm a theory or a model or to determine whether the assumptions of a particular statistical inference procedure are satisfied. As an example of the first case, we may wish to test the randomness of a particular point pattern by using the Poisson probability distribution. Because virtually all classical or parametric statistical inference procedures make the assumption of normality, it is often necessary to check this assumption before one proceeds. This is an example of the second reason for testing hypotheses about the distribution of a random variable. Goodness-of-fit tests are used for these purposes. The label *goodness of fit* is used because the purpose of the test is to see whether a particular probability distribution is a good model for the sampled population. This judgment is made according to whether the probability distribution specified in H_0 provides a good fit for the sample data. The two most commonly used tests are the χ^2 and Kolmogorov-Smirnov tests.

Chi-Square Test

The χ^2 test is a widely used goodness-of-fit test. The basic structure of the test is shown in Table 10-13. The sample data are nominal. Counts are obtained for each of the k classes. Denote the observed number or count of observations of the ith class f_i. The sum of the observed frequency counts over the k classes equals the total sample size n:

$$\sum_{i=1}^{k} f_i = n \tag{10-7}$$

By using the probability distribution specified in the null hypothesis, a set of expected frequencies are determined. Let the expected number of observations in the ith class be denoted F_i. Of course,

$$\sum_{i=1}^{k} F_i = n \tag{10-8}$$

The χ^2 test is based on a comparison of the observed frequencies f_i and the expected frequencies F_i. The closer the set of F_i to f_i, the more likely it is that the distribution of the sample reflects the probability distribution specified in the null

TABLE 10-13
Structure of a χ^2 Goodness-of-Fit Test

Class	Observed frequency	Expected frequency with H_0	Relative squared difference
1	f_1	F_1	$(f_1 - F_1)^2/F_1$
2	f_2	F_2	$(f_2 - F_2)^2/F_2$
....	....	....	....
k	f_k	F_k	$(f_k - F_k)^2/F_k$
Total:	n	n	$X^2 = \sum_{i=1}^{k} (f_i - F_i)^2/F_i$

hypothesis. The farther apart the observed and expected frequencies, the less likely H_0 is true. The test statistic is the sum of the relative squared differences:

$$X^2 = \sum_{i=1}^{k} \frac{(f_i - F_i)^2}{F_i} \tag{10-9}$$

First, note that when $f_i = F_i$ for any class, the contribution of that class to X^2 is zero. Second, note that larger differences are penalized more severely than smaller differences since the difference is squared in the numerator of (10-9). The denominator is used to standardize these errors. Suppose, for example, we expect a count of 40 in some class and we observe 38. This difference of 2 leads to a $(38 - 40)^2/40 = .1$ contribution to X^2. However, a difference of 2 from an observed count of 6 to an expected count of 4 leads to a $(6 - 4)^2/4 = 1$ contribution to X^2. Even though the difference of 2 is the same in each case, the latter situation causes an increase in X^2 ten times as large.

When n is large, the test statistic X^2 is distributed approximately as a χ^2 random variable with $k - m - 1$ degrees of freedom. One degree of freedom is lost because the observed and expected counts both sum to the sample size n. Also, 1 degree of freedom is lost for each parameter of the probability distribution specified in H_0 which must be estimated from sample data. The number of parameters to be estimated m varies with the distribution specified under H_0. For the Poisson distribution, $m = 1$ since only the parameter λ need be estimated. For the normal, two parameters must be estimated, μ and σ. If the null hypothesis includes the values of the required parameters, then no parameters need be estimated from the sample. Thus, there are two cases for the relevant hypotheses:

1. H_0: The distribution is normal. H_A: The distribution is not normal. Degrees of freedom are $k - m - 1$.
2. H_0: The distribution is normal with mean $\mu = a$ and variance $\sigma^2 = b$. H_A: The distribution is not normal with mean a and variance b. Degrees of freedom are $k - 1$.

Case 1 is more common. Our interest lies only in whether we can safely assume the distribution is normal. In case 2, however, we are interested in whether the distribution is normal *and* in the particular parameters for the mean and variance. If we reject the null hypothesis in the second case, we are *not* necessarily saying the distribution is not normal. We are saying only that it is not normal with the given parameters. The distribution may be normal with mean $= c$ and variance $= d$.

The sample size required for the χ^2 approximation depends on the expected frequencies under H_0, or F_i. As a general rule, the expected frequency in each class should exceed 2, and most should exceed 5. If more than 20 percent of the classes have expected frequencies less than 5, the χ^2 approximation is not valid. If there are several classes with $F_i < 5$ or even one with $F_i < 2$, it is usual to group adjacent categories so that the combined expected frequency exceeds 5. For example, if the expected frequencies of the first three classes in a problem are

i	1	2	3
F_i	1.27	3.88	5.99

they can be combined to

i	1	2
F_i	5.15	5.99

so that the required expected frequency in each class exceeds 5. Ordinarily, only adjacent classes are grouped in this way. In almost all goodness-of-fit tests, there is an explicit ordering of the categories according to the value of the variable. It would thus make little sense to group categories 3 and 9 to satisfy the minimum frequency requirement. The principal reason for combining categories to achieve a minimum expected frequency of 5 is the extreme sensitivity of the X^2 test statistic to small

---

TABLE 10-14
Chi-Square Goodness-of-Fit Test

Specification
The χ^2 test compares an observed set of *frequencies* in k classes f_i, $i = 1, 2, \ldots, k$ to a set of expected frequencies F_i, $i = 1, 2, \ldots, k$. It is required that $\Sigma_{i=1}^{k} f_i = \Sigma_{i=1}^{k} F_i = n$. It is assumed that the sample is random and reasonably large.

Hypotheses
H_0: The population has a specific form, for example, normal or normal with mean μ and variance σ^2.
H_A: The population distribution is not of the form specified in H_0.

Note: In some instances H_0 can specify the specific parameters of the population. In other cases, only the family of the distribution is specified, and the parameters are estimated from the sample data.

Test statistic

$$X^2 = \sum_{i=1}^{k} \frac{(f_i - F_i)^2}{F_i}$$

With $k - m - 1$ degrees of freedom, m is the number of parameters estimated from sample data.

Decision rule
If $X^2 \leq \chi^2(1 - \alpha, k - m - 1)$, conclude H_0. If $X^2 > \chi^2(1 - \alpha, k - m - 1)$, conclude H_A.

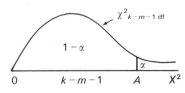

where $A = \chi^2(1 - \alpha, k - m - 1)$

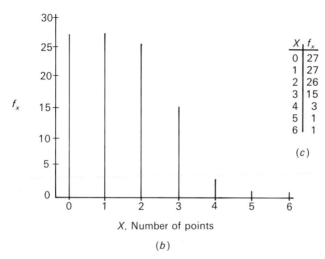

FIGURE 10-3. Maps for Example 10-6.

differences when F_i is small. For example, a difference of $f_i - F_i = 4$ when $F_i = 1$ leads to a contribution of 16 to the statistic X^2. The combination of categories has one drawback. Each time two categories are combined, there is a loss of a degree of freedom.

The χ^2 goodness-of-fit test is summarized in Table 10-14. Two examples illustrate the application of this test. Example 10-6 utilizes a *discrete* probability distribution, the Poisson, in a test of point-pattern randomness. It also illustrates the procedure for pooling classes to yield $F_i > 5$. Example 10-7 represents one of the most common uses of the χ^2 test, testing a variable for normality prior to the application of a parametric statistical test. Also, this problem illustrates a useful method of applying the goodness-of-fit test to continuous probability distribution.

EXAMPLE 10-6. Let us consider a problem in point-pattern analysis. Specifically, let us test whether the distribution of points in the map of Figure 10-3 is random. To do this, we must see whether the frequency distribution of points by quadrat can be considered to be Poisson with $\lambda = 1.5$. The observed and expected numbers of points per quadrat are as follows:

Points per quadrat	Observed number of quadrats	Expected number of quadrats
0	27	22.31
1	27	33.47
2	26	25.10
3	15	12.55
4	3	4.71
5	1	1.41
6+	1	0.45

To apply the χ^2 test, we require frequencies greater than 2. To satisfy this assumption, the final three categories can be pooled into a single class of 4+ points. This pooling is illustrated in the calculations for X^2 of Table 10-15. The original seven classes are now reduced to five by this method. The expected frequency of the pooled class 4+ is now 6.57. The formal test of hypothesis is given in Table 10-16. The calculated test statistic of $X^2 = 3.103$ is not significant at $\alpha = .05$, which leads to the acceptance of H_0. The point pattern is random.

EXAMPLE 10-7. Consider again the 50 DO values given in Table 2-1. Suppose we now wish to test whether this variable is a normal random variable. First, it is necessary to specify a set of classes. Since this is a continuous variable, the specification of appropriate classes is not as easy as it is for a discrete random variable. One efficient way to define classes for continuous random variables is to divide the area under the density function into k intervals, each with a probability of occurrence of $1/k$. Figure 10-4 illustrates this procedure for a normal distribution and $k = 6$ classes. The cumulative z scores defining the class bounds are shown at each class limit. The first class includes all observations with a standardized value less than $z_{.167} = -.97$. The second class includes all observations between $-.97$ and $z_{.333} = -.43$, and so on.

TABLE 10-15
Calculations for χ^2 Test for Example 10-6

Class	Number of points X	Observed frequency f_i	Probability under H_0	Expected frequency F_i	$f_i - F_i$	$(f_i - F_i)^2/F_i$
1	0	27	.2231	22.31	4.69	0.99
2	1	27	.3347	33.47	−6.47	1.25
3	2	26	.2510	25.10	6.90	0.03
4	3	15	.1255	12.55	2.45	0.48
5	4 $\rbrace$ ≥4	3 $\rbrace$ 5	.0657	6.57	−1.57	0.38
	5	1	1.000	100.00	6.000	$X^2 = 3.13$
	6	1				

TABLE 10-16
Chi-Square Test for Example 10-6

Hypotheses
H_0: The distribution of points in the map of Figure 10-3 is Poisson; that is, it is a random point pattern.
H_A: The distribution of points is not Poisson.

Test statistic
A χ^2 goodness-of-fit test is used. Test statistic is X^2 with $k - 2 = 5 - 2 = 3$ degrees of freedom. The parameter $\lambda = 1.5$ is estimated from the map.

Level of significance
$\alpha = .05$.

Decision rule
$A = \chi^2(.95, 3) = 7.82$
If $X^2 < 7.82$, accept H_0; $X^2 \geq 7.82$, accept H_A.

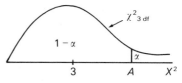

Computation of test statistic
From Table 10-14, $X^2 = 3.13$.

Decision
Since $X^2 < 7.82$, accept H_0.

It is easy to convert these z scores to values of any variable X by using the identity $z = (x - \bar{x})/s$. For the DO lake data we know (from Chapter 2) that $\bar{x} = 5.58$ and $s = .6392$. Therefore, the value of X (or DO value) corresponding to $z = -.97$ can be determined by substitution:

$$-.97 = \frac{x - 5.58}{.6392} \quad \text{or} \quad x = 4.96 \quad (10\text{-}10)$$

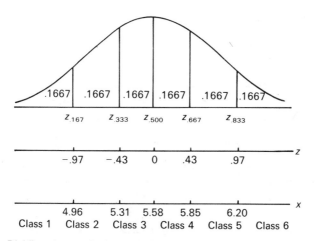

FIGURE 10-4. Dividing the standard normal distribution into six intervals of equal probability.

TABLE 10-17
Testing Lake DO Values for Normality by Using χ^2

Class	Bounds in X	$f_{j'}$ observed frequency	Probability under H_0	$F_{j'}$ expected frequency	$f_i - F_i$	$(f_i - F_i)^2/F_i$
1	< 4.96	9	.1667	8.333	0.667	.05
2	4.96–5.31	6	.1667	8.333	−2.333	.65
3	5.32–5.57	6	.1667	8.333	−2.333	.65
4	5.58–5.85	14	.1667	8.333	5.667	3.85
5	5.86–6.20	7	.1667	8.333	−1.333	.21
6	> 6.20	8	.1667	8.333	−0.333	.01
		50	≈ 1.000	≈ 50.0	0.00	$\chi^2 = 5.43$

All other class limits can be found in the same way. The classes defined by this method are illustrated in Figure 10-4 and Table 10-17. The observed frequencies shown in Table 10-17 are generated by using the DO values of Table 2-1. The probabilities for each class are $1/k = \frac{1}{6} = .1667$, and the expected frequencies are $.1667$ times the sample size of $n = 50$, or 8.333. The calculations for the test statistic X^2 are given in the last two columns.

The procedure for dividing the area under a density function into k classes of equal probability can be utilized for any value of k and any probability density function. How should the value of k be chosen? The greater the value of k, the more degrees of freedom for the statistical test, so it is best to maximize the number of classes. The only constraint is that the expected frequencies exceed 5. The easiest way of meeting the minimum frequency requirement is to choose k so that n/k exceeds 5 by as small an amount as possible. For $n = 20$ observations, it is possible to use 4 classes; for $n = 63$, 12 classes; for $n = 100$, 20 classes; and so on. For the lake DO data, the maximum value of k is $\frac{50}{5} = 10$ (see Problem 10).

Kolmogorov-Smirnov Test

An alternative, and frequently more powerful, goodness-of-fit test is the *Kolmogorov-Smirnov test*. This test can be used for *continuous* random variables. For discrete random variables, the test gives only approximate results. Let us begin by reviewing the concept of a *cumulative probability distribution function* of a random variable. As defined in Chapter 5, the cumulative probability distribution function of a random variable X is defined as

$$F(x) = P(X \leq x) \qquad -\infty < x < +\infty \qquad (10\text{-}11)$$

Figure 10-5 illustrates the cumulative probability distribution functions for the uniform continuous and standard normal distributions. The Kolmogorov-Smirnov test is based on examining the difference between this cumulative probability function $F(x)$ and the *sample cumulative distribution function* $S(x)$. The sample cumulative distribution function $S(x)$ specifies the proportion (or relative frequency) of *sample* values less than or equal to x for each value of x in the sample. It is an unsmoothed

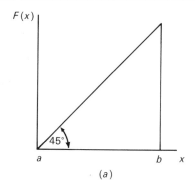

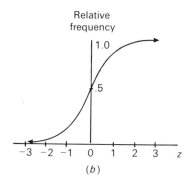

FIGURE 10-5. Cumulative probability functions for the (*a*) uniform continuous and (*b*) standard normal.

ogive. The method for constructing the sample cumulative distribution function for any variable is best illustrated with a small sample. Suppose the sample consists of the following values: 18, 10, 5, 2, 4, 20, 2, 8, 1, 12. First, order the observations: 1, 2, 2, 4, 5, 8, 10, 12, 18, 20. then construct a cumulative relative frequency table of the form

x	1	2	4	5	8	10	12	18	20
$S(x)$	.1	.3	.4	.5	.6	.7	.8	.9	1.0

Of the 10 observations, only one is less than or equal to one, so that $S(1) = .1$. There are three observations in the sample less than or equal to 2, so that $S(2) = .3$. The remainder of the table is constructed in this way. The step-function form of $S(x)$ is clearly seen in Figure 10-6.

The key question in the Kolomogorov-Smirnov test is, Are the differences between the sample $S(x)$ and the cumulative probability distribution function $F(x)$ sufficiently large that they are unlikely to have occurred because of chance fluctuations? The test statistic is

$$D = \max_{x} |S(x) - F(x)| \qquad (10\text{-}12)$$

In other words, D is the largest absolute deviation between the cumulative sample function and the cumulative probability function. Tables of D for $\alpha = .01, .05,$ and

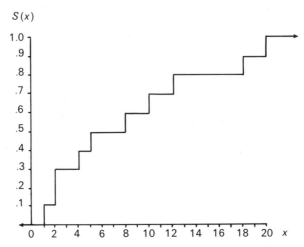

FIGURE 10-6. Graph of cumulative sample distribution function.

.10 levels of significance are given in the Appendix. The Kolomogorov-Smirnov test is summarized in Table 10-18.

EXAMPLE 10-8. A cartographer records the times taken by 10 subjects to complete an experiment in which subjects are asked to estimate data values illustrated as proportional circles. The times in seconds are

| 65 | 62 | 50 | 80 | 83 | 64 | 67 | 62 | 66 | 68 |

Past research indicates that these times tend to be normally distributed with a mean of 60 s and a standard deviation of 10 s. Use the Kolmogorov-Smirnov test to determine whether it is likely that these times could be a random sample from a normal distribution with these parameters.

TABLE 10-18
Kolmogorov-Smirnov Test

Specification
A cumulative distribution function $S(x)$ is generated from the random sample of n observations. This cumulative distribution function is compared to the cumulative probability distribution function of some hypothesized distribution, $F(x)$, with known parameters.

Hypotheses
H_0: $S(x) = F(x)$, the sample is from the hypothesized probability distribution with known parameters.
H_A: $S(x) \neq F(x)$, it is not.

Test statistic
$D = \max_{x} |S(x) - F(x)|$ sample size $= n$

Decision rule
The table of critical values for D is given in the Appendix. Reject H_0 if D exceeds the tabulated value at the chosen level of significance. Otherwise, do not reject H_0. There are n degrees of freedom.

TABLE 10-19
Calculation of D for Example 10-8

z	$S(z)$	$F(z)$	$|S(z) - F(z)|$
−1.0	.1000	.1587	.0587
0.2	.3000	.5793	.2793
0.4	.4000	.6554	.2254
0.5	.5000	.6915	.1915
0.6	.6000	.7257	.1257
0.7	.7000	.7580	.0580
0.8	.8000	.7881	.0119
2.0	.9000	.9772	.0772
2.3	1.0000	.9893	.0107

$$D = \max_z |S(z) - F(z)| = .2793$$

First, standardize the values, using $z_i = (x_i - \mu)/\sigma$. The 10 z scores are

0.5   0.2   −1.0   2.0   2.3   0.4   0.7   0.2   0.6   0.8

and when they are reordered by value, they are

−1.0   0.2   0.2   0.4   0.5   0.6   0.7   0.8   2.0   2.3

The cumulative sample distribution function is therefore

z	−1.0	0.2	0.4	0.5	0.6	0.7	0.8	2.0	2.3
$S(z)$	.1	.3	.4	.5	.6	.7	.8	.9	1.0

The calculations required for the test statistic D are given in Table 10-19. The values for $F(z)$ are extracted from the table of the standard normal distribution. The value of $D = .2793$ for $n = 10$ degrees of freedom is not significant at $\alpha = .05$. From the table of the Kolmogorov-Smirnov test statistic in the Appendix, we note that a value of $D = .409$ is requried to reject H_0 at this level of significance. We conclude that the times taken by this sample of 10 subjects could be a random sample from a normal distribution with mean 60 and standard deviation 10. See Figure 10-7.

The Kolmogorov-Smirnov test can also be used with grouped data from any cumulative frequency table of a continuous variable. The values of $S(x)$ are the relative cumulative frequency values at the endpoints of the intervals, and the value of x is simply the upper limit of the interval. The principal disadvantage of using grouped data for a continuous variable is the loss of degrees of freedom. There is only 1 degree of freedom for each class or category rather than each observation. Of course, this must be weighed against the reduced computational load necessary to calculate D. Except perhaps when the computations are being completed by hand, it is advisable to use the original grouped data.

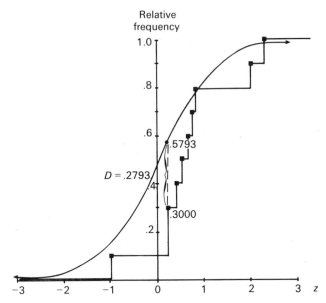

FIGURE 10-7. $F(z)$ and $S(z)$ for Example 10-8.

10.4. Contingency Tables

A *contingency table* is a cross-classified set of frequency counts for two nominally or ordinally scaled variables, each having two or more classes. In Chapter 5, we introduced the use of contingency tables within a discussion of *bivariate random variables* and *bivariate probability functions*. Let us reconsider the relationship between household size and car ownership used as an example in Section 5.5. The contingency table and bivariate probability function relating these two variables are repeated in Table 10-20(*a*) and (*b*). The *test for statistical independence*, or *contingency table test*, is a nonparametric test utilizing the χ^2 distribution to determine whether two variables are statistically independent or a relationship exists between them. Recall that two random variables X and Y are said to be statistically independent if

$$P(x, y) = P(x)P(y) \tag{10-13}$$

holds for all (x, y) pairs. For the variables called car ownership and household size, this equation is violated for at least one (x, y) pair. Note that

$$P(x = 3) = .26 \quad\quad P(y = 1) = .25 \quad\quad P(x = 3, y = 1) = .10$$

Therefore $P(x = 3, y = 1) \neq P(x = 3)P(y = 1)$ since $.10 \neq (.26)(.25)$. This result presupposes that Table 10-20(*b*) is the bivariate probability function for a *population*. Suppose we consider the data to be a sample from a *bivariate multinomial population*. It is bivariate since it concerns two different variables. It is multinomial since each nominal variable has more than two possible classes or values. The test for statistical independence of two variables in a bivariate multinomial population uses the χ^2

TABLE 10-20
(a) Contingency Table and (b) Bivariate Probability Function for Household Size and Car Ownership Data

		Car ownership				
		0	1	2	3	Total
Household size	2	10	8	3	2	23
	3	7	10	6	3	26
	4	4	5	12	6	27
	5	1	2	6	15	24
	Total	22	25	27	26	100

(a)

		Car ownership				
		0	1	2	3	Marginal
Household size	2	.10	.08	.03	.02	.23
	3	.07	.10	.06	.03	.26
	4	.04	.05	.12	.06	.27
	5	.01	.02	.06	.15	.24
	Marginal	.22	.25	.27	.26	1.00

(b)

distribution. It is a simple extension of the goodness-of-fit test described in Section 10.3.

To fix ideas, consider Table 10-21(a). Let π_{ij} denote the population proportion in the ijth cell, and let π_i^r and π_j^c be the marginal probability distributions for the random variables X and Y, respectively. The null hypothesis in this test is that the two variables X and Y are statistically independent, or

$$H_0: \pi_{ij} = \pi_i^r \pi_j^c \qquad i = 1, 2, \ldots, r; \; j = 1, 2, \ldots, c \qquad (10\text{-}14)$$

where the variable X has r categories or classes and the variable Y has c categories. The alternate hypothesis is

$$H_A: \pi_{ij} \neq \pi_i^r \pi_j^c \qquad \text{for at least one } i, j \text{ pair} \qquad (10\text{-}15)$$

As usual, this population is unobserved. From a sample, such as the contingency table shown in Table 10-20(a), the *sample* proportions p_{ij} for each cell are determined as well as the row and column marginal probability functions p_i^r and p_j^c. The form of the sample tableau is illustrated in Table 10-21(b). Now, it is easy to check for statistical independence in the sample by comparing the value p_{ij} to $p_i^r p_j^c$. Any discrepancies between these values suggests that the two categories may not be independent in the population. But since this is only a sample and so is subject to random or chance fluctuations, the equation may not hold *exactly* for all i, j pairs in the sample. The inferential question boils down to this: How much deviation from (10-14) can

TABLE 10-21
(a) Bivariate Multinomial Population and (b) Sample

X \\ Y	1	2	$\cdots$	C	Total
1	π_{11}	π_{12}	$\cdots$	π_{1c}	π_1^r
2	π_{21}	π_{22}	$\cdots$	π_{2c}	π_2^r
r	π_{r1}	π_{r2}	$\cdots$	π_{rc}	π^r
Total	π_1^c	π_2^c	$\cdots$	π_c^c	1.00

(a)

X \\ Y	1	2	$\cdots$	C	Total
1	p_{11}	p_{12}	$\cdots$	p_{1c}	p_1^r
2	p_{21}	p_{22}	$\cdots$	p_{2c}	p_2^r
r	p_{r1}	p_{r2}	$\cdots$	p_{rc}	p_r^r
Total	p_1^c	p_2^c	$\cdots$	p_c^c	1.00

(b)

there be in the sample before we can say, at a certain level of significance, that the two variables are not statistically independent?

The χ^2 test for contingency tables does not directly utilize the sample proportions given in Table 10-21(b) in the calculation of the test statistic X^2. Instead, the test is based on a comparison of the *observed frequencies* [as in Table 10-20(a)] and a set of *expected frequencies* which should occur if the null hypothesis is true. The underlying idea behind the test is the same as in the goodness-of-fit test. If H_0 is true, then the observed and expected frequencies will be quite similar; if it is not true, then there will be substantial differences between the sets of frequencies.

The data used in the contingency table test are the raw frequencies in a sample or observed contingency table:

Variable Y
class

		1	2	$\cdots$	c	Total
	1	f_{11}	f_{12}	$\cdots$	f_{1c}	R_1
	2	f_{21}	f_{22}	$\cdots$	f_{2c}	R_2
Variable X						
class	r	f_{r1}	f_{r2}	$\cdots$	f_{rc}	R_r
	Total	C_1	C_2	$\cdots$	C_c	n

Here f_{ij} is the observed frequency in cell i, j, R_i is the sum of the frequencies in row i, C_j is the sum of the frequencies in column j, and n is the number of observations. The sum of the row frequencies $\Sigma_{i=1}^{r} R_i$ equals the sum of the column frequencies $\Sigma_{j=1}^{c} C_j$ which equals the total number of observations n.

To compute the expected frequencies F_{ij} under H_0, use the relation

$$F_{ij} = \frac{R_i C_j}{n} \qquad (10\text{-}16)$$

This shortcut formula uses the respective row and column totals for cell i, j and the total number of observations n. For example $F_{11} = R_1 C_1/n$. For cell 1, 1 of Table 10-20(a), we compute $F_{11} = 23(22)/100 = 5.06$. It is instructive to see how this simple formula relates to the null hypothesis of statistical independence. First, note that

$$p_{ij} = \frac{f_{ij}}{n} \qquad p_i^r = \frac{R_i}{n} \qquad p_j^c = \frac{C_j}{n}$$

Therefore, if the null hypothesis is true and $p_{ij} = p_i^r p_j^c$, then

$$p_{ij} = \frac{R_i}{n} \frac{C_j}{n} \qquad (10\text{-}17)$$

To compute the *expected frequency* F_{ij}, we multiply the joint probability by the sample size n:

$$F_{ij} = n \frac{R_i}{n} \frac{C_j}{n} = \frac{R_i C_j}{n} \qquad (10\text{-}18)$$

This is the same as (10-16).

The test statistic is

$$X^2 = \sum_{i=1}^{r} \sum_{j=1}^{c} \frac{(f_{ij} - F_{ij})^2}{F_{ij}} \qquad (10\text{-}19)$$

which is obtained by summing over *all* classes. The double summation is required since each class is identified by a row and column subscript. When H_0 is true, the sampling distribution of X^2 is approximately χ^2 with $(r-1)(c-1)$ degrees of freedom. The degrees of freedom of the distribution are obtained as follows. There are rc classes. The number of estimated parameters required to compute F_{ij} is $r - 1 + c - 1$. Since both sets of marginal probabilities must sum to 1.00, only $r - 1$ of the row probabilities and $c - 1$ of the column probabilities are needed to compute F_{ij}. The available degrees of freedom are $rc - $ (number of estimated parameters) $- 1$, or $rc - (r - 1 + c - 1) - 1 = (r-1)(c-1)$. Just as in the goodness-of-fit test, n must be reasonably large and all $F_{ij} > 2$ and most $F_{ij} \geq 5$. The test is summarized in Table 10-22.

EXAMPLE 10-9. Let us complete the example of car ownership and household size. First, compute the expected frequencies F_{ij}, by using (10-16). These results, along with the calculation of the test statistic X^2, are given in Table 10-23. Note that all expected frequencies exceed 5, so that the χ^2 approximation is valid for these data. If this condition is violated, then it is necessary to collapse categories

TABLE 10-22
Chi-Square Test for Independence in Contingency Tables

Specification
A total of n observations of two nominal/ordinal variables are cross-classified in a contingency table of dimensions $r \times c$. The observed frequency in the ijth cell is denoted f_{ij}. Expected frequencies for these cells are calculated from

$$F_{ij} = \frac{R_i C_j}{n}$$

Here R_i is the observed frequency count of the ith row, and C_j is the observed frequency count of the jth column.

Hypotheses
H_0: $\pi_{ij} = \pi_i^r \pi_j^c$ $i = 1, 2, \ldots, r;\ j = 1, 2, \ldots, c$
The variables are statistically independent.

H_A: $\pi_{ij} \neq \pi_i^r \pi_j^c$ for at least one ij pair
The variables are not statistically independent.

Test statistic

$$X^2 = \sum_{i=1}^{r} \sum_{j=1}^{c} \frac{(f_{ij} - F_{ij})^2}{F_{ij}} \quad \text{with } (r-1)(c-1) \text{ degrees of freedom}$$

Decision rule
Reject H_0 if $X^2 > A$, where $A = \chi^2[1 - \alpha, (r-1)(c-1)]$.

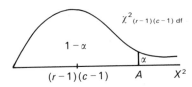

so that the approximation can be used. Since $r = 4$ and $c = 4$, there are $(r-1)(c-1) = (4-1)(4-1) = 9$ df for χ^2 when H_0 holds. For $\alpha = .05$ we require $\chi^2(.95, 9) = 16.92$ in order to reject H_0. Because $X^2 = 37.17$ exceeds 16.92, we conclude H_A—that household size and car ownership are statistically dependent. A listing of the SAS program and computer output for this problem is provided in Appendix 10A.

EXAMPLE 10-10. Four candidates A, B, C, and D are running for office in a city divided into four major electoral districts, North (N), South (S), East (E), and West (W). A pollster randomly samples 100 voters in the city and asks them to express their first preference among the candidates. The results are as follows:

		Candidate				
		A	B	C	D	Total
	N	5	9	5	10	29
	S	8	6	7	1	22
Electoral district	E	0	11	10	4	25
	W	4	8	3	9	24
	Total	17	34	25	24	100

Do these data suggest that support for the candidates depends on electoral district? The expected frequencies are calculated by using (10-16):

		Candidate			
		A	B	C	D
	N	4.93	9.86	7.25	6.96
Electoral district	S	3.74	7.48	5.50	5.28
	E	4.25	8.50	6.25	6.00
	W	4.08	8.16	6.00	5.76

Unfortunately, $\frac{4}{16}$, or 25 percent, of the cells in the table have expected frequencies less than 5. There are three options. First, one is tempted to ignore the problem. Cell 1, 1 has an expected frequency of almost 5, and $\frac{3}{16}$ of the cells represents only 18.75 percent of the total. This is inadvisable and, as we shall see, could lead to a false rejection of H_0. The second option is to collapse the categories so that the requirement of $F_{ij} \geq 5$ is met. It makes little sense to group any of the candidates, although we might be able to group them on the basis of their perceived location along some political spectrum such as left wing–right wing. Or we could group the electoral districts, say into two categories of north-south and east-west. Although this would solve the technical problem, it is a poor alternative since we know geographical aggregation often masks data variation.

The last and probably best alternative is to drop candidate A. It is not a perfect

TABLE 10-23

Calculation of Expected Frequencies and Test Statistic for Example 10-9

Expected frequency			
$\frac{23(22)}{100} = 5.06$	$\frac{23(25)}{100} = 5.75$	$\frac{23(27)}{100} = 6.21$	$\frac{23(26)}{100} = 5.98$
$\frac{26(22)}{100} = 5.72$	$\frac{26(25)}{100} = 6.50$	$\frac{26(27)}{100} = 7.02$	$\frac{26(26)}{100} = 6.76$
$\frac{27(22)}{100} = 5.94$	$\frac{27(25)}{100} = 6.75$	$\frac{27(27)}{100} = 7.29$	$\frac{27(26)}{100} = 7.02$
$\frac{24(22)}{100} = 5.28$	$\frac{24(25)}{100} = 6.00$	$\frac{24(27)}{100} = 6.48$	$\frac{24(26)}{100} = 6.24$

$$X^2 = \frac{(10 - 5.06)^2}{5.06} + \frac{(8 - 5.75)^2}{5.75} + \cdots + \frac{(6 - 6.48)^2}{6.48} + \frac{(15 - 6.24)^2}{6.24}$$

$$= 4.82 + 0.88 + \cdots + 0.04 + 12.30 = 37.17$$

solution, but is solves the technical problem without any arbitrary judgmental grouping. The new observed and expected frequencies are as follows:

		Observed					Expected		
	B	C	D	Total		B	C	D	Total
N	9	5	10	24	N	9.83	7.22	6.94	24
S	6	7	1	14	S	5.73	4.22	4.05	14
E	11	10	4	25	E	10.24	7.53	7.23	25
W	8	3	9	20	W	8.19	6.02	5.78	20
Total	34	25	24	83	Total	34	25	24	83

Only two-twelfths of the cells have expected frequencies less than 5. To complete the test, we compute the test statistic X^2 from (10-19):

$$X^2 = \frac{(9-9.83)^2}{9.83} + \frac{(5-7.22)^2}{7.22} + \cdots + \frac{(9-5.78)^2}{5.78}$$

$$= .07 + .68 + \cdots + 1.79 = 11.93$$

Since $X^2 = 11.93$ does not exceed $\chi^2(.95, 6) = 12.59$, we conclude H_0—voter preference for candidates B, C, and D is statistically independent of electoral districts. Interestingly, this conclusion is at odds with what would have been concluded had we ignored the minimum frequency requirement for the complete data set with candidates A, B, C, and D.

10.5. Tests for Randomness

Randomness is a key assumption for all the statistical tests we have discussed. Whether this assumption is reasonable for any set of data may be very difficult to decide—particularly when a random sampling procedure has not been used to generate the data. For example, suppose we wish to ascertain whether the sequence of annual rainfall totals over the past 30 years at some location has been generated by a random process or whether the sequence contains a trend. Even if the location was chosen randomly, the data may still not be random since the individual rainfall values are not selected from a random number table or similar device. Nonparametric methods can be used to test for randomness of a sample, even after it has been collected and even if it has not been collected by using a randomization device. As an example of a test for randomness we present the number-of-runs test.

Number-of-Runs Test

The number-of-runs test is based on examining the number of *runs* in a sequence of items. A *run* is defined as an unbroken sequence of like items surrounded by unlike items. Let us use the convention of denoting the two types of items in the sequence by + and −. The sequence of items [+ + + + − − − −] contains two runs. The first run is the first four pluses, and the second run is the four minuses. As another example, the sequence [+ − + − − − + −] contains six runs. Each of the first

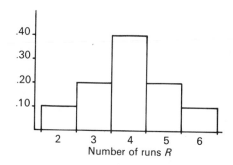

FIGURE 10-8. Probability distribution for number of runs R for $n = 6$, $n_1 = n_2 = 3$.

three signs is a run of length 1, the middle three minuses represent a fourth run, and the last two signs are the fifth and sixth runs. The total number of runs in such a sequence is a good indicator of randomness. If there are too few runs, it is possible there is some clustering pattern and the series is probably not random. If, however, there are too many runs, then there may be a repeating or alternating pattern uncharacteristic of random series.

The principle behind the number-of-runs test is best illustrated by using a short series as an example. Suppose the entire series consists of $n = 6$ observations of which $n_1 = 3$ are pluses and $n_2 = 3$ are minuses. Table 10-24 lists the 20 different arrangements of these pluses and minuses. Four runs is the most likely outcome, occurring in 8 of the 20 sequences. Suppose the observed sequence is $[+ + + - - -]$. This

TABLE 10-24
Arrangements of Signs for $n_1 = n_2 = 3$

Sequence	Number of runs	Number of realizations
+ + + − − − − − − + + +	2	2
+ − − − + + + + − − − + − + + + − − − − + + + −	3	4
+ − + + − − − + − − + + − + + − − + + − − + + − − − + + − + + + − − + − − − + − + + + + − + − −	4	8
+ − + − − + + − − + − + − + + − + − − + − + + −	5	4
+ − + − + − − + − + − +	6	2

series has 2 runs. Is the number of runs in this sequence indicative of nonrandomness—or can the small number of runs be attributed to chance? The sampling distribution for the number of runs for a series of $n = 6$ observations with $n_1 = n_2 = 3$ is illustrated in Figure 10-8. This sampling distribution is compiled from Table 10-24. Although the number of runs is really a discrete distribution, it has been illustrated as a conventional histogram. From this distribution we see there is a 10 percent chance (2 out of 20) of two runs occurring in a random sequence under these conditions.

For small values of n_1, n_2, and n it is feasible to construct the sampling distribution of the number of runs in this way. For longer series this is impractical. The sampling distribution of the number of runs R has mean and variance given by, respectively,

$$E(R) = \frac{2n_1n_2}{n_1 + n_2} + 1 \tag{10-20}$$

and

$$\sigma^2(R) = \frac{2n_1n_2(2n_1n_2 - n_1 - n_2)}{(n_1 + n_2)^2(n_1 + n_2 - 1)} \tag{10-21}$$

Moreover, this sampling distribution is closely approximated by the normal distribution, provided n_1, $n_2 \geq 10$. The various hypotheses are summarized in Table 10-25. A one-tail test can be used if we assert there is clustering in the pattern (for example, $[+ + + + - - - -]$) or if there is a repeating pattern ($[+ - + - + - + -]$). Each of these patterns if found at one extreme of the sampling distribution. A two-tail test can also be used.

As an example, consider the 30-year sequence of annual rainfall at some hypothetical climatic station. The annual totals are listed in Table 10-26. Does the process underlying annual rainfall have a decreasing trend? The first step is to divide the observations into two categories, plus and minus. For an interval-scaled variable, it is easiest to define the two classes with respect to the median rainfall. The median rainfall over the 30-yr period is 58 in. All observations above 58 are given a plus and all those below a minus. What is to be done with the observations with value 58? It turns out that the number of runs in this example is not affected by the sign assigned to these three observations. The runs in the 30-yr sequence are illustrated in Table 10-27. Consider first the observation of year 1961 with a rainfall of 58 in. If we assign it a plus sign, the net effect is to make run 1 longer and run 2 shorter. There is no change in the number of runs. Any "tied" value which is surrounded by two observations of opposite sign is *noncritical*. No matter which sign is applied to the observation, the number of runs remains the same.

Suppose, however, that observation 1954 had a rainfall of 58 in. If it is assigned a plus sign, then run 1 remains as it is and the entire series has 4 runs. If it is assigned a minus sign, then run 1 is broken into three separate runs and the entire series has 6 runs! This may be a significant change. The best method of handling such *critical* ties is to calculate the test statistic R twice: once by assigning the sign most conducive

TABLE 10-25
Number-of-Runs Test

Specification
The observations are taken in the order in which they are generated. Each observation below the median is assigned a minus sign, and each observation above the median is assigned a plus sign. This procedure generates a new series of n signs, n_1 of which are positive and n_2 of which are negative. A run is defined as an unbroken series of plus or minus signs.

Hypotheses
H_0: The series of observations is generated by a random process.
H_A: (a) The series is not generated by a random process (two-tail).
　　(b) The series is increasing (decreasing) (one-tail).
　　(c) The series has a repeating cyclical pattern (one-tail).

Test statistic
The total number of runs of like items in the observation series is R.

Decision rule
Reject H_0 if
$$R > \left(\frac{2n_1 n_2}{n_1 + n_2} + 1\right) + z_{\alpha/2}\sqrt{\frac{2n_1 n_2(2n_1 n_2 - n_1 - n_2)}{(n_1 + n_2)^2(n_1 + n_2 - 1)}}$$
or
$$R < \left(\frac{2n_1 n_2}{n_1 + n_2} + 1\right) - z_{\alpha/2}\sqrt{\frac{2n_1 n_2(2n_1 n_2 - n_1 - n_2)}{(n_1 + n_2)^2(n_1 + n_2 - 1)}}$$
for a two-tail test. For a one-tail test, reject H_0 if
$$R < \left(\frac{2n_1 n_2}{n_1 + n_2} + 1\right) - z_\alpha\sqrt{\frac{2n_1 n_2(2n_1 n_2 - n_1 - n_2)}{(n_1 + n_2)^2(n_1 + n_2 - 1)}}$$
for $H_A(b)$ and reject H_0 if
$$R > \left(\frac{2n_1 n_2}{n_1 + n_2} + 1\right) + z_\alpha\sqrt{\frac{2n_1 n_2(2n_1 n_2 - n_1 - n_2)}{(n_1 + n_2)^2(n_1 + n_2 - 1)}}$$
for $H_A(c)$.

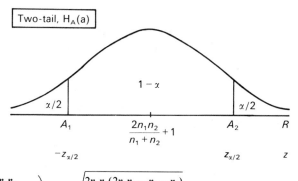

where
$$A_1 = \left(\frac{2n_1 n_2}{n_1 + n_2} + 1\right) - z_{\alpha/2}\sqrt{\frac{2n_1 n_2(2n_1 n_2 - n_1 - n_2)}{(n_1 + n_2)^2(n_1 + n_2 - 1)}}$$

$$A_2 = \left(\frac{2n_1 n_2}{n_1 + n_2} + 1\right) + z_{\alpha/2}\sqrt{\frac{2n_1 n_2(2n_1 n_2 - n_1 - n_2)}{(n_1 + n_2)^2(n_1 + n_2 - 1)}}$$

(*continued*)

TABLE 10-25 (continued)

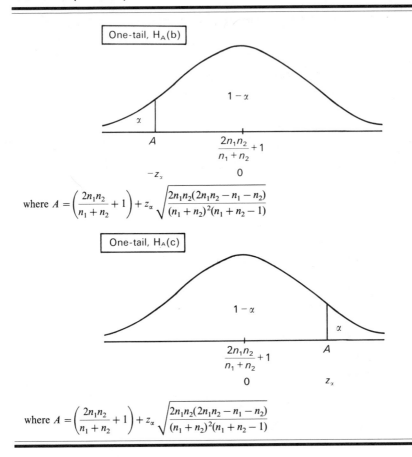

One-tail, $H_A(b)$

$1 - \alpha$

α

A

$\dfrac{2n_1n_2}{n_1+n_2}+1$

$-z_\alpha$ 0

where $A = \left(\dfrac{2n_1n_2}{n_1+n_2}+1\right) + z_\alpha \sqrt{\dfrac{2n_1n_2(2n_1n_2-n_1-n_2)}{(n_1+n_2)^2(n_1+n_2-1)}}$

One-tail, $H_A(c)$

$1 - \alpha$

α

$\dfrac{2n_1n_2}{n_1+n_2}+1$ A

0 z_α

where $A = \left(\dfrac{2n_1n_2}{n_1+n_2}+1\right) + z_\alpha \sqrt{\dfrac{2n_1n_2(2n_1n_2-n_1-n_2)}{(n_1+n_2)^2(n_1+n_2-1)}}$

TABLE 10-26
Annual Rainfall Levels at a Climatic Station

Year	Annual rainfall, in	Year	Annual rainfall, in
1951	66	1966	59
1952	65	1967	58
1953	64	1968	56
1954	65	1969	55
1955	66	1970	53
1956	64	1971	54
1957	63	1972	52
1958	62	1973	51
1959	60	1974	50
1960	59	1975	51
1961	58	1976	49
1962	57	1977	48
1963	58	1978	49
1964	59	1979	47
1965	60	1980	48

TABLE 10-27
Number-of-Runs Test for Climatic Data of Table 10-26

Year	Rainfall, in	Run		Year	Rainfall, in	Run
1951	66	+ ⎤		1966	59	+ ⎤ Run 3
1952	65	+		1967	58	0 ⎦
1953	64	+		1968	56	− ⎤
1954	65	+		1969	55	−
1955	66	+ ⎬ Run 1		1970	53	−
1956	64	+		1971	54	−
1957	63	+		1972	52	−
1958	62	+		1973	51	−
1959	60	+		1974	50	− ⎬ Run 4
1960	59	+ ⎦		1975	51	−
1961	58	0 ⎤		1976	49	−
1962	57	− ⎬ Run 2		1977	48	−
1963	58	0 ⎦		1978	49	−
1964	59	+ ⎤		1979	47	−
1965	60	+ ⎦		1980	48	− ⎦

$$R = 4 \qquad n = 30$$
$$n_1 = 14 \qquad n_2 = 16$$

$$E(R) = \frac{2(14)(16)}{14 + 16} + 1 = 15.93$$

$$\sigma^2(R) = \frac{2(14)(16)[2(14)(16) - 14 - 16]}{(14 + 16)^2(14 + 16 - 1)}$$

$$\sigma(R) = 2.68$$

to rejection of H_0 and the second time by assigning the sign least conducive to rejection of H_0. The PROB-VALUES obtained by these two cases must bound the PROB-VALUE for H_0.

The calculations for the number-of-runs test for the rainfall data are given in Table 10-27. The series has $R = 4$ runs with $n_1 = 14$, $n_2 = 16$, and $n = 30$. For this size problem we calculate $E(R) = 15.93$ and $\sigma(R) = 2.68$, using (10-20) and (10-21). To test the hypothesis that rainfall has been decreasing, a one-tail test is applied. A small value of R sustains this hypothesis. For $\alpha = .05$, $z_{.05} = -1.645$. Hence $A = 15.93 + (-1.645)(2.68) = 11.52$. Since $R = 4 < 11.52$, we conclude $H_A(b)$, that rainfall has been decreasing over the 1951–1980 period at this climatic station.

10.6. Summary

When the standard parametric procedures are not appropriate for analyzing a set of data, nonparametric tests can often be used. These tests are most useful when (1) the level of measurement of the variables concerned is nominal or ordinal and (2) the assumptions of the equivalent parametric procedure are appreciably violated. Violating a parametric test's assumptions causes the test to be inexact. For most, but not all, parametric tests there are some conditions for which even a fairly large violation of an assumption causes only a minor disturbance in the sampling distribution of the test statistic. However, for other assumptions, even a small violation may lead to a significant degree of error in test results. Nonparametric tests are at a significant disadvantage compared with parametric tests when a data set satisfies the assumptions of the appropriate statistical test.

Four types of nonparametric procedures have been presented in this chapter. The nonparametric analogs to the one- and two-sample parametric tests discussed in Chapter 9 are the median test, the Mann-Whitney test, and the Wilcoxon signed-ranks test. The Kruskal-Wallis test is the nonparametric equivalent to one form of the analysis of variance, a powerful parametric procedure. Goodness-of-fit tests are used to test the hypothesis that a random variable follows some specified theoretical probability distribution. The χ^2 and Kolmogorov-Smirnov tests are the most widely used goodness-of-fit tests. The χ^2 test can also be extended to the analysis of dependence between two nominal and/or ordinal variables, summarized in a contingency table.

Nonparametric tests are becoming more and more popular in geographical research. They are decidedly easier to apply than parametric tests. Moreover, these methods are applicable to the sorts of variables collected in questionnaire-based surveys. Such surveys are increasingly common in geographical research. Unfortunately, the simplicity of application of these techniques has made them available to unsophisticated users who sometimes apply them inappropriately or in an uncomprehending, cookbook way. Understanding when to use these methods also requires an understanding of when *not* to use them.

It has not been possible to describe all the nonparametric tests used by geographers in this brief overview of nonparametric methods. Other tests are described in the references cited at the end of this chapter under Further Reading.

Appendix 10A. Contingency Table Analysis Using SAS

SAS Program

```
DATA CAR;
INPUT SIZE CAR @ @;
CARDS;
2 0 2 0 2 0 2 0 2 0 2 0 2 0 2 0 2 0 2 0
2 1 2 1 2 1 2 1 2 1 2 1 2 1 2 1 2 2 2 2
2 2 2 3 2 3 3 0 3 0 3 0 3 0 3 0 3 0 3 0
3 1 3 1 3 1 3 1 3 1 3 1 3 1 3 1 3 1 3 1
3 2 3 2 3 2 3 2 3 2 3 2 3 3 3 3 3 3 4 0
4 0 4 0 4 0 4 0 4 1 4 1 4 1 4 1 4 1 4 2 4 2
4 2 4 2 4 2 4 2 4 2 4 2 4 2 4 2 4 2 4 2
4 3 4 3 4 3 4 3 4 3 4 3 5 0 5 1 5 1 5 2
5 2 5 2 5 2 5 2 5 2 5 3 5 3 5 3 5 3 5 3
5 3 5 3 5 3 5 3 5 3 5 3 5 3 5 3 5 3 5 3
;
PROC PRINT;
OPTIONS LINESIZE=70;
PROC FREQ;
   TABLES SIZE*CAR/ EXPECTED CELLCHI2 CHISQ;
```

SAS Output

```
SAS

TABLE OF SIZE BY CAR

SIZE        CAR

FREQUENCY|
EXPECTED |
CELL CHI2|
 PERCENT |
 ROW PCT |
 COL PCT |       0|       1|       2|       3|  TOTAL
---------+--------+--------+--------+--------+
       2 |     10 |      8 |      3 |      2 |     23
         |    5.1 |    5.8 |    6.2 |    6.0 |
         |4.82285 |.880435 |1.65928 | 2.6489 |
         |  10.00 |   8.00 |   3.00 |   2.00 |  23.00
         |  43.48 |  34.78 |  13.04 |   8.70 |
         |  45.45 |  32.00 |  11.11 |   7.69 |
---------+--------+--------+--------+--------+
       3 |      7 |     10 |      6 |      3 |     26
         |    5.7 |    6.5 |    7.0 |    6.8 |
         |.286434 |1.88462 |.148205 |2.09136 |
         |   7.00 |  10.00 |   6.00 |   3.00 |  26.00
         |  26.92 |  38.46 |  23.08 |  11.54 |
         |  31.82 |  40.00 |  22.22 |  11.54 |
---------+--------+--------+--------+--------+
       4 |      4 |      5 |     12 |      6 |     27
         |    5.9 |    6.8 |    7.3 |    7.0 |
         |.633603 |.453704 |3.04309 |.148205 |
         |   4.00 |   5.00 |  12.00 |   6.00 |  27.00
         |  14.81 |  18.52 |  44.44 |  22.22 |
         |  18.18 |  20.00 |  44.44 |  23.08 |
---------+--------+--------+--------+--------+
       5 |      1 |      2 |      6 |     15 |     24
         |    5.3 |    6.0 |    6.5 |    6.2 |
         |3.46939 |2.66667 |.035556 |12.2977 |
         |   1.00 |   2.00 |   6.00 |  15.00 |  24.00
         |   4.17 |   8.33 |  25.00 |  62.50 |
         |   4.55 |   8.00 |  22.22 |  57.69 |
---------+--------+--------+--------+--------+
TOTAL          22       25       27       26        100
            22.00    25.00    27.00    26.00     100.00

STATISTICS FOR TABLE OF SIZE BY CAR

STATISTIC                        DF     VALUE      PROB
---------------------------------------------------------
CHI-SQUARE                        9     37.170     0.000
LIKELIHOOD RATIO CHI-SQUARE       9     36.426     0.000
MANTEL-HAENSZEL CHI-SQUARE        1     27.723     0.000
PHI                                     0.610
CONTINGENCY COEFFICIENT                 0.521
CRAMER'S V                              0.352

SAMPLE SIZE = 100
```

FURTHER READING

Extensions and refinements to all the tests described in this chapter can be found in most textbooks focusing on nonparametric statistics. Examples of situations in which nonparametric statistics can be applied are described in the three textbooks below as well as the problems at the end of this chapter.

W. J. Conover, *Practical Nonparametric Statistics*, 2d ed. (New York: Wiley, 1980).

W. W. Daniel, *Applied Nonparametric Statistics* (Boston: Houghton Mifflin, 1978).

S. Siegel, *Nonparametric Statistics for the Behavioral Sciences* (New York: McGraw-Hill, 1956).

PROBLEMS

1. Explain the meaning of (a) nonparametric statistics, (b) robustness, (c) goodness-of-fit test, and (d) contingency table.

2. Briefly compare the advantages and disadvantages of parametric and nonparametric tests.

3. An urban geographer randomly samples 20 new residents of a neighborhood to determine their ratings of local shopping facilities. The measurement scale is as follows: strongly dislike, 0; dislike, 1; neutral, 2; like, 3; and strongly like, 4. The 20 responses are 0, 4, 3, 2, 2, 1, 1, 2, 1, 0, 0, 1, 2, 1, 3, 4, 2, 0, 4, 1. Use the median test to see whether the population median is 2.

4. A course in statistical methods for geographers is team-taught by two instructors, Professor Jovita Fontanez and Professor Clarence Old. In the student evaluations for the course, students were asked to indicate their preference for the two professors. Of the 13 students in the class, 8 preferred Professor Fontanez and 2 preferred Professor Old. The remaining students were unable to express a preference for either instructor. Test the hypothesis that students prefer Professor Fontanez. (*Hint*: Use the median test.)

5. Consider the data given in Table 10-11. Use two-sample Mann-Whitney tests to test whether each pair of intersection types has different carbon monoxide levels.

6. Consider the data of Problem 4 in Chapter 9. Using the Mann-Whitney test at $\alpha = .05$, test the null hypothesis that the two samples come from the same population, that is, fertilizer XLC does not increase corn yields.

7. Solid-waste generation rates measured in metric tons per household per year are collected in randomly selected areas of a city. These areas are classified as high-density, low-density, or sparsely settled. It is thought that generation rates are probably higher in the areas of highest residential density because they are likely to be served by more efficient waste collection systems and there is very little opportunity for on-site disposal or storage, particularly in comparison to sparsely settled areas. Do the following data support this hypothesis?

High density	Low density	Sparsely settled
1.84	2.04	1.07
3.06	2.28	2.31
3.62	4.01	0.91
4.91	1.86	3.28
3.49	1.42	1.31

8. Use the Wilcoxon signed-ranks test to verify the conclusions of the matched-pairs test for Problem 5 of Chapter 9.

9. The distances traveled by a random sample of 12 people to their places of work in 1970 and again in 1980 are shown in the following table:

Person	Distance traveled, km		Person	Distance traveled, km	
	1970	1980		1970	1980
1	8.6	8.8	7	7.7	6.5
2	7.7	7.1	8	9.1	9.0
3	7.7	7.6	9	8.0	7.1
4	6.8	6.4	10	8.1	8.8
5	9.6	9.1	11	8.7	7.2
6	7.2	7.2	12	7.3	6.4

Has the length of the journey to work changed over the decade?

10. Test the normality of the DO data of Table 2-1, (a) using the χ^2 test with $k = 10$ classes, (b) using the χ^2 test with $k = 6$ classes defined in Table 2-3, and (c) using the Kolmogorov-Smirnov test for the data in the form of both (a) and (b).

11. One hundred randomly sampled residents of a city subject to periodic flooding are classified according to whether they are on the floodplain of the major river bisecting the city or off the floodplain. These households are then surveyed to determine whether they currently had flood insurance of any kind. The results are shown in the following table:

	On the floodplain	Off the floodplain
Purchased insurance	50	10
Did not purchase insurance	15	25

Test a relevant hypothesis.

12. In a study researching the effects of sociocultural environments on children's evaluation of other family members, 40 subjects in Nairobi, Kenya, Oslo, Norway, and Seattle, Washington, were instructed to draw a picture of a family. The following table classifies the leading figure in the drawings made by each group of children:

	Parent	Self	Relatives
Nairobi, Kenya	5	10	25
Oslo, Norway	15	10	15
Seattle, Washington	10	25	5

Test the hypothesis that these three populations are homogeneous with respect to the relationship of the leading figure in the drawings.

13. The actual daily occurrence of sunshine in a given city over a 30-day period is calculated as the percentage of possible time the sun could have shone, had it not been for the existence of cloudy skies. If we define a sunny day as one with over 50 percent sunshine, determine whether the pattern of occurrence of sunny days is random for the following data:

Day	Percentage of sunshine	Day	Percentage of sunshine	Day	Percentage of sunshine
1	75	11	21	21	77
2	95	12	96	22	100
3	89	13	90	23	90
4	80	14	10	24	98
5	7	15	100	25	60
6	84	16	90	26	90
7	90	17	56	27	100
8	18	18	0	28	90
9	90	19	22	29	58
10	100	20	44	30	0

14. In a certain city, the mortality rate due to household fire per 100,000 inhabitants for the last 15 years is as follows: 3.65, 3.41, 3.53, 3.23, 3.50, 3.43, 3.73, 3.40, 3.70, 3.58, 3.80, 3.40, 3.38, 3.36, 3.21. Is there a trend in these rates?

III

STATISTICAL RELATIONSHIPS BETWEEN TWO VARIABLES

11

Correlation Analysis

Quite often in statistical analysis we are interested in studying the relationship *between* variables. For example, a geomorphologist might wish to explore the relationship between the annual discharge of a stream and its width downstream from the source. Also urban geographers are sometimes interested in the association between the selling price of a residential property and other structural and locational characteristics such as house size and accessibility. *Multivariate statistical techniques* are methods which can be applied in situations where the analysis involves two or more different variables. The simplest class of multivariate statistical techniques is *bivariate*, involving only two variables at a time. An investigation of the *strength* of the association between two variables is called *simple (or bivariate) correlation analysis*, and a study of the *nature* of the relationship between two variables is termed *simple regression analysis*. In simple regression analysis, we are concerned with *estimating* the value of one variable by considering its relationship with one other variable. But, as we shall see in Chapter 12, regression and correlation are intimately related. This chapter is limited to simple linear correlation. Our purpose is to illustrate the use and interpretation of various quantitative indices that express the relative strength of the *linear* relationship between two variables. Clearly this is not the only way in which two variables might be related. For example, we might find that two variables are not linearly related at all, but are perfectly or almost perfectly related in some nonlinear way. This issue is explored further in Section 11.1.

The most commonly used measure of correlation for interval or ratio variables is known as *Pearson's product-moment correlation coefficient*, or simply as *Pearson's r*. The calculation and interpretation of Pearson's *r* are discussed in Section 11.1. In fact, this measure of the relationship between two variables is one of the parameters of the bivariate normal distribution discussed in Section 5.4. Before beginning Section 11.1, you may find it useful to review the material in Section 5.4, particularly the presentation of the bivariate normal distribution.

It is also possible to derive indices of correlation for variables measured at the nominal or ordinal scales. These measures are discussed in Section 11.3. The last two sections of this chapter describe particular problems and applications of correlation analysis in map association studies in geography.

11.1. Pearson's Product-Moment Correlation Coefficient

In Chapter 5, the bivariate normal distribution is introduced along with the concepts of the *covariance* between two random variables $C(X, Y)$ and the correlation ρ_{XY}. All the examples utilized in Chapter 5 describe situations in which the parameters of the population distribution are known. In most practical situations, the parameters of the bivariate normal distribution, and therefore the correlation coefficient, are unknown. As is usual in statistical analysis, the correlation coefficient is then estimated from sample data. The best point estimate of rho is Pearson's r, known also, for reasons outlined below, as the product-moment correlation coefficient.

Scatter Diagrams and the Calculation of r

It is possible to calculate a correlation coefficient for any set of n paired values of two variables X and Y. Label these observations (X_i, Y_i), $i = 1, 2, \ldots, n$. Since these observations may or may not be random variables, we drop the convention of distinguishing values of the variable by a lowercase designation. Of course, if we are interested in testing the statistical significance of a correlation coefficient, it is necessary to make the additional assumption that the n observations are a true random sample for some statistical population.

First, the observations are graphed in a *scatter plot* (or *scattergram* or *scatter diagram*). Each observation or pair (X_i, Y_i) represents one dot in a scatter plot. Examples of scatter plots with corresponding values of r are illustrated in Figure 11-1. Where all points in the scatter plot lie on a positively sloped line, $r = 1$, as in

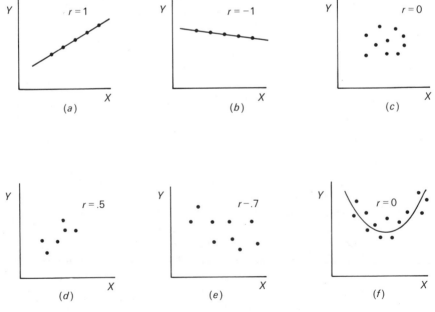

FIGURE 11-1. Scatter plots.

Figure 11-1(a). If the line has a negative slope, then $r = -1$, as in (b). When $r = 0$, the scatter of points is near circular, as in (c). The general elliptical shape to the scatter plots occurs for intermediate values of r, as in (d) and (e). Note especially Figure 11-1(f). This is an example of a scatter of points with a strong *nonlinear* association, but $r = 0$. This is because Pearson's r measures only the strength of the *linear* association between variables.

The sample correlation coefficient r is obtained by substituting the appropriate point estimators for $C(X, Y)$, or σ_X and σ_Y, into Equation (5-40). The best point estimate for $C(X, Y)$ is

$$S_{XY} = \frac{1}{n-1} \sum_{i=1}^{n} (X_i - \bar{X})(Y_i - \bar{Y})$$
(11-1)

The sample covariance S_{XY} measures the tendency for the two variables X and Y to covary. Consider, for example, the scatter plot of Figure 11-2. The plot suggests that the two variables X and Y are positively related. We would expect S_{XY} to be positive. The calculation of S_{XY} can be given a simple geometric explanation by using Figure 11-2. The two means $\bar{X}$ and $\bar{Y}$ divide the scatter plot into four quadrants labeled I, II, III, and IV. Let us examine the sign and magnitude of $(X_i - \bar{X})(Y_i - \bar{Y})$ in each of these four quadrants. If both X_i and Y_i are *greater* than their respective means, then the observation is in quadrant I. The product of the deviations from the mean is positive, since both deviations are positive. This holds true for all observations in quadrant I. In quadrant III, both deviations are negative since the values of X_i and Y_i are below their means. However, the product of the two deviations $(X_i - \bar{X})(Y_i - \bar{Y})$ is positive since it is the product of two negative values.

In quadrants II and IV, the deviations are of opposite signs, and therefore the product of the deviations must be negative. The outcome for S_{XY} is based on the sum of these products for all observations. Where there is a positive relation between the two variables, as in Figure 11-2, $S_{XY} > 0$. Why? Because there are many points in quadrants I and III with large positive values for $(X_i - \bar{X})(Y_i - \bar{Y})$, but only a few points in quadrants II and IV where the product is negative. If, however, the scatter

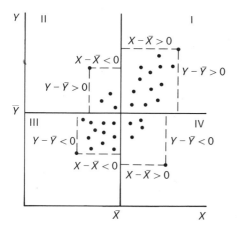

FIGURE 11-2. Geometric illustration of correlation coefficient derivation.

suggested a negative relation between X and Y, there would be many points in quadrants II and IV and few in quadrants I and III. In this case, the covariance turns out to be negative. If the scatter reveals a virtual equal scatter in all quadrants, the covariance S_{XY} is near zero.

The sample covariance S_{XY} has the same disadvantage as $C(X, Y)$ for measuring the strength of the relation between the two variables. It is highly influenced by the units in which the two variables are measured. Just as $C(X, Y)$ is standardized by σ_X and σ_Y in the correlation coefficient ρ_{XY}, the sample covariance is standardized by S_X and S_Y, so that

$$r = \frac{S_{XY}}{S_X S_Y} = \frac{[1/(n-1)] \sum_{i=1}^{n} (X_i - \bar{X})(Y_i - \bar{Y})}{\sqrt{[1/(n-1)] \sum_{i=1}^{n} (X_i - \bar{X})^2} \sqrt{[1/(n-1)] \sum_{i=1}^{n} (X_i - \bar{X})^2}} \qquad (11\text{-}2)$$

Equation (11-2) can be simplified by canceling the term $1/(n-1)$ which appears in the numerator and the denominator. This yields

$$r = \frac{\sum_{i=1}^{n} (X_i - \bar{X})(Y_i - \bar{Y})}{\sqrt{\sum_{i=1}^{n} (X_i - \bar{X})^2} \sqrt{\sum_{i=1}^{n} (Y_i - \bar{Y})^2}} \qquad (11\text{-}3)$$

For computing purposes, especially when the calculations are performed by hand or a nonprogrammable hand calculator is used, it is helpful to rearrange (11-3) to yield

$$r = \frac{\sum_{i=1}^{n} X_i Y_i - (\sum_{i=1}^{n} X_i)(\sum_{i=1}^{n} Y_i)/n}{\sqrt{\sum_{i=1}^{n} X_i^2 - (\sum_{i=1}^{n} X_i)^2/n} \sqrt{\sum Y_i^2 - (\sum_{i=1}^{n} Y_i)^2/n}} \qquad (11\text{-}4)$$

This is the most convenient formula for calculating Pearson's r.

To illustrate the calculation of the correlation coefficient by using these alternative formulas, consider the following three examples:

EXAMPLE 11-1. The four points $(0, 3)$, $(1, 5)$, $(2, 7)$, and $(3, 9)$ fall on the line $Y = 2X + 3$. As is evident in Figure 11-3, there is a perfectly linear relationship

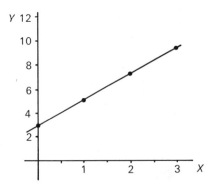

FIGURE 11-3. Graph for Example 11-1.

TABLE 11-1

Calculation of Pearson's r for Data of Figure 11-3

Point (X_i, Y_i)	$X_i - \bar{X}$	$(X_i - \bar{X})^2$	$Y_i - \bar{Y}$	$(Y_i - \bar{Y})^2$	$(X_i - \bar{X})(Y_i - \bar{Y})$
(0, 3)	−1.5	2.25	−3	9	4.50
(1, 5)	−0.5	0.25	−1	1	0.50
(2, 7)	0.5	0.25	1	1	0.50
(3, 9)	1.5	2.25	3	9	4.50
	0	5.00	0	20	10.00

$$r = \frac{S_{XY}}{S_X S_Y} = \frac{\frac{10}{3}}{\sqrt{\frac{20}{3}}\sqrt{\frac{5}{3}}} = \frac{10}{\sqrt{100}} = 1$$

between X and Y for these four points. The sample correlation coefficient should be $r = 1$. First, calculate $\bar{X} = (0 + 1 + 2 + 3)/4 = 1.5$ and $\bar{Y} = (3 + 5 + 7 + 9)/4 = 6$. The necessary calculations for Pearson's r obtained by using the covariance formulation in (11-3), are summarized in Table 11-1. As expected, $r = 1$.

EXAMPLE 11-2. The six points in Figure 11-4 do not fall on a straight line. It appears that the two variables are not linearly associated. For these data let us utilize Equation (11-4). It is useful to organize the calculations for the correlation coefficient by using the tabular format of Table 11-2. As the scatter diagram in Figure 11-4 suggests, these two variables have no linear association whatever and $r = 0$.

TABLE 11-2

Tabular Format for Calculation of Pearson's r from Data of Figure 11-4

X_i	Y_i	X_i^2	Y_i^2	$X_i Y_i$
0	2	0	4	0
0	4	0	16	0
2	2	4	4	4
2	4	4	16	8
4	2	16	4	8
4	4	16	16	16
$\sum_i X_i = 12$	$\sum_i Y_i = 18$	$\sum_i X_i^2 = 40$	$\sum_i Y_1^2 = 60$	$\sum_i X_i Y_i = 36$

$$r = \frac{36 - (12)(18)/6}{\sqrt{40 - 12^2/6}\sqrt{60 - 18^2/6}} = \frac{0}{\sqrt{16}\sqrt{6}} = 0$$

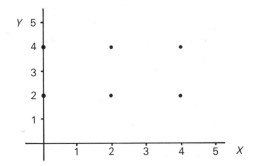

FIGURE 11-4. Graph for Example 11-2.

EXAMPLE 11-3. In practice, most correlation coefficients lie between the two extremes of perfect correlation ($r = \pm 1$) and no linear correlation ($r = 0$). Consider the following data consisting of the selling price of a sample of 10 homes:

Selling price Y, $000	Size X, 000 ft^2
45	1.0
60	1.3
55	1.6
70	1.8
75	1.9
80	2.0
100	2.2
90	2.4
105	2.8
120	3.0

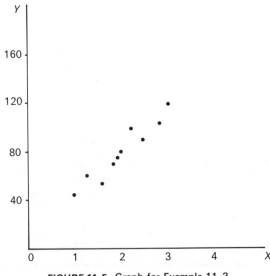

FIGURE 11-5. Graph for Example 11-3.

TABLE 11-3
Calculation of Pearson's r for Example 11-3

X_i	Y_i	X_i^2	Y_i^2	X_iY_i
45	1.0	2,025	1.00	45
60	1.3	3,600	1.69	78
55	1.6	3,025	2.56	88
70	1.8	4,900	3.24	126
75	1.9	5,625	3.61	142.5
80	2.0	6,400	4.00	160
100	2.2	10,000	4.84	220
90	2.4	8,100	5.76	216
105	2.8	11,025	7.84	294
120	3.0	14,400	9.00	360
800	20.0	69,100	43.54	1729.5

$$r = \frac{1729.5 - (800)(20.0)/10}{\sqrt{69,100 - 800^2/10} - \sqrt{43.54 - 20.0^2/10}} = \frac{129.5}{134.4} = .96$$

When illustrated in the scatter diagram of Figure 11-5, the data reveal a strong positive correlation. The necessary calculations are summarized in Table 11-3 and confirm that the selling price of a house is strongly positively associated ($r = .96$) with the square footage of the house. Put simply, larger houses tend to be more expensive.

Frequently in geographical research the data collected in a single study include observations for many different variables. In a study of house price variations in an urban area, for example, let us suppose that the following six variables are available from a sample of $n = 100$ homes:

1. Selling price, thousands of dollars (Price)
2. Size of house, thousands of square feet (Size)
3. Number of bedrooms (Bed)
4. Lot frontage, feet (Lot)
5. Age, years since constructed (Age)
6. Number of bathrooms (Bath)

Correlation coefficients can be calculated between each pair of variables in this set. These correlation coefficients are frequently summarized in a *correlation matrix* such as Table 11-4. There are several important features of such matrices. First, the correlation between a variable and itself is always 1, so that the *diagonal* elements of the matrix always contain the value $r = 1.0$. To see why this is so, examine Equation (11-4). If we substitute X for Y in this equation, the numerator and the denominator are identical; thus the ratio must be 1.0. Second, correlation matrices are symmetric. The correlation between two variables X and Y necessarily equals the correlation between variables Y and X. Note in Equation (11-4) that the identical numerical result is obtained if variable X is substituted for variable Y and vice versa.

What does Table 11-4 reveal? First, note the strong positive correlations between Price and the variables Size, Bed, Lot, and Bath. For this sample of 100

TABLE 11-4
Correlation Matrix for Six Housing Variables

	Price	Size	Bed	Lot	Age	Bath
Price	1.0	.96	.83	.91	−.53	.92
Size	.96	1.0	.98	.80	.23	.91
Bed	.83	.98	1.0	.62	−.71	.94
Lot	.91	.80	.62	1.0	.07	.21
Age	−.53	.23	−.71	.07	1.0	−.77
Bath	.92	.91	.94	.21	−.77	1.0

homes, this indicates that higher-price homes tend to be larger and have more bed-rooms, wider lot frontages, and more bathrooms. Also, these variables tend to co-vary. For example, larger homes generally have more bedrooms, more bathrooms, and wider lots. There are also a few strong negative correlations. Older homes are generally *cheaper* and have *fewer* bedrooms and bathrooms. Finally, there are several weak correlations. There seems to be little relation between Size and Age ($r = .23$), Age and Front ($r = .07$), and Front and Bed ($r = .21$). Besides using the correlation coefficient to describe the *degree* of association between two variables, it is also possible to test the statistical significance of any sample correlation coefficient.

Significance Testing of a Correlation Coefficient

The sample correlation coefficient r calculated by using (11-2), (11-3), or (11-4) is the best point estimator of ρ, the population correlation coefficient. Figure 11-6 illustrates the sampling procedure which underlies the significance test of r. The statistical population consists of the two random variables X and Y known to be bivariate normally distributed. The population correlation coefficient is given by

$$\rho = \frac{C(X, Y)}{\sigma_X \sigma_Y} \tag{11-5}$$

If the population cannot be observed, then ρ is unknown and the null hypothesis to be tested is H_0: $\rho = 0$. From the population, a random sample of n pairs (X, Y) is drawn, and the sample correlation coefficient r is calculated. The sampling distribution of r, under the null hypothesis that $\rho = 0$, (1) is t-distributed with $n - 2$ degrees of freedom and (2) has an *estimated* standard error of

$$s_r = \sqrt{\frac{1 - r^2}{n - 2}}$$

Therefore, the required test statistic is

$$t = \frac{r}{s_r} = \frac{r}{\sqrt{(1 - r^2)/(n - 2)}} = \frac{r\sqrt{n - 2}}{\sqrt{1 - r^2}} \tag{11-6}$$

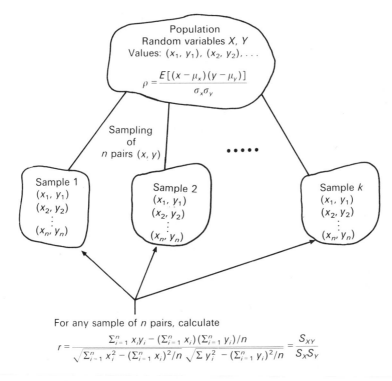

For any sample of n pairs, calculate

$$r = \frac{\sum_{i=1}^{n} x_i y_i - (\sum_{i=1}^{n} x_i)(\sum_{i=1}^{n} y_i)/n}{\sqrt{\sum_{i=1}^{n} x_i^2 - (\sum_{i=1}^{n} x_i)^2/n} \sqrt{\sum y_i^2 - (\sum_{i=1}^{n} y_i)^2/n}} = \frac{S_{XY}}{S_X S_Y}$$

FIGURE 11-6. Sampling procedure underlying significance test of correlation coefficient.

For example, consider the value of $r = .96$ obtained for the variables Price and Size for $n = 10$ houses listed in Example 11-3. Using a two-tail test at $\alpha = .05$ yields

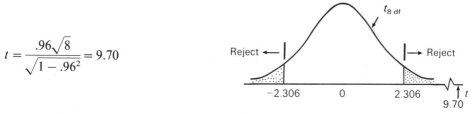

$$t = \frac{.96\sqrt{8}}{\sqrt{1 - .96^2}} = 9.70$$

Since $t > t_{.025,\,8\text{ df}}$, we must reject H_0: $\rho = 0$.

To test a hypothesis where $\rho \neq 0$ or to construct a confidence interval for ρ, the procedure becomes more complicated. This is because the sampling distribution of r is symmetric only for the case in which the sampling distribution is centered at 0. When $E(r) \neq 0$ (that is, we cannot assume that $\rho = 0$), the sampling distribution of r is skewed. It is increasingly skewed as $E(r)$ approaches ± 1. This situation is depicted in Figure 11-7. Fortunately, it is possible to use Fisher's Z transformation as an approximate method for constructing confidence intervals in this case.

The transformation $Z = \frac{1}{2} \ln[(1 + r)/(1 - r)]$ is approximately normally distributed with mean $\mu_Z = \frac{1}{2} \ln[(1 + \rho)/(1 - \rho)]$ and standard error $\sigma_Z = \sqrt{1/(n - 3)}$ for $n > 50$. With this transformation of the correlation coefficient it is possible to

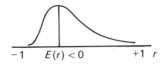

$-1 \quad E(r) < 0 \qquad +1 \quad r$

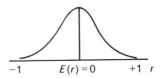

$-1 \qquad E(r) = 0 \qquad +1 \quad r$

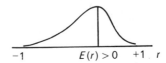

$-1 \qquad E(r) > 0 \quad +1 \quad r$

FIGURE 11-7. Sampling distributions of the sample correlation coefficient r.

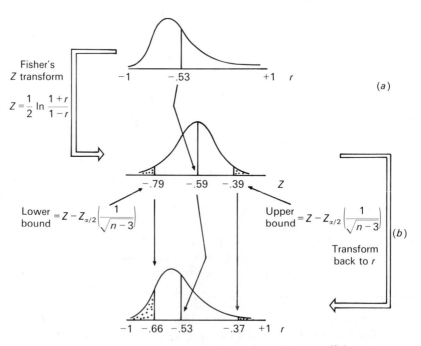

FIGURE 11-8. Confidence interval estimate of correlation coefficient ρ.

derive confidence intervals in the usual way. The method is best illustrated with an example.

EXAMPLE 11-4. Consider the correlation of $r = -.53$ between house selling price (Price) and the number of years since the house was constructed (Age) for $n = 100$ homes. In Figure 11-8(a) the sampling distribution of r for the case $r = -.53$ illustrates the problem in deriving a confidence interval. The transformation $Z = \frac{1}{2} \ln (.47/1.53) = -.59$ leads to the approximately normal distribution in Figure 11-8(b). From this distribution, confidence limits can be constructed in the usual way. For $\alpha = .05$, or 95 percent, confidence

$$\text{Upper bound} = -.59 + 1.96 \left(\frac{1}{\sqrt{97}} \right) = -.39$$

$$\text{Lower bound} = -.59 - 1.96 \left(\frac{1}{\sqrt{97}} \right) = -.79$$

As usual, these bounds are symmetric around $-.59$. To express these confidence limits in terms of r, it is necessary to convert these values from Z to r by using the normal table in the Appendix. From this table, we see the limits on ρ are $-0.66 \le \rho \le -0.37$. The process is illustrated in Figure 11-8. Because of the asymmetry of the interval, the lower bound is $.66 - .53 = .13$ below r, and the upper bound is $.53 - .37 = .16$ unit above $r = .53$. The asymmetric nature of the interval becomes increasingly pronounced as it approaches ± 1.

11.2. Nonparametric Correlation Coefficients

Pearson's product-moment correlation coefficient is the most widely used measure of association for interval-scaled variables. Nonparametric measures of association can also be applied to pairs of variables measured at the ordinal or nominal scale. Nonparametric measures are useful in cases in which the available data are limited to these scales, or in cases in which interval variables do not meet the exacting requirements sometimes necessary for parametric measures such as Pearson's r. For example, any statistical inference concerning Pearson's r requires the assumption that the population joint probability distribution of X and Y is bivariate normal. Just as nonparametric statistical tests sometimes prove useful in univariate inferential statistics, so, too, do nonparametric measures of association in bivariate situations. Nonparametric measures can be conveniently divided into two classes: *rank correlation coefficients* for ordinal data and *nominal scale measures*.

Rank Correlation Coefficients

Ordinal measures of association, or rank correlation coefficients, are appropriate whenever the relationship between two variables X and Y is *monotonic increasing* or *decreasing*. A monotonic increasing function for Y is one that either remains constant or increases as X increases. Analogously, a monotonic decreasing function for Y is one that either remains constant or decreases as X increases. We frequently encounter such functions in many propositions in geography and the social sciences. For example, it is common to say, "The greater the X, the greater the Y" or "The greater the X, the less the Y." Both statements imply monotonicity in the relationship between X and Y. Whereas Pearson's r is a measure of the *linear* association between two variables, rank correlation coefficients cannot distinguish linear from nonlinear monotonic functions. This is entirely due to the nature of ordinal data. Recall that the ordinal level of measurement requires only that we are able to *order* the observations of a variable, and the notion of *distance* between observations is inappropriate. Although we may be able to say X_1 is greater than X_2, and X_2 is greater than X_3, we are unable to compare the differences $X_1 - X_2$ and $X_2 - X_3$. Therefore we have no way of knowing whether the relationship between X and Y is linear, quadratic, or of some other monotonic form.

Because it is a parametric measure of correlation based on continuous interval data, Pearson's r is the most powerful measure of association presented in this text. Rank correlation measures should be used only where (1) the level of measurement of the two variables X and Y is ordinal or (2) the assumptions required for a test of statistical significance for Pearson's r cannot be met (that is, the data are not bivariate normal). At the same time, Pearson's r is a robust measure of correlation, and the statistical test of significance is valid under modest departures from the assumption of bivariate normality.

Two of the more commonly used measures of ordinal association are *Spearman's* ρ *(rho)* and *Kendall's* τ *(tau)*. They share three properties with Pearson's r:

1. Both are limited to the range $[-1, +1]$.
2. Both coefficients are positive (negative) when an increase (a decrease) in X corresponds to an increase (a decrease) in Y.
3. A value near zero indicates that the values of X are uncorrelated with the values of Y.

Consider n observations of two variables X and Y measured on at least the ordinal scale. Assume that there are no tied values among either the values of X or those of Y. Each of the X values is replaced by its rank, the lowest rank 1 is given to the smallest value of X, and rank n is given to the highest value of X. Each of the Y values is ranked in the same way. Now, arrange the X values in ascending order with the rank of the Y values in the same sequence. Table 11-5 illustrates this procedure for the two variables Size (house size in thousands of square feet) and Price (selling price in thousands of dollars) examined in Section 11.2. The values of X are listed in column 2 and the Y values in column 3. Note that the X variable has been placed in ascending order. Let $r(X_i)$ be the rank of the ith observation X_i and $r(Y_i)$ be the rank of the ith observation Y_i. For example, observation $i = 1$ is the lowest X and lowest Y, so that $r(X_1) = r(Y_1) = 1$. Observation $i = 2$ is the second lowest X and third lowest Y, so that $r(X_2) = 2$ and $r(Y_2) = 3$. If the two variables are perfectly positively correlated, then the rank orders of the two variables will be identical. If there is no correlation, then the distribution of Y ranks is random with respect to the ranks of the paired X values. If there is a perfect negative correlation, the ranks of the Y values

TABLE 11-5
Calculation of r_S

i	X_i	Y_i	$r(X_i)$	$r(Y_i)$	$[r(X_i) - r(Y_i)]^2$
1	1.0	45	1	1	0
2	1.3	60	2	3	1
3	1.6	55	3	2	1
4	1.8	70	4	4	0
5	1.9	75	5	5	0
6	2.0	80	6	6	0
7	2.2	100	7	8	1
8	2.4	90	8	7	1
9	2.8	105	9	9	0
10	3.0	120	10	10	0
					4

$$r_S = 1 - 6 \sum_i \frac{[r(X_i) - r(Y_i)]^2}{n(n^2 - 1)}$$

$$= 1 - 6 \frac{(1-1)^2 + (2-3)^2 + (3-2)^2 + (4-4)^2 + (5-5)^2 + (6-6)^2 + (7-8)^2 + (8-7)^2 + (9-9)^2 + (10-10)^2}{10(10^2 - 1)}$$

$$= 1 - 6(\tfrac{4}{990}) = .976$$

will decrease perfectly with the increase in the ranks of X. Rank correlation coefficients assess the degree of association between the ranks. In Table 11-5 the ranks of the X's and Y's are almost perfectly correlated. Six of the ten rankings are identical, and the other four differ only by one ranking. Both Spearman's ρ and Kendall's τ are measures which tell us the precise degree of association between these two sets of ranks.

SPEARMAN'S RANK CORRELATION COEFFICIENT

The idea behind Spearman's ρ is quite simple. The two sets of rankings are compared by taking the differences between the ranks, squaring these differences, summing the squares over all n rankings, and then standardizing the result so that the bounds are $[-1, +1]$. An estimate of Spearman's ρ, denoted r_S, can be computed from

$$r_S = 1 - 6 \sum_{i=1}^{n} \frac{[r(X_i) - r(Y_i)]^2}{n(n^2 - 1)} = 1 - 6 \sum_{i=1}^{n} \frac{d_i^2}{n(n^2 - 1)} \qquad (11\text{-}7)$$

where $d_i = r(X_i) - r(Y_i)$. This formula can be derived directly from Pearson's r simply by substituting $r(X_i)$ for X_i and $r(Y_i)$ for Y_i. Whenever the ranks are identical, $d_i = 0$ for all i and $r_S = +1$. Whenever the ranks are in perfect reverse order, the second part of (11-7) simplifies to -2 and $r = 1 - 2 = -1$. Suppose the ranks for X are 1, 2, 3 and the corresponding ranks for Y are 3, 2, 1. Using (11-7) to calculate r_S, we find $r_S = 1 - 6(4 + 0 + 4)/[3(8)] = 1 - \frac{48}{24} = 1 - 2 = -1$. The calculation of r_S for Price and Size is summarized in Table 11-5. As is apparent from the two rankings, Price and Size are indeed highly positively correlated with $r_S = .976$. Note that this result is almost identical to the value of $r = .96$ obtained for Pearson's r. If we were to take the ranks and substitute them directly into (11-4), the result for r_S would also be .976. It is for this reason that Spearman's r_S can be interpreted in the same way as the product-moment correlation coefficient.

If there are tied values for any of the ranks, then it is still possible to calculate Spearman's r_S by using the convention for ties suggested in Chapter 10. Each tied value is assigned the average of the ranks that would have been assigned had no ties existed. If the number of ties is small, then Equation (11-7) can be used to compute r_S. Whenever there are a substantial number of tied values, a correction factor should be used. Details are available in most advanced textbooks on nonparametric texts such as Siegel (1956). As a rule of thumb, the correction factor can be ignored when the number of pairs of ties is less than one-fourth of the sample.

If the n pairs of observations represent a random sample from a bivariate population, then it is possible to test the statistical significance of r_S. Since it is a nonparametric measure, Spearman's coefficient makes no assumption whatsoever about the nature of the underlying population distribution. For $n < 10$ the significance of r_S can be determined directly from tables. When $n \geq 10$, it is usual to test Spearman's r_S by using a standard normal deviate. In this case the sampling distribution of r_S is approximately normal with a standard error of $1/\sqrt{n-1}$. Under the null hypothesis that there is no relationship between the two variables X and Y in the population, the test statistic

$$Z = \frac{r_S - 0}{1/\sqrt{n-1}} = r_S\sqrt{n-1} \tag{11-8}$$

can be used to test the significance of r_S. For Price and Size we have $r_S = .976$ and $n = 10$, so that $Z = .976\sqrt{9} = 2.93$. This is significant at $\alpha = .05$ since it exceeds $Z = 1.96$. Confidence limits on ρ_S can be computed from

$$r_S - Z_{\alpha/2}\left(\frac{1}{\sqrt{n-1}}\right) \leq \rho_S \leq r_S + Z_{\alpha/2}\left(\frac{1}{\sqrt{n-1}}\right) \tag{11-9}$$

The 95 percent limits on ρ_S are $.976 - 1.96(1/\sqrt{10-1}) \leq \rho_S \leq .976 + 1.96(1/\sqrt{10-1})$ which leads to $.323 \leq \rho_S \leq 1.629$ (or 1.0 since this is the upper bound on Spearman's rank correlation coefficient).

KENDALL'S τ

Spearman's r_S is the most widely used measure of ordinal correlation. It can be derived directly from Pearson's r and is extremely easy to calculate. Kendall's τ is an alternative correlation coefficient based on an entirely different computational procedure. The first step is to rank the observations in the same way as for Spearman's r_S.

TABLE 11-6
Calculation of Kendall's τ

i	X_i	Y_i	$r(X_i)$	$r(Y_i)$	N_C	N_D
1	1.0	45	1	1	9	0
2	1.3	60	2	3	7	1
3	1.6	55	3	2	7	0
4	1.8	70	4	4	6	0
5	1.9	75	5	5	5	0
6	2.0	80	6	6	4	0
7	2.2	100	7	8	2	1
8	2.4	90	8	7	2	0
9	2.8	105	9	9	1	0
10	3.0	120	10	10	0	0
					43	2

$$\tau = \frac{43 - 2}{10(9)/2} = \frac{41}{45} = .91$$

Table 11-6 contains the rankings for the variables Size and Price. Note that the first five columns are identical to the first five columns of Table 11-5. The calculation of Kendall's τ is based on the comparison of each pair of observations. For example, compare observations $i = 1$ and $i = 2$. Note that $r(X_1) = 1 < r(X_2) = 2$ and $r(Y_1) = 1 < r(Y_2) = 3$. This implies that the rankings of variables X and Y are in agreement or ordinally correct for this pair of observations. Pairs of observations which have this property are said to be *concordant*. Consider now observations $i = 2$ and $i = 3$. Note that $r(X_2) = 2 < r(X_3) = 3$ but $r(Y_2) = 3 > r(Y_3) = 2$. This pair is said to be *discordant* since the rankings of variable X are not in agreement with those of variable Y. To compute Kendall's τ, we must compare all pairs of observations and record the number of discordant and concordant pairs. If there are n observations, then there are $\binom{n}{2} = [n(n - 1)]/2$ comparisons to be made. Of these N_C are concordant and N_D are discordant. Kendall's τ uses the function

$$\tau = \frac{N_C - N_D}{n(n - 1)/2} \tag{11-10}$$

as the measure of association between variables X and Y. If all the pairs are concordant, then $N_C = [n(n - 1)]/2$ and $N_D = 0$, so that $\tau = +1$. When $N_C = N_D$, $\tau = 0$ and there is no correlation. When $N_D = [n(n - 1)]/2$ and $N_C = 0$, we know $\tau = -1$.

The most confusing part of the calculation of τ is the enumeration of the pairs of observations. This procedure is simplified by organizing the calculations in the form of Table 11-6. Consider any observation i. Since the X's are in ascending order, we know that $r(X_i) < r(X_{i+1})$. But since the Y's are not in ascending order, we must count all the values of $r(Y_k)$ that are larger than $r(Y_i)$ but farther down the list than *i*. These will be the discordant values. All other comparisons must be concordant. This procedure is repeated, beginning at row $i = 1$ and continuing down the list until row $i = n$. To see this operation, let us begin in Table 11-6 at row $i = 1$. Since $r(X_1) = r(Y_1) = 1$ comparisons with all other nine observations must be concordant, they must have ranks greater than 1. In the last column we record nine concordant pairs and zero discordant pairs. Continuing with observation $i = 2$, we note $r(X_2) = 2$ and $r(Y_2) = 3$. Therefore any rank in the Y column below $i = 2$ and less than 3 must be discordant. Looking down the list of the Y ranks, we see that only one observation has a Y rank of less than 3. Note that the pair $i = 2, 3$ is discordant since $r(X_2) < r(X_3)$ but $r(Y_2) > r(Y_3)$. There are seven concordant pairs and one discordant pair. If we continue down the list in this way, all $10(9)/2 = 45$ pairs will be enumerated and no double counting occurs. In this case there are 43 concordant pairs and 2 discordant pairs. Kendall's $\tau = .91$, quite close to $r_S = .976$.

The test of significance of Kendall's τ is based on examining the difference $N_C - N_D$. Kendall has shown that for $n \geq 10$ the sampling distribution of $N_C - N_D$ is approximately normal with mean 0 and standard error $\sigma_{N_C - N_D} = \sqrt{\frac{1}{18}n(n-1)(2n+5)}$. To test the null hypothesis H_0: $\tau = 0$, the appropriate test statistic is therefore

$$Z = \frac{N_C - N_D}{\sqrt{\frac{1}{18}n(n-1)(2n+5)}} \tag{11-11}$$

For example, for Price and Size, $n = 10$, $\tau = .91$, $N_C = 43$, and $N_D = 2$,

$$Z = \frac{43 - 2}{\sqrt{\frac{1}{18}(10)(9)(25)}} = \frac{41}{\sqrt{125}} = 3.67$$

This is significant at $\alpha = .05$ since it exceeds 1.96.

Nominal Scale Measures of Association

In Chapter 10 we introduced the χ^2 test as a method for determining whether a statistically significant association exists between two nominally scaled variables expressed in a contingency table. Although a χ^2 test can be used to verify the existence of a *significant statistical association* between the two variables, it does not necessarily tell us much about the *strength* of the association. The level of significance is not a good indicator of strength since it depends, as we found in both parametric and nonparametric testing, on the size of the sample. So once we have used χ^2 to verify the existence of a statistically significant association, the next step is to measure the strength of this relationship. As usual, we seek single summarizing measures with bounds $[-1, +1]$ or $[0, 1]$. Such measures can be used to compare several different relationships and select the strongest or weakest relationship. There are several measures for contingency tables.

Traditional Measures Based on χ^2

The difficulty with χ^2 is that the value of χ^2 in any contingency table is directly proportional to the total sample size n. Two tables with identically proportional cell frequencies will have different χ^2 values. One way of overcoming this problem is to use the ϕ (phi) coefficient:

$$\phi = \sqrt{\frac{\chi^2}{n}} \tag{11-12}$$

This measure has many desirable properties. It can also be derived independently from a 2×2 contingency table of the form

		Variable Y		
		Category 1	Category 2	Total
Variable X	Category 1	f_{11}	f_{12}	$f_{11} + f_{12}$
	Category 2	f_{21}	f_{22}	$f_{21} + f_{22}$
	Total	$f_{11} + f_{21}$	$f_{12} + f_{22}$	$f_{11} + f_{12} + f_{21} + f_{22}$

by using the formula

$$\phi = \frac{f_{11}f_{22} - f_{21}f_{12}}{\sqrt{(f_{11} + f_{12})(f_{21} + f_{22})(f_{12} + f_{22})(f_{11} + f_{21})}} \tag{11-13}$$

where f_{ij} is the observed count in cell ij.

When Equation (11-12) is used to calculate ϕ, it has a range $[0, 1]$. By using (11-13) the sign of the coefficient is retained. If $f_{11}f_{22} > f_{21}f_{12}$, then ϕ is positive; and if $f_{21}f_{12} > f_{11}f_{22}$, then it is negative. However, the sign of the coefficient is not particularly meaningful since we could change the sign simply by interchanging the order of the rows and columns of the contingency table. Since the data are purely nominal, row and column order is of no significance. Note that if the cross-products are equal, then the numerator of ϕ is zero and so ϕ must be zero. If both the off-diagonal elements f_{21} and f_{12} are zero, then $\phi = 1$ and the relationship is perfect. To see this, simply substitute the values $f_{12} = f_{21} = 0$ into (11-13). The numerator and denominator become equal, and $\phi = 1$. If, however, f_{11} and f_{22} are both zero, then $\phi = -1$. The range of ϕ is thus $[-1, +1]$.

In the general $r \times c$ contingency table, ϕ can attain a value greater than 1. To correct for this, three other measures of χ^2 are often used for contingency tables larger than 2×2:

$$\text{Tschuprow's } T = \sqrt{\frac{\chi^2}{n\sqrt{(r-1)(c-1)}}} \qquad (11\text{-}14)$$

$$\text{Cramer's } V = \sqrt{\frac{\chi^2}{n \min(r-1, c-1)}} \qquad (11\text{-}15)$$

$$\text{Pearson's } C = \sqrt{\frac{\chi^2}{\chi^2 + n}} \qquad (11\text{-}16)$$

All three measures attempt to standardize χ^2 by the maximum attainable value of χ^2 so that the coefficient has a true range $[0, 1]$. For an $r \times c$ contingency table, the maximum value of χ^2 is $n \min(r-1, c-1)$. This the denominator of the radical in Cramer's V. Whenever the contingency table is square and $r = c$, it is easily shown that Tschuprow's $T =$ Cramer's V. In fact, for 2×2 tables $T = V = \phi$. Another commonly used measure is Pearson's contingency coefficient C. Like ϕ, T, and V, it has a value of zero whenever there is no correlation between the two variables. The maximum value of C is always less than 1 since χ^2 can never be as large as $\chi^2 + n$. Since the maximum value of χ^2 is $n \min(r-1, c-1)$, the maximum value of C is, by substitution for χ^2 into (11-16),

$$\min\left(\sqrt{\frac{r-1}{r}}, \sqrt{\frac{c-1}{c}}\right)$$

For a 2×2 table this maximum is .707, for a 3×3 table it is .816, and so on. Since the maximum value of C depends on the dimensions of the contingency table, Pearson's C is not as easily interpretable as T or V. All three contingency coefficients are functions of χ^2, and a test of significance for χ^2 is always equivalent to a test of significance of the contingency coefficient.

EXAMPLE 11-5.   A rural geographer samples 100 farmers of the rural-urban fringe of a large metropolitan area. Farmers in the sample are classified according to whether they are a hobby farmer, part-time commercial farmer, or full-time commercial farmer. Also, the principal farming activity can be classified as mixed cash crops,

TABLE 11-7
Calculation of Nominal-Scale Association Measures

Observed

	Mixed cash crop	Mixed livestock	Dairy	Total
Hobby	8	12	5	25
Part-time	10	9	11	30
Full-time	7	14	24	45
Total	25	35	40	100

Expected

	Mixed cash crop	Mixed livestock	Dairy	Total
Hobby	6.25	8.75	10	25
Part-time	7.50	10.50	12	30
Full-time	11.25	15.75	18	45
Total	25	35	40	100

$$\chi^2 = \frac{(8-6.25)^2}{6.25} + \frac{(12-8.75)^2}{8.75} + \cdots + \frac{(14-15.75)^2}{15.75} + \frac{(24-18)^2}{18}$$

$$= 9.13$$

$$T = \sqrt{\frac{9.13}{100\sqrt{(2)(2)}}} = .214$$

$$V = \sqrt{\frac{9.13}{100(2)}} = .214$$

$$C = \sqrt{\frac{9.13}{9.13 + 100}} = .289$$

mixed livestock, or dairy. The results of the survey are summarized in the following table:

		Principal farm activity			
		Mixed cash crops	Mixed livestock	Dairy	Total
	Hobby	8	12	5	25
Ownership type	Part-time	10	9	11	30
	Full-time	7	14	24	45
	Total	25	35	40	100

How strongly associated are the variables ownership type and crop choice?

The strength of the association is measured by any of the three contingency coefficients T, V, or C. In Table 11-7 the calculations commence with the computation of χ^2. A value of $\chi^2 = 9.13$ has a PROB-VALUE in the range $.05 \leq$ PROB-VALUE $\leq .10$ with 4 degrees of freedom. This value is then substituted into Equations (11-14) to (11-16) to estimate T, V, and C. All three coefficients indicate a rather weak association between ownership type and crop choice. Note the equality of coefficients V and T. This is to be expected since the observed contingency table has dimensions 3×3.

OTHER MEASURES FOR NOMINAL VARIABLES

The handicap of all coefficients based on χ^2 is the lack of a meaningful interpretation of the coefficient except at the extremes of 0 and $+1$. To get around this, Goodman and Kruskal have devised a coefficient which is easily interpretable. The calculation and interpretation of Goodman and Kruskal's τ_b are best explained in the context of an empirical example. Consider once again the relationship between farm ownership type and crop choice examined above. Suppose that we are asked to predict the ownership type of each of the 100 farmers as they are presented to us one by one. The only information known to us is that there are 25 hobby farmers, 30 part-time farmers, and 45 full-time farmers. That is, we know the marginal totals by ownership type. How many errors would we expect to make in classifying these 100 farmers?

First, consider the hobby farmers. We know that 75 of the 100 farmers are not hobby farmers. Therefore, of the 25 farmers we decide to place in this category, we should expect to make (in the long run) $25(\frac{75}{100}) = 18.75$ errors or incorrect classifications. This product is simply the total number of farmers assigned to the hobby class multiplied by the probability that any one of them is not a hobby farmer. Similarly, we can expect to incorrectly assign $30(\frac{70}{100}) = 21$ part-time farmers and $45(\frac{55}{100}) = 24.75$ full-time farmers. In total there are $18.75 + 21 + 24.75 = 64.5$ likely classification errors. Only about one-third of the farmers would be correctly classified.

Now, suppose we are given some additional information about the farmers—their principal farming activity. Can this additional information be used to reduce our classification errors? First, let us suppose the farmer is introduced as one whose principal farming activity is mixed cash crops. It is necessary to place 8 of the 25 cash crop farmers into the hobby farm category. Since 17 of the 25 mixed cash crop farmers are not hobby farmers, we can expect to make $8(\frac{17}{25}) = 5.44$ errors. For part-time farmers the expected number of errors is $10(\frac{15}{25}) = 6$, and for full-time farmers we expect to make $7(\frac{18}{25}) = 5.04$ errors. Consider now the second column. These are mixed livestock farmers. Using the same logic, we would expect to make $12(\frac{23}{35}) = 7.89$ hobby farmer misclassifications, $9(\frac{26}{35}) = 6.69$ part-time farmer errors, and $14(\frac{21}{35}) = 8.40$ full-time farmer errors. For the last category of dairy farmers, the respective errors are $5(\frac{35}{40}) = 4.375$, $11(\frac{29}{40}) = 7.98$, and $24(\frac{16}{40}) = 9.6$. Adding these quantities leads to a total of 61.42 errors.

Goodman and Kruskal's τ_b is defined as the proportional reduction in assignment errors for the first variable (ownership type) gained by the knowledge of the second variable (principal farming activity). Specifically, define

$$\text{Goodman and Kruskal's } \tau_b = \frac{\begin{array}{c}\text{Number of} \\ \text{original errors}\end{array} - \begin{array}{c}\text{Errors with} \\ \text{additional variable}\end{array}}{\text{Number of original errors}} \qquad (11\text{-}17)$$

In the example above, $\tau_b = (64.5 - 61.42)/64.5 = .048$. Put simply, the assignment errors have been reduced by 4.8 percent. Suppose that the contingency table relating ownership type and principal farming activity is given by the following idealized table:

		Principal farm activity			
		Mixed cash crops	Mixed livestock	Dairy	Total
	Hobby	100	0	0	100
Ownership type	Part-time	0	100	0	100
	Full-time	0	0	100	100
	Total	100	100	100	300

If we tried to predict ownership type without utilizing crop choice, we would misclassify two-thirds of the 300 farmers. However, once we were given the crop choice of the farmer, we would make no misclassifications since the relationship between the two variables is perfect. For this table $\tau_b = 1$.

There is one problem with Goodman and Kruskal's τ_b. It is asymmetric. To calculate τ_b, we examined the error reduction in terms of the variable specified by the rows of the contingency table, using the additional information given by the variable specified in the columns of the table. If the process were reversed and the error reduction analyzed for the variable at the top of the table, a different value for τ_b would result. It is common to differentiate these coefficients by labeling the first τ_b and the second τ_a.

11.3. Correlation Coefficients and Areal Association

As we found in Chapter 2, geographers are frequently concerned with spatial distributions. The transfer of most statistical concepts to a spatial context is relatively straightforward, although it can occasionally raise some thorny conceptual issues. For correlation analysis, all that is required is that the observations refer to a set of locations expressed as areal units or as discrete points in space. Measurements of any two variables at the nominal, ordinal, or interval scales can be used in the calculation of one of the correlation coefficients described in Sections 11.1 and 11.2. As usual, the resulting coefficient indicates the direction and strength of the relationship between the pair of variables over the set of locations. When the observations refer to areal units, the correlation coefficient can be interpreted as *a standardized measure of areal association*. A correlation coefficient is a measure of the degree of similarity of the two maps of the individual variables.

Maps of nominal variables with two categories are known as *two-color maps*. Figure 11-9 illustrates two-color maps for five different nominal variables X, Y_1, Y_2, Y_3, and Y_4. Is the map of variable X similar to the maps of variables Y_1 to Y_4? The degree of similarity can be measured by any of the nominal-scale measures of

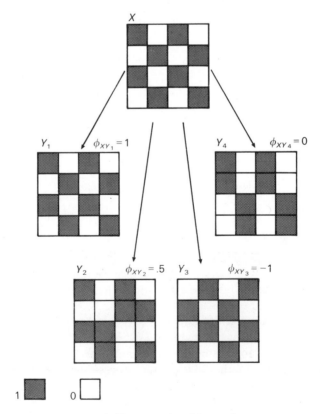

FIGURE 11-9. Five examples of two-color maps.

association. Consider first the maps of variables X and Y_1. These two maps are identical. Each corresponding cell of the maps has an identical value and thus color. The ϕ coefficient for these two variables is $+1$. This can be verified by a quick examination of the contingency table for X and Y_1.

Variable Y_1

		Black = 1	White = 0	Total
	Black = 1	8	0	8
Variable X	White = 0	0	8	8
	Total	8	8	16

The zeros on the off-diagonal area indicate the perfect association between these two variables. For variables X and Y_2, the relationship is not perfect. Six of the eight black cells of the map of variable X are black in the map of variable Y_2, and two are white. Also, six of the eight white cells of map X are white in map Y_2, and two are black. The contingency table

Variable Y_2

		Black = 1	White = 0	Total
	Black = 1	6	2	8
Variable X	White = 0	2	6	8
	Total	8	8	16

reveals a moderately strong relationship and $\phi = .5$.

Variable X is perfectly negatively correlated with variable Y_3. Each cell that is black in one map is white in the other, and vice versa. The ϕ coefficient of -1 verifies the perfect inverse relationship between these two variables. Although the maps of variables X and Y_4 both depict alternating patterns, the ϕ coefficient of zero indicates there is no relation between these two variables. The contingency table

Variable Y_A

		Black = 1	White = 0	Total
	Black = 1	4	4	8
Variable X	White = 0	4	4	8
	Total	8	8	16

verifies this. Although it is easy to visualize the areal correspondence of these relatively small and simple maps, it is not as easy to visually interpret the degree of areal association between complicated maps with many cells. Quantitative measures of association are particularly useful in these cases.

Ordinal or interval areal data are usually displayed on a *k-color map*. By dividing the data of a variable into three or more classes, a k-color map can be used to give a visual portrait of its areal distribution. For ordinal data one can assign a single color (or pattern or shade) to each of the four quartiles of the ranks, or 10 colors to the deciles. For interval data, the data can be divided into classes in several different ways. For example, a three-color map with the classes

$$Z > 1.0 \qquad \text{(black)}$$
$$-1.0 \le Z \le 1.0 \qquad \text{(gray)}$$
$$Z < -1.0 \qquad \text{(white)}$$

could be used to portray the distribution of an interval scale variable standardized by its standard deviation through Equation (5-33). In other cases more ad hoc classes may be defined. The bounds defining color categories should be carefully chosen. It is important that the visual impression of the maps of two variables accurately reflect the degree of association between them. Just as the selection of an improper interval width can mask important characteristics of the frequency distribution of a variable, so, too, can poorly chosen classes mask the pattern of a k-color map.

Let us consider a simple example drawn from political geography. The percentage support of a particular political party is known for the 16 electoral districts of the hypothetical city illustrated in Figure 11-10. These percentages are compiled in Table 11-8. It is hypothesized that support for this particular party is closely associated with certain demographic characteristics of the individual electoral districts. In this case,

TABLE 11-8
Electoral District Data

Zone	Voter support for party, %	Electoral district population ages 25 to 45, %
1	10	20
2	50	60
3	10	10
4	60	60
5	20	10
6	50	50
7	20	20
8	50	40
9	20	30
10	40	30
11	20	40
12	40	40
13	30	30
14	40	50
15	30	40
16	30	50

1	2	3	4
5	6	7	8
9	10	11	12
13	14	15	16

FIGURE 11-10. Sixteen electoral districts of a hypothetical city.

let us suppose that the party appeals mainly to the age group of 25- to 45-year-olds. The percentage of each electoral district's population in this age group is also listed in Table 11-8. A quick visual check of the areal association of these two variables can be made by comparing the k-color maps for each. To simplify matters, identical class bounds are used in each map and only three colors. As is clearly illustrated in Figure 11-11, these two maps are very similar. Eleven of the 16 zones have identical colors. One of the problems with k-color maps is that it is sometimes difficult to get an accurate visual impression of the degree of similarity. Too many colors can lead to unnecessary complexity, and too few colors can mask the real variation portrayed by the data. Whenever the data are grouped into classes and then translated into color categories, there is an inevitable loss of information.

There are two important problems in the use and interpretation of correlation coefficients for areal data: the *scale problem* and the *modifiable areal units* problem. All correlation coefficients measure the areal association between two variables only for the particular areal units used in the calculation. As we noted in Chapter 3, when areal data are aggregated into fewer and larger areal units, there can be substantial variations in the value of most statistics. This is the *scale or aggregation problem*. And

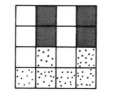

Percentage of party support

Percentage of population aged 25 to 45 years

0–20% ☐   21–40% ⬚   41–60% ■

FIGURE 11-11. k-color maps for variables in Table 11-8.

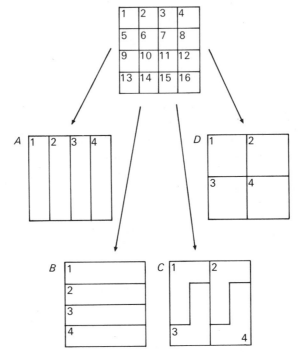

FIGURE 11-12. Alternative spatial aggregations.

TABLE 11-9
Zonal Aggregation of Electoral District Data

	Zone	Party support, %	Zone aged 25 to 45, %
Aggregation *A*	1	20	22.5
	2	45	47.5
	3	20	27.5
	4	45	47.5
Aggregation *B*	1	32.5	37.5
	2	35	30
	3	30	35
	4	32.5	42.5
Aggregation *C*	1	25	30
	2	27.5	32.5
	3	40	40
	4	37.5	42.5
Aggregation *D*	1	32.5	35
	2	35	32.5
	3	32.5	35
	4	30	42.5

even if we use the same number, size, and shape of areal units, different partitions of space can yield different values for the correlation coefficients for the same data. This is the *modifiable areal units problem*. Neither problem invalidates the use of correlation coefficients for areal data, but each presents specific interpretation problems.

First, let us illustrate the potential problems by using the voting behavior data. Figure 11-12 illustrates four possible aggregations in the data of the original 16 districts. In each of cases *A*, *B*, *C*, and *D* the original data are aggregated to four larger equal-area zones. The values for *X* and *Y* for zone 1 of aggregation *A* are *averages* of the four electoral districts numbered 1, 5, 9, and 13. The (X, Y) pair of $(20, 22.5) = ((10 + 20 + 20 + 30)/4, (20 + 10 + 30 + 30)/4)$. The values for all zones in each of the four aggregations are computed in the same way and are summarized in Table 11-9. To see the effects of the aggregation, it is useful to compare the scatter diagrams of each of the aggregations to the original data set. As is illustrated in Figure 11-13, the data of the original 16 districts exhibit the expected strong positive correlation between these two variables. The effects of aggregation are dramatic.

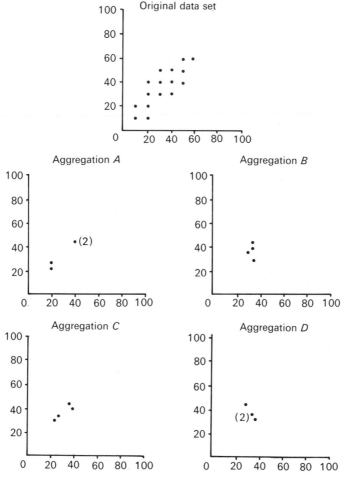

FIGURE 11-13. Scatter diagrams for aggregations of Figure 11-12.

TABLE 11-10
Correlation Coefficients for Alternative Spatial Aggregations
of Electoral District Data

Aggregation	Pearson's r
Original 16 zones	.81
Aggregation A	.99
Aggregation B	−.39
Aggregation C	.95
Aggregation D	−.94

Notice that there is a substantial reduction in the total variation of both variables. Aggregation D is the most extreme example. The range of the variable party support declines from $60 - 10 = 50$ in the original data to $37.5 - 32.5 = 5$. A similar reduction occurs for the percentage of the population aged 25 to 45 years.

The correlation coefficients between party support and percentage of the population aged 25 to 45 years are summarized in Table 11-10. Note the large discrepancies between the 16-district correlation coefficient $r = .81$ and the correlations for the four aggregations. Partly because the data have been selected in order to dramatize the possibilities, the correlations range from the almost perfect positive to the almost perfect negative. This is not uncommon when there are only a small number of observations. Even where the *scale* of the observations is held constant, there can also be large variations in the value of r. Notice that all four aggregations involve *contiguous* districts and have exactly the same areas. Despite this, the range of the correlation coefficient for these modified units has the extreme range of $+.99$ to $-.94$.

In many situations in the social sciences, the effects of data aggregation are neither as dramatic nor as random as occurs in the voting behavior case. A systematic *increase* in the value of the correlation coefficient appears to be common, particularly for cases in which contiguous grouping procedures are used. Different scales of aggregation produce different degrees of association between variables. When we attempt to interpret correlation coefficients or test hypotheses about the relationships between variables, using areal data, it is important to keep in mind what sociologists term *the ecological fallacy*. In short, it is fallacious to transfer conclusions concerning the relationship between variables obtained from data at one level of aggregation to other levels of aggregation. The most obvious fallacies occur when we attempt to make statements about the behavior of individuals, using data based on aggregates or groups of these individuals.

Consider the voting behavior data in this light. Of course, we are really interested in whether individuals aged 25 to 45 years are likely to support the party. Because of confidentiality, all that is available is data aggregated to the level of the electoral district. The fact that these two variables are strongly associated at the electoral district level does not necessarily mean that the same degree of association applies to *individual* voters. All we can legitimately say is that supporters of the party *tend to live in electoral districts with a high proportion of residents from 25 to 45 years*. The relationship holds between *areas*, not necessarily between *individuals*. In fact, it is possible that the people who support the party are of entirely different age groups. They may vote for this party in response to the 25- to 45-year-olds who populate their electoral districts! In many instances the solution to the scale problem is simply to employ individual or household data. At other times, one of the variables in which we are interested might necessarily require an areal definition. For example, individual house selling prices depend on many characteristics of the individual house and lot, but also on neighborhood variables which are based on areal aggregations of individual houses. Neighborhood stability, land-use mix, and other such variables are known to have significant effects on residential property values. Changes in scale may mask important associations between variables or overemphasize others.

Areal association studies represent an initial attempt to develop explanations for the areal variations of many phenomena. Unfortunately, the lack of geographical theory concerning the areal distribution of many phenomena hinders our ability to satisfactorily interpret the results of many correlation analyses. It is often difficult to explain observed scale and modifiable areal unit effects. Nevertheless, statistical correlation analysis represents one of the most widely used techniques of map comparison and areal association.

11.4. Spatial Autocorrelation

Correlation analysis is not limited to the comparison of the maps of two different variables. It can also be used as a measure of order or pattern in the spatial distribution of a single variable depicted on a two- or k-color choropleth map. The term *spatial autocorrelation* refers to the correlation of a variable with itself through space. If there is any systematic pattern in the spatial distribution of a variable, it is said to be spatially autocorrelated. If the pattern on the map is such that nearby or neighboring areas are more similar than more distant areas, the pattern is said to be *positively spatially autocorrelated*. This is a common occurrence in geography. The distribution of house prices or household income in a city is usually positively spatially autocorrelated. This is because wealthy households tend to live in exclusive neighborhoods, segregating themselves from lower-income households, relegating these households to other areas in the city. Many land-use activity types are agglomerated in space and also exhibit positive spatial autocorrelation. Variables collected by physical geographers can also display this pattern. Climatologists find similar temperature and precipitation levels in adjacent areas. *Negative spatial autocorrelation* describes patterns in which neighboring areas are unlike. This is not a common pattern in geography. Between the two extremes of positive and negative autocorrelation are *random* patterns which exhibit *no spatial autocorrelation.*

There are two reasons why autocorrelation is of concern to geographers. First and foremost, the search for spatial pattern is one of the dominating themes of geographical research. Areal distributions are rarely random. If there are any "laws" in geography, then surely the first is "Nearby locations are more alike than more distant ones." In statistical jargon this means the areal distributions of many phenomena are positively spatially autocorrelated. The second reason for studying spatial autocorrelation is that the inferential techniques presented in Chapters 8 to 10 are

Map *A*   Map *B*

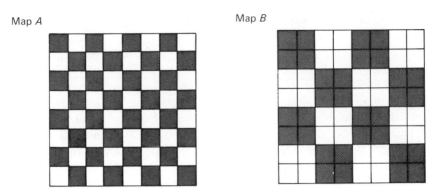

FIGURE 11-14. Map patterns.

based on the assumption that the values of the observations in each sample are
independent of one another. One of the ways in which spatial data might violate this
assumption is if the data reveal a pattern of spatial autocorrelation. If two of the
observations in a sample are from nearby locations, they are not usually independent.
Our interest in spatial autocorrelation is twofold: to measure the strength of spatial
autocorrelation in a map and to test the assumption of independence or randomness.

One difficulty in devising measures of spatial autocorrelation is that many
different patterns could exist in a map. Each measure and test is useful in detecting
only one of these patterns. To see the problem, consider the two patterns in Figure
11-14. Both exhibit a pattern of negative spatial autocorrelation when they are examined at different scales. However, any test of the pattern is scale-dependent. Note that
no black cell in map *A* is contiguous with another black cell. White cells have the
same property. In map *B* each black cell is contiguous with two other black cells. A
test to uncover the simple pattern in map *A* may fail to uncover the pattern in map
B and vice versa. To illustrate the procedures of spatial autocorrelation, we restrict
ourselves to simple contiguity tests which examine the correlation between a cell and
those neighboring cells with which it shares a boundary. First, let us consider a simple
test for nominal variables with two categories based on a two-color map. This will
introduce the basic rationale behind the more complicated methods used to analyze
k-color maps.

A Contiguity Test for Two-Color Maps

Measures of spatial autocorrelation for nominal variables mapped in two colors are
based on the pattern of joins between contiguous areas or cells of the map. Each cell
is colored either black (B) or white (W). The *join*, or border, between adjacent cells
can be classed as white-white (WW), black-black (BB), or (BW) depending on the
colors of the two cells. Only joins of nonzero length are considered, and diagonal cells
are by definition noncontiguous. Different patterns of autocorrelation have different
numbers of joins in each of the BB, WW, and BW categories. Compare the three
maps of Figure11-15. Map *A* is negatively autocorrelated. Adjacent cells are unlike
since they are of different colors. All joins are BW. Map *C* represents the other

Map A 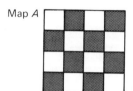 Map B 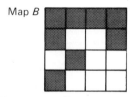 Map C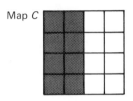

FIGURE 11-15. Two-color maps.

extreme, positive autocorrelation. Most of the joins are BB or WW. Only the four along the middle boundary are BW. In between these two extremes lies map B. The pattern is a mix between the alternating pattern of map A and the clustering pattern of map C. There is almost an equal number of each type of join. A summary of the join counts for these three maps confirms these observations:

		Join counts			
		BB	WW	BW	Total
	A	0	0	24	24
Map	B	5	7	12	24
	C	10	10	4	24

The simplest tests for autocorrelation use the numbers of joins of each type as an indicator of the type of autocorrelation displayed on the map.

To illustrate the two-color test for contiguity, let us analyze map B. To begin, we must enumerate the number and type of joins on the map. There are $n = 16$ cells numbered consecutively, beginning in the northwest corner and proceeding by rows. There are $n_B = 8$ black cells and $n_W = 8$ white cells. Table 11-11 is a convenient format for enumerating the information required for a contiguity test. Let L_i be the number of joins of cell i. Cell 1 has $L_1 = 2$ joins, cell 2 has $L_2 = 3$ joins, and so on. The total number of joins on the map is $L = \sum_{i=1}^{n} L_i/2$. The sum of L_i is divided by 2 to avoid double counting of joins. In map B there are $\frac{48}{2} = 24$ joins. Now, if we know a map has n_B black cells and $n_W = n - n_B$ white cells, the probability that a randomly selected cell is black is $p = n_B/n$. The probability it is white is $q = n_W/n$, and $p + q = 1$. If it is selected randomly, cell color is a *binomial* random variable.

If the areal pattern on the map is random, how many joins of each type would we expect in an n-cell region with L total joins? Begin by selecting a color for the first cell. The probability it is black is p, and the probability it is white is q. Now, consider a second cell contiguous with zone 1. The probability it is black is also p. The probability that the join between the two cells is BB is $p \cdot p = p^2$. To get a WW join, the required probability is $q \cdot q = q^2$. BW joins occur in two ways. Either a black cell can be followed by a white cell, or vice versa. Both probabilities are equal to pq, so that the probability of a BW join is $pq + pq = 2pq$. If there are L joins on the map, the expected number of cells of each type is simply each probability multiplied by L:

$$E(BB) = \mu(BB) = p^2 L$$

$$E(WW) = \mu(WW) = q^2 L \qquad (11\text{-}18)$$

$$E(BW) = \mu(BW) = 2pqL$$

TABLE 11-11
Join Counts for Map B of Figure 11-12

Cell	Color	Joins			L_i	$L_i(L_i - 1)$
		BB	WW	BW		
1	B	2	0	0	2	2
2	B	2	0	1	3	6
3	B	2	0	1	3	6
4	B	2	0	0	2	2
5	B	1	0	2	3	6
6	W	0	1	3	4	12
7	W	0	2	2	4	12
8	B	1	0	2	3	6
9	W	0	0	3	3	6
10	B	0	0	4	4	12
11	W	0	3	1	4	12
12	W	0	2	1	3	6
13	B	0	0	2	2	2
14	W	0	1	2	3	6
15	W	0	3	0	3	6
16	W	0	2	0	2	2
Totals	$n_B = 8$	10	14	24	48	104
	$n_W = 8$					

$$p = \tfrac{8}{16} = .5 \qquad L = \tfrac{1}{2}(48) = 24$$
$$q = \tfrac{8}{16} = .5 \qquad K = \tfrac{1}{2}(104) = 52$$

$$BB = \tfrac{1}{2}(10) = 5 \qquad WW = \tfrac{1}{2}(14) = 7 \qquad BW = \tfrac{1}{2}(24) = 12$$

where $\mu(BB)$ is the expected or mean number of black-black joins, on a map with a random pattern. The expected number of joins adds to L since $p^2 L + q^2 L + 2pqL = (p + q)^2 L = L$ (since $p + q = 1$, by definition).

Of course, these expectations will not occur on each random two-color map, and it is necessary to specify the standard deviations of the sampling distributions of BB, WW, and BW joins. Since the details are quite complicated, the standard deviations are presented without comment:

$$\sigma(BB) = \sqrt{p^2 L + p^3 K - p^4(L + K)}$$
$$\sigma(WW) = \sqrt{q^2 L + q^3 K - q^4(L + K)} \qquad (11\text{-}19)$$
$$\sigma(BW) = \sqrt{2pqL + pqK - 4p^2 q^2(L + K)}$$

where $K = \tfrac{1}{2} \Sigma_{i=1}^{n} L_i(L_i - 1)$ is a constant for any map. It turns out that the sampling distributions of the number of joins of each type are normal under the following general conditions:

1. The number of cells is large, that is, greater than approximately 30.
2. Both $p, q > .2$.
3. The region is not elongated.
4. The join structure is not dominated by a few cells. The normal approximation is not close if one cell (or a small number of cells) has many joins and most others have only 1 or 2.

When these conditions are satisfied, the appropriate test statistics for each of the join counts are

$$Z(BB) = \frac{BB - \mu(BB)}{\sigma(BB)}$$

$$Z(WW) = \frac{WW - \mu(WW)}{\sigma(WW)} \tag{11-20}$$

$$Z(BW) = \frac{BW - \mu(BW)}{\sigma(BW)}$$

where BB, WW, and BW are the *observed* number of joins of each type.

Since it contains only 16 cells, map B does not satisfy the requirements for normality Z_{BB}, Z_{WW}, and Z_{BW}. The computational steps required in the test can still be illustrated with this example, if we recognize the fact that it is a poor approximation in this case. Using the information contained in Table 11-11, we first calculate $p = q = .5$, $L = 24$, and $K = 52$. These values are then substituted to obtain

$$\mu(BB) = 24(.5)^2 = 6$$

$$\mu(WW) = 24(.5)^2 = 6$$

$$\mu(BW) = 2(24)(.5)(.5) = 12$$

Notice that these expected numbers of joins are $6 + 6 + 12 = 24$. The standard errors of the sampling distributions are calculated from (11-9):

$$\sigma(BB) = \sqrt{(.5)^2(24) + (.5)^3(52) - (.5)^4(24 + 52)} = 2.78$$

$$\sigma(WW) = \sqrt{(.5)^2(24) + (.5)^3(52) - (.5)^4(24 + 52)} = 2.78$$

$$\sigma(BW) = \sqrt{2(.5)(.5)(24) + (.5)(.5)(52) - 4(.5)^2(.5)^2(24 + 52)} = 2.45$$

Substituting this information into (11-20) yields the following values for $Z(BB)$, $Z(WW)$, and $Z(BW)$:

$$Z(BB) = \frac{5 - 6}{2.78} = -.36$$

$$Z(WW) = \frac{7 - 6}{2.78} = .36$$

$$Z(BW) = \frac{12 - 12}{2.45} = 0$$

Taking these values of Z to the standard normal table, we find that the PROB-VALUES for each type of join are $P(Z(BB) \geq -.36) = P(Z(WW) \geq .36) = .3594$ and $P(Z(BW) \geq 0) = .5000$. These PROB-VALUES suggest that the numbers of joins in map B frequently occur in random patterns with 24 joins. We cannot reject the null hypothesis H_0: "There is no autocorrelation of contiguous cells in map B" at any common level of significance. We must conclude that the pattern on the map is *not* spatially autocorrelated. Another example of the use of this test is provided in Chapter 13 where it is used to test one of the key assumptions of bivariate regression analysis.

This test can also be used for nominal variables with more than two classes expressed as k-color maps. The extension is quite straightforward. Details can be obtained in Cliff and Ord (1973, 1981).

A Contiguity Measure and Test for Ordinal and Interval Data

In general, it is inadvisable to group ordinal and interval data into classes and use the join-count statistical test described above to measure autocorrelation. There is a significant loss of information when the level of measurement of the variables is reduced in this way. Instead it is advisable to use a measure of spatial autocorrelation which employs the actual ordinal or interval value of a cell in its calculations. One such measure is *Geary's contiguity ratio c*. This statistic relates the difference between contiguous cells to the differences in the entire set of cells which comprise the map.

The first step is to construct a contiguity matrix. Each entry in this matrix δ_{ij} is assigned the value 1 if cells i and j are contiguous and 0 otherwise. By definition, $\delta_{ij} = 0$ for all i. For a map with n cells the contiguity matrix is square of order $n \times n$ and symmetric. The contiguity matrix for the variable percentage party support (see Table 11-8 and Figure 11-10) is given in Table 11-12. Since each of the 24 joins appears twice in the matrix, there are 48 entries of 1. Geary's contiguity ratio c uses the information in this contiguity matrix and the actual ordinal or interval cell values in the formula

$$c = \frac{(n - 1) \sum_{i=1}^{n} \sum_{j=1}^{n} \delta_{ij}(X_i - X_j)^2}{4L \sum_{i=1}^{n} (X_i - \bar{X})^2} \tag{11-21}$$

where n is the number of cells, L is the total number of joins, and X_i is the value of

TABLE 11-12
Contiguity Matrix for Map of Figure 11-10

Cell	1	2	3	4	5	6	7	8	9	10	11	12	13	14	15	16
1	1	1			1											
2	1	1	1			1										
3		1	1	1			1									
4			1	1				1								
5	1				1	1			1							
6		1			1	1	1			1						
7			1			1	1	1			1					
8				1			1	1				1				
9					1				1	1			1			
10						1			1	1	1			1		
11							1			1	1	1			1	
12								1			1	1				1
13									1				1	1		
14										1			1	1	1	
15											1			1	1	1
16												1			1	1

Note. Only 1s are shown.

cell i. This statistic takes the ratio of the sum of the squared differences between contiguous cells to the sum of the squared differences of each cell value to the mean cell value on the map. In the numerator the squared difference between cells enters into the summation only when the two cells are contiguous and $\delta_{ij} = 1$. The coefficient $(n - 1)/(4L)$ is used to scale the ratio, so that the expected value of c, $E(c)$, for a random map pattern is 1. Of course, we cannot expect all maps without autocorrelation to have $c = 1$. The sampling distribution of c for a map with no autocorrelation is normal with $E(c) = 1$ and a standard deviation given by

$$\sigma(c) = \sqrt{\frac{(2L + K)(n - 1) - 2L^2}{(n + 1)^2}} \tag{11-22}$$

where K is defined as $\frac{1}{2}\Sigma_{i=1}^{n} L_i(L_i - 1)$. This leads to the following test statistic:

$$Z_c = \frac{E(c) - c}{\sigma(c)} \tag{11-23}$$

The preliminary calculations for Geary's c for the party support data are given in Table 11-13. Consider cell 1. It has a value of 10 and is contiguous with two cells— cell 2 with a value of 50 and cell 5 with a value of 20. Two terms enter the numerator of c, the values of which are listed in the last column. The mean party support in the city is 32.5, so that $(X_1 - \bar{X})^2 = (10 - 32.5)^2 = 506.25$. The 16 terms in the summation for the denominator are listed in the second column, one for each cell. Table 11-13 is a convenient format for organizing the calculations for Geary's c. The values of L and K are 24 and 52, respectively. Note that these values are the same as those obtained in the two-color join-count test for map B of Figure 11-15. This is because both maps are regular 4×4 lattices.

Substituting these calculated values into (11-21) yields

$$c = \frac{15(21{,}200)}{4(24)(3500)} = .946$$

We now calculate $\sigma(c)$:

$$\sigma(c) = \sqrt{\frac{[2(24) + 52]15 - 2(24)^2}{17^2}} = 1.097$$

The test statistic

$$Z_c = \frac{1 - .946}{1.097} = .049$$

leads us to accept the null hypothesis of no autocorrelation at $\alpha = .05$ since the calculated test statistic is less than 1.96.

TABLE 11-13
Work Table for Geary's Contiguity Ratio for Party Support Data

Cell	$(X_i - \bar{X})^2$	Joins	$X_i - X_j$	$(X_i - X_j)^2$
1	506.25	1–2	-40	1600
		1–5	-10	100
2	306.25	2–1	40	1600
		2–3	40	1600
		2–6	0	0
3	506.25	3–2	-40	1600
		3–4	-50	2500
		3–7	-10	100
4	756.25	4–3	50	2500
		4–8	10	100
5	156.25	5–1	10	100
		5–6	-30	900
		5–9	0	0
6	306.25	6–2	0	0
		6–5	30	900
		6–7	30	900
		6–10	10	100
7	156.25	7–3	10	100
		7–6	-30	900
		7–8	-30	900
		7–11	0	0
8	306.25	8–4	-10	100
		8–7	30	900
		8–12	10	100
9	156.25	9–5	0	0
		9–10	-20	400
		9–13	-10	100
10	56.25	10–6	-10	100
		10–9	20	400
		10–11	20	400
		10–14	0	0
11	156.25	11–7	0	0
		11–10	-20	400
		11–12	-20	400
		11–15	-10	100
12	56.25	12–8	-10	100
		12–11	20	400
		12–16	10	100
13	6.25	13–9	10	100
		13–14	-10	100

(*continued*)

TABLE 11-13 (continued)

Cell	$(X_i - \bar{X})^2$	Joins	$X_i - X_j$	$(X_i - X_j)^2$
14	56.25	14–10	0	0
		14–13	10	100
		14–15	10	100
15	6.25	15–11	10	100
		15–14	−10	100
		15–16	0	0
16	6.25	16–12	−10	100
		16–15	0	0

$$\sum_{i=1}^{n}(X_i - \bar{X})^2 = 3500.00 \qquad \sum_{i=1}^{n}\sum_{j=1}^{n}(X_i - X_j)^2 = 21{,}200$$

Limitations of Autocorrelation Measures

The join-count two-color test and Geary's contiguity ratio have two distinct limitations. First, many different partitions of space have identical join structures. Figure 11-16 illustrates three different cell patterns, each of which has the contiguity matrix

$$\begin{array}{c c c c c}
 & A & B & C & D \\
A & 0 & 1 & 1 & 0 \\
B & 1 & 0 & 0 & 1 \\
C & 1 & 0 & 0 & 1 \\
D & 0 & 1 & 1 & 0 \\
\end{array}$$

In (*a*), there are four equal-sized cells, in (*b*) there is a single cell of dominating size, and in (*c*) the map does not completely encompass the area inside its exterior

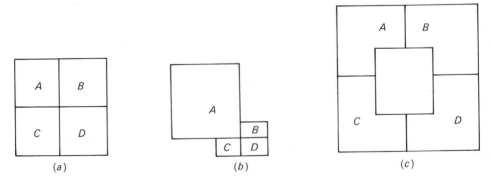

FIGURE 11-16. Different partitions of space with identical join structures.

boundary. These are quite distinct patterns in themselves, and there appear to be few reasons to expect similar autocorrelation patterns.

The second problem with these measures is that they are capable of evaluating only the simplest forms of spatial autocorrelation—between contiguous zones. What about more complex patterns of areal dependency? If we use a better measure of cell separation, say intercell distance, we might be able to evaluate many more patterns than is possible with contiguity-based measures. In-depth treatment of the problem of spatial autocorrelation may be found in Cliff and Ord (1973, 1981).

11.5. Summary

This chapter has introduced the notion of relations between variables and several aspects of bivariate statistical analysis. Of particular importance is the notion of correlation, which is used to measure the degree of association between two variables measured over the same set of observations. Correlation coefficients can be defined for interval variables (Pearson's r), ordinal variables (Kendall's τ and Spearman's ρ), and for nominal variables (for example, ϕ). These measures can be used to gauge the strength of the association and as input to tests of statistical significance. Geographic applications of correlation analysis incorporate the notion of areal association. The scale and modifiable areal units problems, first described in Chapter 3, surface again. Difficulties in interpretation of correlation coefficients often arise. Correlation analysis used to uncover the spatial pattern on a single map is known as spatial autocorrelation analysis. Measures and tests of autocorrelation for two- and k-color maps are indicators of certain systematic map patterns.

REFERENCES

A. D. Cliff and J. K. Ord, *Spatial Autocorrelation* (London: Pion, 1973).
A. D. Cliff and J. K. Ord, *Spatial Processes* (London: Pion, 1981).
S. Siegel, *Nonparametric Statistics for the Behavioral Sciences* (New York: McGraw-Hill, 1956).

FURTHER READING

Examples of the application of correlation analysis in social science research are contained in virtually all introductory textbooks for statistical methods in the various disciplines. Several textbooks for geographers are cited at the end of Chapter 2. Extensions to, and applications of, nonparametric methods of correlation can be found in the textbooks cited at the end of Chapter 10. Advanced material on spatial autocorrelation can be found in the two texts by Cliff and Ord (1973, 1981) listed in the References. The problems which follow also include some examples relevant to geographers.

PROBLEMS

1. Explain the meaning of the following terms or concepts:
   a. Correlation coefficient
   b. Covariance
   c. Correlation matrix
   d. Monotonic increasing function
   e. Concordant observations
   f. Spatial autocorrelation
   g. Contiguity
   h. Ecological fallacy
   i. Two-color map
   j. k-color map

2. Explain what is meant by the concept of a correlation coefficient between (a) two ordinal variables and (b) two nominal variables.

3. What pattern would appear on a map characterized by positive spatial autocorrelation? Negative spatial autocorrelation?

4. What is the value of the correlation coefficient r between two variables X and Y when (a) variable X always takes on some constant value C and, (b) the value of variable Y is always 3 times the value of variable X? (*Hint*: Draw a graph.)

5. In each of the following cases, state whether you would expect the correlation coefficient r to be positive, negative, or near zero.
   a. X = stream discharge
      Y = stream depth
   b. X = distance from a neighborhood to the central business district
      Y = residential density
   c. X = sound level, in decibels
      Y = distance from a major urban expressway
   d. X = number of patrons of a day care center
      Y = distance traveled to the center
   e. X = percentage of neighborhood residents who are black
      Y = percentage of neighborhood residents who are Caucasian
   f. X = percentage of neighborhood residents who are Australian
      Y = percentage of neighborhood residents who are Icelandic

6. A behavioral geographer is interested in whether two different methods of estimating familiarity with a set of neighborhoods, method A and method B, give similar results. Each method uses questions concerning the knowledge of subjects of neighborhood landmarks, streets, activities, and residents. Method A uses an entirely different set of items from method B. Ten subjects are given both tests. An interval between the tests of 1 month is used to eliminate possible bias. Five subjects are given method A first, and five subjects are given method B first. These subjects are chosen randomly. The results are summarized as follows:

	Scores	
Subject	Method A	Method B
1	58	72
2	49	63
3	60	73
4	81	82
5	92	81
6	34	50
7	14	32
8	68	74
9	25	41
10	72	80

a. Plot a scattergram of these data.
b. Calculate Pearson's r.
c. Rank each of the variables and calculate Spearman's r_S.
d. What do the results suggest?
e. Test the significance of the correlations.

7. Calculate Kendall's τ for the data in Problem 6.

8. Consider the following two-color map:

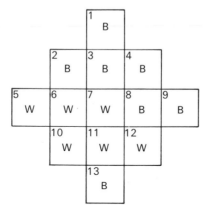

Note: Cell numbers are located in the upper left corner of each cell; B = black, W = white.

a. Use the contiguity test for two-color maps to test for possible spatial autocorrelation in this map.
b. Suppose that the values for the variable depicted on this map are, in order of cell number, 3, 4, 7, 6, 3, 5, 11, 14, 18, 21, 17, 16, 5. Calculate Geary's contiguity ratio and test it for significance.
c. Group 12 of the cells in pairs, leaving one cell ungrouped. Replace each cell with the mean of the two cells from which it has been formed. Construct a two-color map of this new configuration with black assigned to values below the mean and white to values above the mean. Test the new map for autocorrelation.
d. Calculate Geary's contiguity ratio for the new map created in part (c).

9. Residents of a city were surveyed by telephone to determine their support for a proposal to construct a multipurpose stadium on a large tract of previously undeveloped land on the outskirts of the city. When cross-tabulated with age of the respondent, the results of the survey are as follows:

	Age of respondent		
	< 30	31–44	45+
Approve	43	13	10
Neutral or no opinion	27	7	60
Disapprove	17	72	12

a. Calculate Tschuprow's T, Cramer's V, Pearson's C, and Goodman and Kruskal's τ_b for these data.
b. Does age of respondent correlate with support for the stadium?
c. What other ways might one analyze these data?

12

Introduction to Regression Analysis

Suppose now that we wish to go beyond measuring the mere strength of association between two variables through a correlation coefficient. In particular, we might be interested in the precise mathematical form of the function linking variable X to variable Y. We might even wish to take this further and use our derived mathematical function to predict the value of variable Y from our knowledge of variable X. This is a problem in regression analysis.

Regression analysis can be broadly defined as the analysis of statistical relationships among variables. There are two types of variables in regression problems. First, there are the *dependent variables*, which are also often termed *response*, or *endogenous*, variables. By convention, these variables are normally labeled $Y_1, Y_2, \ldots, Y_r$. Second, there are the *independent*, or *predictor*, or *exogenous*, *variables*. These are usually labeled $X_1, X_2, \ldots, X_s$. In this chapter we confine ourselves to simple regression problems. As the name suggests, simple regression concerns itself with the case in which we have one dependent variable Y and one independent variable X. For purposes of clarity, we now suppress the subscript on these two variables since there is only one variable of each type.

Although the terms *dependent* (or response) and *independent* (or predictor) *variables* are quite conventional, no implication of causality is necessarily implied in any given case. This is true no matter how strong the statistical relationship might be. In some instances there may be strong a priori grounds for the specification of a cause-and-effect relationship and the selection of a dependent variable and one or more independent variables. In many other situations this may not be easy. Geography, and social science generally, is characterized by the recognition of relationships between variables which are unclear, ambiguous, or possibly even exhibiting two-way dependence. The difficulty in establishing clear-cut cause-and-effect relations most often derives from the nature of theory in social science. In the physical sciences, however, the specification of regression relations is often straightforward. For example, the rate at which some chemical reaction occurs can be made a function of the ambient temperature. Cause and effect are clear. The temperature is not caused by the rate of the chemical reaction! In this chapter let us assume that causal and functional dependence go hand in hand. Much of the material is easier to comprehend if we make this assumption.

Section 12.1 describes the situations in which we employ regression analysis and introduces the simple linear regression model. In Section 12.2 we discuss the procedure used to estimate the parameters of the simple regression model. Section 12.3 details various methods of assessing the goodness of fit of the regression equation to the observed data. Here we will see the intimate relationship which exists between correlation and regression. In Section 12.4 the use of a regression equation in the context of prediction is examined, as well as several issues concerning the interpretation of regression parameters and regression equations. Finally, in Section 12.5 we briefly introduce some technical and methodological issues which sometimes arise in empirical applications.

12.1. Simple Linear Regression Model

Relations between Variables

The concept of a relation between variables is a familiar one to geographers. We speak, for example, of the relationship between land rent and distance to the central business district (CBD) in cities or of the relationship between stream discharge and sediment load. However, let us now distinguish a *functional* relationship from a *statistical* one.

FUNCTIONAL RELATIONS

A functional relation between two variables X and Y is always expressed by a mathematical formula of the form $Y = f(X)$. Given any value of X, the function f indicates the corresponding value of Y. The function f can take on any form whatsoever, but let us consider one particular example involving a linear function (see Appendix 12A for a review of the elementary geometry of straight lines).

EXAMPLE 12-1.   A shipper charges prospective customers according to the following schedule for a full truckload delivery

$$Y = 50.00 + 0.40X$$

Here Y is the total shipment cost in dollars, 0.40 is the per-mile rate in dollars, and X is the distance to the destination in miles. The bills of lading for four recent customers provide the following data:

Shipment No.	Total cost, $	Miles
1	70.00	50
2	90.00	100
3	100.00	125
4	124.00	185

These values are plotted in Figure 12-1. Note that all points lie directly on the line. There is no error. All functional relations share this characteristic.

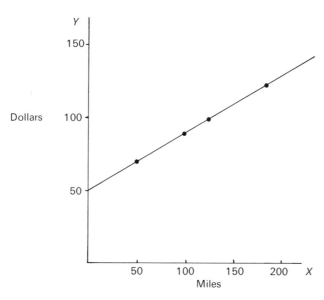

FIGURE 12-1. Graph for Example 12-1.

STATISTICAL RELATIONS

A statistical relation is not necessarily a perfect one. Except in the most unlikely cases, the observations of a statistical relation do not fall on the line or curve which expresses the functional relationship.

EXAMPLE 12-2. Scientists at an experimental agricultural station applied varying amounts of fertilizer to homogeneous 2-acre plots and recorded the yield from each plot at the end of the harvest. The graph of corn yield versus the amount of fertilizer applied is shown in Figure 12-2. The figure clearly suggests there is a relation between these two variables. As increasing amounts of fertilizer are applied, there is a corresponding increase in the yield of corn. The relation is not a perfect one, and the scatter of points suggests that some of, but not all, the variation in corn yield of these plots can be accounted for by the variations in the amount of fertilizer applied. For example, two of the plots are treated with 300 lb of fertilizer per acre, but yields are somewhat different. Because of the scattering of points in a statistical relation, we often also call plots such as Figure 12-2 *scatter diagrams* or *scattergrams*. Each point in the scattergram represents one of our observations.

In Figure 12-2 we have also plotted a line which describes this statistical relation reasonably well. Most of the points fall close to the line, but the scatter suggests there is some variation in corn yield not accounted for by the amount of fertilizer applied. If the experimental agricultural station irrigated each plot equally, and if the underlying soil and drainage are homogeneous, we might think of this error as being random. Notice that even though we do not have an exact functional relationship, we still have what appears to be a highly useful statistical one.

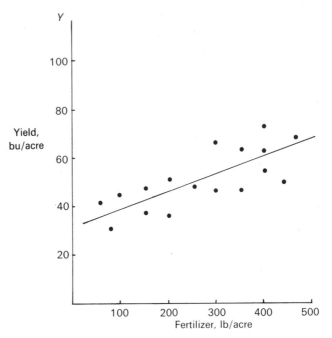

FIGURE 12-2. Graph for Example 12-2.

Linear Regression Model

A linear regression model is a formal means of expressing the two characteristic ingredients of a statistical relationship. First, the dependent variable Y varies systematically with the independent variable or variables. Second, a scattering of the observations occurs around this systematic component. Although the systematic component can take on any form and we can have as many variables in the relation as is necessary, we restrict ourselves to the simple two-variable regression model. We conceptualize the problem in the following way.

FIXED-X MODEL

Suppose we are able to control the value of variable X at some particular value, say at $X = X_1$. At $X = X_1$ we have a distribution of Y values and a mean value of Y for this value of X. There are similar distributions of Y and mean responses at each other fixed value of X, $X = X_2$, $X_3 = X_4$, ..., $X = X_n$. The regression relationship of Y on X is derived by tracing the path of the mean value of Y for these fixed values of X. Figure 12-3 illustrates a regression relation for a fixed-X model. Now the regression function which describes the *systematic* part of this statistical relation may be of any form whatsoever. We restrict ourselves to linear regression equations such as the one graphed in Figure 12-4(a). The function which passes through the means has the equation

$$Y = a + bX \tag{12-1}$$

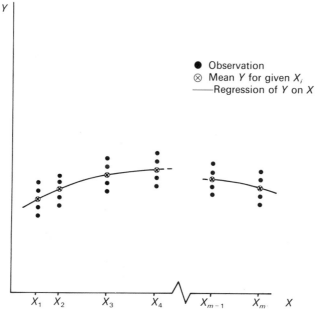

FIGURE 12-3. Regression relationship.

which is the familiar slope-intercept form of a straight line. Although we limit ourselves to linear regression equations in this chapter, regression analysis can be utilized in situations for which the path of the means follows many different functional forms. Figures 12-4(*b*) to (*e*) illustrate several potential alternatives and the appropriate functional forms.

EXAMPLE 12-3. To examine the rate of noise attenuation with distance from a major transport facility, sound recorders are placed at intervals of 50 ft over a range of 50 to 500 ft from the centerline of an expressway. Sound recordings are then analyzed, and the sound level in decibels which is exceeded 10 percent of the day is calculated. This measure, known as L_{10}, is a general indicator of the highest typical sound levels at a location. Recordings are then repeated for five consecutive weekdays by using the same sound recorders placed at the identical locations. A plot of the data generated from this experiment is shown in Figure 12-5.

The graph reveals the typical pattern of distance decay. The variation in decibel levels at any fixed location either is due to the effects of other variables—perhaps variations in traffic volumes—or can be thought to represent random error. Let us presume that all other variables which might have caused sound-level variations are more or less constant and that the variations do represent random error. Note that the path of the mean sound levels with distance follows the linear function superimposed on the scattergram. One would feel quite justified in fitting a linear regression equation to this scatter of points.

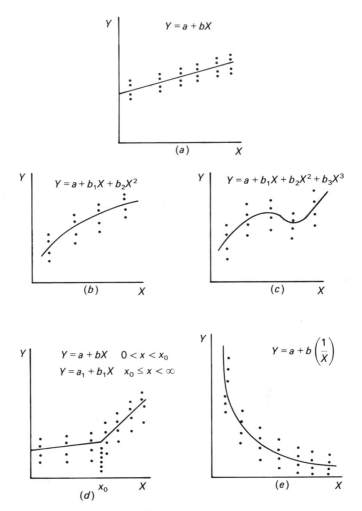

FIGURE 12-4. (*a*) Linear; (*b*) quadratic; (*c*) cubic; (*d*) piecewise linear; (*e*) reciprocal.

RANDOM-X MODEL

If situations in which we can control the values of X so that they are limited to certain predetermined levels were the only conditions in which regression analysis could be applied, it would be of limited value in geographical research. Only rarely do geographers gather data of this form. They are more commonly collected in experimental situations. *All* the *mechanical* operations presented in this chapter can be performed on *any* set of paired observations of two variables X and Y. And, as we shall see in Chapter 13, virtually all the inferential results of regression analysis apply to data characteristic of the fixed-X model *and* data in which *both* X and Y can be considered random variables. This greatly generalizes the applicability of the regression model. Let us consider an example of a random-X model.

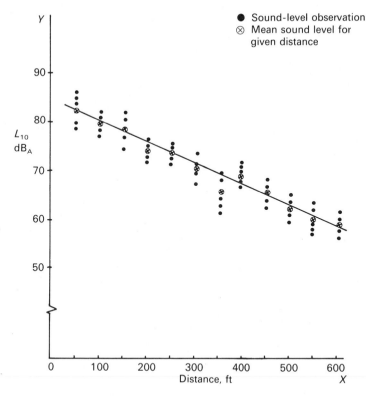

FIGURE 12-5. Graph for Example 12-3.

EXAMPLE 12-4. One of the key tasks in metropolitan area transportation planning is to make estimates of the total number of trips made by households during a typical day. Data collected from individual households are usually aggregated to the level of the *traffic zone*. Traffic zones are small subareas of the city with relatively homogeneous characteristics. The dependent variable is the number of trips made by a household per day, known as the *trip generation rate*. One of the most important determinants of variations in household trip generation rates is household income. We would expect the total number of trips made by a household in a day to be positively related to the income of the household.

Consider the hypothetical city of Figure 12-6 which has been divided into 12 traffic zones. Values of the average trip generation rates and household income for each of 12 zones are listed in Table 12-1. The data are typical for a North American city. Inner-city traffic zones 1 to 4 generally are composed of lower-income households who make few trips per day. Suburban zones 5 to 12 are populated with higher-income households with high trip generation rates. The scattergram of Figure 12-7 verifies the positive relation between trip generation and household income.

Notice how these data differ from the data of a fixed-X model. First, we do not have more than one observation of trip generation rates for any single level of household income. Second, we do not have observations of trip generation rates for

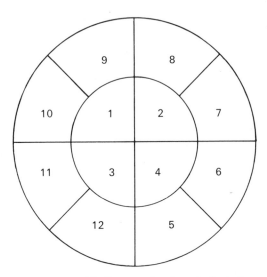

FIGURE 12-6. Traffic zones in a hypothetical city.

systematic values of household income, say every $500 or $1000 of income. Neverthe-less, it appears that the simple linear regression of household trip generation rates on income represents a useful statistical relation. We are going to utilize this example throughout this chapter as well as in Chapter 13.

But first let us examine the nature of these data. In what sense can we think of variable X as being a random variable? We might think of household income as a random variable since it is derived from one particular spatial aggregation of house-holds into traffic zones. The traffic zones of Figure 12-6 can be thought of as one random choice among all the possible ways in which we might construct a set of

TABLE 12-1
Trip Generation Rates and Household Income in a Hypothetical City

Traffic zone	Trips per household per day	Average household income, $000
1	3	10
2	5	12
3	5	11
4	4	9
5	6	14
6	7	16
7	9	21
8	7	18
9	8	22
10	8	24
11	6	17
12	5	15

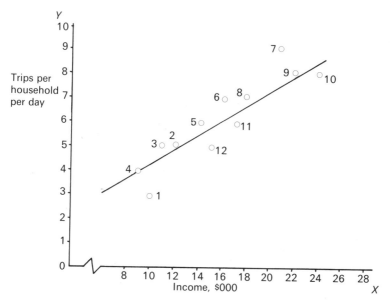

FIGURE 12-7. Scattergram for Example 12-4.

traffic zones from the household data. Other aggregations to different traffic zones would yield different average incomes and different trip generation rates. This is not a very persuasive argument. We would think of income as being a random variable if the original observations from which we generated Table 12-1 were a random sample taken from all the households in the city. If this were the case, then we might have some justification in treating income as a random variable. But what if the aggregated data represent all the households in the city? Can we make a case for treating income as a random variable? As you see, many applications of regression analysis in geography raise thorny theoretical issues. Most of these issues arise in situations in which we utilize regression analysis in an inferential mode. We discuss them further in Chapter 13.

12.2. Estimation of a Linear Regression Function

Ordinarily, we do not know the values of the regression parameters a and b, and we must *estimate* their values from our data. Once we know a and b, we can precisely locate our regression line within the scattergram. The method used to derive a and b from our set of observations does not depend on the type of input data at our disposal. It can be consistent with either the fixed-X or the random-X model. We utilize the data from our trip generation example, given in Table 12-1.

Let us denote (X_i, Y_i) as the values of our two variables for the ith observation. There are n different observations, and thus i can take on the values $1, 2, \ldots, n$. Now if the scatter of points lies quite close to falling on a single line, we could probably do a fairly reasonable job of fitting the line by eye. We could then simply estimate the values of a and b from the graph. This is seldom the case in reality. For example, in Figure 12-7, it is not at all clear where our regression line would be drawn. What we need is an objective method of fitting the line which can be easily operationalized. Fortunately, a simple algebraic solution to this problem exists.

Possible Criteria for Fitting a Regression Line

First, we must choose the criterion to assess the success of our line of best fit. An approach based on the deviations of the points to the line is ideal. Since we are interested in explaining the variation in variable Y, it seems natural to measure the *vertical* deviations from the observed Y to the line. We do this because the regression line should pass through the mean value of Y for any given X. Our error is the difference between the mean value of Y predicted by our line and the observed Y_i. This measure of error is illustrated in Figure 12-8.

First, we locate the point on the line corresponding to the observed X by defining $\hat{Y}_i$ (read "Y hat i") $= a + bX_i$. The deviation or error is therefore $e_i = Y_i - \hat{Y}_i$. Note that when the observed Y lies above the line, this error is positive, and when the observed Y_i lies below the line, it is negative. Now we might think that a good criterion for fitting the regression line is to minimize the sum of these errors, taking into account all our observations:

$$\min \sum_{i=1}^{n} (Y_i - \hat{Y}_i) \qquad (12\text{-}2)$$

Unfortunately, the positive deviations offset the negative deviations, and it is often possible to draw several lines through the set of points—each with a zero sum of deviations. An example of this problem is illustrated in Figure 12-9.

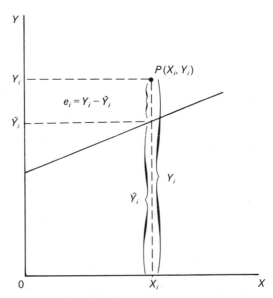

FIGURE 12-8. Error in fitting a line to a typical point.

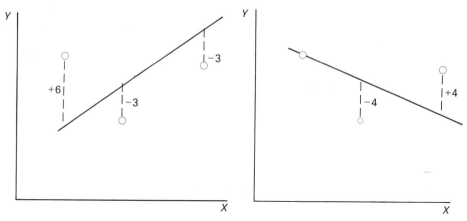

FIGURE 12-9. Both regression lines have $\Sigma_{i=1}^{3} (Y_i - \hat{Y}_i) = 0$.

To overcome this problem, we utilize the *least squares criterion*:

$$\min \sum_{i=1}^{n} (Y_i - \hat{Y}_i)^2 \qquad (12\text{-}3)$$

which selects a and b to minimize the sum of squared deviations. If the line is very close to the points, then this sum is very small; as the points become increasingly spread about the line, this index increases. If all the observed points lie on the line, then the sum equals zero. The use of squared deviations overcomes the "sign problem" inherent in the use of Equation (12-2). Moreover, we see in Chapter 13 that there are important theoretical justifications for using this criterion.

Estimation through Least Squares

It is now time to ask, "How can we find the values of a and b that minimize the sum of squared deviations?" Let us substitute $\hat{Y}_i = a + bX_i$ into (12-3). The problem now becomes one of choosing a and b to

$$\min \sum_{i=1}^{n} (Y_i - a - bX_i)^2 \qquad (12\text{-}4)$$

Since all the X and Y values are known, the value of this expression clearly depends on the values we select for a and b. As we change the values of these two parameters, we change the location of the line and also the value of this minimand. Although the details need not concern us, the solution requires that the following two "normal" equations be satisfied (students who have studied elementary calculus can refer to Appendix 12B for the derivation of these two normal equations):

$$na + b \sum X = \sum Y \qquad (12\text{-}5)$$

$$a \sum X + b \sum X^2 = \sum XY \qquad (12\text{-}6)$$

For ease of presentation we have omitted the limits of the various summations. All summations extend from $i = 1$ to $i = n$.

Let us take the first equation, (12-5). If we solve for a by dividing by n and simplifying, we get

$$a = \frac{\Sigma\, Y}{n} - b\, \frac{\Sigma\, X}{n} \tag{12-7}$$

which, from the identities $\bar{Y} = \Sigma\, Y/n$ and $\bar{X} = \Sigma\, X/n$, leads to

$$a = \bar{Y} - b\bar{X} \tag{12-8}$$

Since $\bar{X}$ and $\bar{Y}$ are calculated from the data, we can easily determine the value of our intercept a once we know b. There is also one important implication which is derived from (12-8). If we rewrite (12-8) as $\bar{Y} = a + b\bar{X}$, we see that the least squares line passes through the point $(\bar{X},\ \bar{Y})$. Note that there is no error at this point since $\bar{Y} = \hat{Y}$.

How do we solve for b? Now that we have a, we can substitute (12-7) into our second normal equation (12-6):

$$-\Sigma\, XY + \left(\frac{\Sigma\, Y}{n} - b\, \frac{\Sigma\, X}{n} \right)\left(\Sigma\, X \right) + b\, \Sigma\, X^2 = 0 \tag{12-9}$$

This can be multiplied to yield

$$-\Sigma\, XY + \frac{\Sigma\, X\, \Sigma\, Y}{n} - \frac{b(\Sigma\, X)^2}{n} + b\, \Sigma\, X^2 = 0 \tag{12-10}$$

If we multiply by n and rearrange this equation, we arrive at

$$nb\, \Sigma\, X^2 - b\left(\Sigma\, X \right)^2 = n\, \Sigma\, XY - \Sigma\, X\, \Sigma\, Y \tag{12-11}$$

Finally, we solve for b:

$$b = \frac{n\, \Sigma\, XY - \Sigma\, X\, \Sigma\, Y}{n\, \Sigma\, X^2 - (\Sigma\, X)^2} \tag{12-12}$$

Equations (12-8) and (12-12) are the most straightforward computational formulas for solving linear regression problems by using hand calculators. Of course, we utilize computers for large-scale applications of linear regression analysis. Normally, these programs use other computational techniques. By convention, the parameter a is known as the *regression intercept* or just *intercept*, and the parameter b is termed the *regression coefficient*.

Let us now return to our example of household trip generation. Table 12-2 contains the required calculations. It is an extremely useful tableau fromat for organizing the calculations required for the linear regression problem. Strictly speaking, we do not need the first column which contains $\Sigma\, Y^2$, but we need this term to calculate the correlation coefficient between X and Y. As we shall see in Section 12.3,

TABLE 12-2
Calculations for a Simple Linear Regression Problem

Observation No.	X	Y	XY	X^2	Y^2
1	10	3	30	100	9
2	12	5	60	144	25
3	11	5	55	121	25
4	9	4	36	81	16
5	14	6	84	196	36
6	16	7	112	256	49
7	21	9	189	441	81
8	18	7	126	324	49
9	22	8	176	484	64
10	24	8	192	576	64
11	17	6	102	289	36
12	15	5	75	225	25
$n = 12$	$\sum X = 189$	$\sum Y = 73$	$\sum XY = 1237$	$\sum X^2 = 3237$	$\sum Y^2 = 479$

$$\bar{X} = \frac{189}{12} = 15.75 \qquad \bar{Y} = \frac{73}{12} = 6.083$$

$$a = 6.083 - .335(15.75) \qquad b = \frac{12(1237) - (189)(73)}{12(3237) - 189^2}$$

$$= .807 \qquad\qquad\qquad = .335$$

the correlation coefficient is a useful measure for assessing the goodness of fit of our regression line.

From Table 12-2 we see that the regression of household trip generation rates on household income yields

$$\hat{Y}_i = .807 + .335X_i \tag{12-13}$$

where $\hat{Y}_i$ is the estimated or predicted *mean* number of trips generated for a given income level. Recall that a regression line follows the path of the means. It is relatively straightforward to graph our regression line within the scatter of observations. To precisely locate the line, we require two points. Although we can determine any two points simply by substituting two different values of X into (12-13), it is often much simpler to use the points $(\bar{X}, \bar{Y})$ and $(0, a)$. We know $(\bar{X}, \bar{Y})$ is on the line because of (12-8), and by definition the parameter a is the intercept on the Y axis. We can simply connect these points by a straight line. If the intercept a is not close to the range of our data (this is a normal occurrence), we can utilize any other convenient point. In the example of household trip generation, the income variable is limited to the range 9 to 24. We estimate $\hat{Y}_i = .807 + .335(20) = 7.507$ and obtain the point $(20, 7.507)$. We use this point and our bivariate mean $(15.75, 6.083)$ to draw the regression line in Figure 12-10.

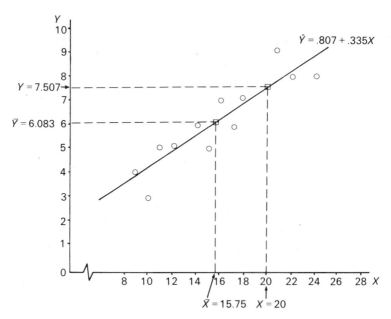

FIGURE 12-10. Drawing the regression line.

12.3. Assessing the Goodness of Fit

We now know that the least squares estimates of a and b given by (12-8) and (12-12) provide the *best* possible fit of a straight line to the data according to criterion (12-4). Since we can calculate these two parameters from any set of observations having at least two different values of X, the question remains: How well does the regression line fit the original data? We cannot use the total sum of squared deviations (12-4) as our index since it clearly depends on the scale of the numbers involved as well as on the number of observations. What we require are indices capable of assessing the goodness of fit of regression lines drawn from different sets of data. With such indices we could compare the fit of several different regressions drawn from different sets of data. Let us examine three different measures which can be used to assess the goodness of fit of a regression line: the simple *correlation coefficient r*, the *coefficient of determination* r^2, and the *standard error of estimate* $s_{Y \cdot X}$.

Correlation Coefficient r

As we found in Chapter 11, the product-moment correlation coefficient called Pearson's r is a dimensionless measure of the degree of linear association between two variables X and Y. It has a range $-1 \le r \le +1$, and it can be used to compare the association between two sets of variables drawn from the same set of observations or of the same two variables drawn from different data sets. The closer r is to ± 1, the greater the strength of the linear association between the two variables. It thus represents a useful summary measure in regression analysis. If our two variables are positively related, then $r > 0$ and $b > 0$. If they are negatively related, then r, $b < 0$. We can thus say that a regression line with a correlation coefficient higher than that of another equation fits the data better.

As you might suspect, there is an important algebraic relation between b and r which shows the intimate relationship between correlation and regression. A quick comparison of (11-4) and (12-12) should convince you of this fact. To illustrate this relationship, let us recall the definition of the correlation coefficient

$$r = \frac{\Sigma (X_i - \bar{X})(Y_i - \bar{Y})}{\sqrt{\Sigma (X_i - \bar{X})^2}\sqrt{\Sigma (Y_i - \bar{Y})^2}} \qquad (12\text{-}14)$$

Let us define $x_i = X_i - \bar{X}$ and $y_i = Y_i - \bar{Y}$. With this pair of definitions we can now rewrite (12-14) as

$$r = \frac{\Sigma x_i y_i}{\sqrt{\Sigma x_i^2}\sqrt{\Sigma y_i^2}} \qquad (12\text{-}15)$$

This is often termed the *deviations form* of the correlation coefficient. It is used in many situations because it makes certain formulas such as the correlation coefficient notationally simpler. We can make a similar simplification of the regression co-efficient:

$$b = \frac{\Sigma x_i y_i}{\Sigma x_i^2} \qquad (12\text{-}16)$$

The similarity of (12-15) and (12-16) again suggests the relation between correlation and regression. To formalize this relationship, let us take the ratio of the regression coefficient to the correlation coefficient:

$$\frac{b}{r} = \frac{\Sigma x_i y_i / \Sigma x_i^2}{\Sigma x_i y_i / (\sqrt{\Sigma x_i^2}\sqrt{\Sigma y_i^2})}$$

$$= \frac{\sqrt{\Sigma x_i^2}\sqrt{\Sigma y_i^2}}{\Sigma x_i^2} = \frac{\sqrt{\Sigma y_i^2}}{\sqrt{\Sigma x_i^2}} \qquad (12\text{-}17)$$

Now if we divide both the numerator and denominator of the right-hand side of (12-17) by $1/\sqrt{n-1}$, we obtain

$$\frac{b}{r} = \frac{\sqrt{\Sigma y_i^2 / (n-1)}}{\sqrt{\Sigma x_i^2 / (n-1)}} = \frac{s_Y}{s_X} \qquad (12\text{-}18)$$

by using the familiar formula for the standard deviation. We now express (12-18) as

$$b = r\frac{s_Y}{s_X} \qquad (12\text{-}19)$$

which indicates the direct algebraic link between correlation and regression analysis.

TABLE 12-3
Work Table for Illustrating Equation (12-19)

The simple correlation coefficient for the household trip generation data, obtained by using computational formula (12-4) with the data from Table 12-2, is

$$r = \frac{1237 - [(189)(73)]/12}{\sqrt{3237 - 189^2/12}\sqrt{479 - 73^2/12}}$$

$$= .9153$$

Using the computational formula for the standard deviation, we find

$$s_X = \sqrt{\frac{3237}{11} - \frac{189^2}{12(11)}} \qquad s_Y = \sqrt{\frac{479}{11} - \frac{73^2}{12(11)}}$$

$$= 4.864 \qquad\qquad = 1.7816$$

$$b = .9153\left(\frac{1.7816}{4.864}\right)$$

$$= .335$$

This agrees with the calculation of b in Table 12-2.

This equation implies $b = 0$ if $r = 0$, and vice versa. Also, since s_Y and s_X must always be positive, b and r must have identical signs.

Let us now return to our example of trip generation. In Table 12-3 we calculate the simple correlation coefficient between income and household trip generation rates to be $r = .9153$. This indicates a very strong linear association between these two variables. Also, in this table we illustrate the numerical validation of the formal relation between correlation expressed by (12-19).

Coefficient of Determination r^2

The *residuals* from a regression line are defined by $e_i = (Y_i - \hat{Y}_i)$. There is one residual for each of the n observations. They provide extremely useful information on the extent to which the calculated regression line fits the data. Generally, a good regression is one which has small residuals, and a poor regression is one with large residuals. What is needed is a measure of goodness of fit which utilizes these residuals and is unit-free. That is, our measure must be independent of the units in which our dependent variable Y is measured. The *coefficient of determination* is such a measure.

Let us begin by defining the *variation* of variable Y about its mean as

$$\sum (Y_i - \bar{Y})^2 \tag{12-20}$$

We now wish to divide this total variation into two distinct parts. The first part we term the part *due to regression*, or *explained by regression*, or *explained by variable X*. The second part is the *residual variation* which is *unexplained by regression*, or the *error variation*.

Let us assume that we fit a regression model to our sample of observations and found the regression coefficient b to be zero. Then our best prediction for Y given any value of X is

$$\hat{Y}_1 = a + 0X_i = \bar{Y} \qquad (12\text{-}21)$$

and our knowledge of the value of variable X is of no use in improving our prediction of Y. Without knowing the value of X we would predict $\bar{Y}$ since it lies in the center of the distribution. In this case, the variation of Y about its mean (12-20) represents the sum of the square of the differences between our observed Y and the predicted values $\hat{Y}_i = \bar{Y}$. Now suppose that the regression coefficient or slope of the fitted line is nonzero. We can improve the prediction of Y by accounting for the fact that the predicted values of Y_i are dependent on our observation X_i and given by $\hat{Y}_i = a + bX_i$.

To see this improvement, consider the following identity, geometrically illustrated in Figure 12-11:

$$Y_i - \bar{Y} = (Y_i - \hat{Y}_i) + (\hat{Y}_i - \bar{Y}) \qquad (12\text{-}22)$$

On the left-hand side is the difference between the observed Y_i and the mean $\bar{Y}$. This is then decomposed into two parts. The first term on the right-hand side is the residual, or error e_i, and the second part represents the difference between the predicted values of Y and $\bar{Y}$. This second part represents the improvement "due to the

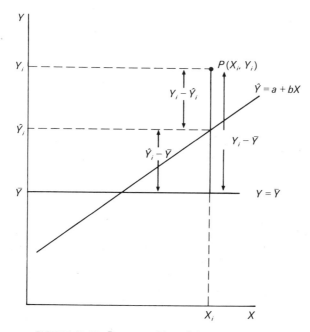

FIGURE 12-11. Decomposition of the total variation.

regression of Y on variable X." We now expand this relation by summing over all n observations and squaring both sides:

$$\sum (Y - \bar{Y})^2 = \sum (Y_i - \hat{Y}_i)^2 + \sum (Y_i - \bar{Y})^2 + 2 \sum (Y_i - \hat{Y}_i)(Y_i - \bar{Y}) \qquad (12\text{-}23)$$

To square the right-hand side of (12-22), we have used the identity $(a + b)^2 = a^2 + b^2 + 2ab$. Although the details need not concern us, the last term on the right-hand side of (12-23) always equals zero, leaving

$$\sum (Y_i - \bar{Y})^2 = \sum (Y_i - \hat{Y})^2 + \sum (\hat{Y}_i - \bar{Y})^2$$

or

$$\text{TSS} = \text{ESS} + \text{RSS} \qquad (12\text{-}24)$$

where TSS is the total variation of Y, or total sum of squares; ESS is the residual or error variation, or error sum of squares; and RSS is the explained variation of Y, or regression sum of squares.

We have still not developed a unit-free measure of goodness of fit since all our components of variation depend on the units in which the dependent variable Y is measured. To eliminate this difficulty, let us divide both sides of (12-24) by TSS to get

$$1 = \frac{\text{ESS}}{\text{TSS}} + \frac{\text{RSS}}{\text{TSS}} \qquad (12\text{-}25)$$

and we then define the *coefficient of determination* r^2 as

$$r^2 = 1 - \frac{\text{ESS}}{\text{TSS}} = \frac{\text{RSS}}{\text{TSS}} \qquad (12\text{-}26)$$

From this definition we see that r^2 can be interpreted as *the proportion of the total variation in variable Y that is explained by the regression of variable Y on variable X*. Since $0 \le \text{ESS} \le \text{TSS}$, we see that $0 \le r^2 \le 1$. When $\text{ESS} = 0$, we have a perfect fit, $r^2 = 1$, and our statistic tells us that we have explained 100 percent of the variation in Y. When $\text{ESS} = \text{TSS}$, $r^2 = 0$, and variable X is of no value in improving our prediction of Y. It explains 0 percent of the total variation. Most often, r^2 lies between these extremes.

In general, a high value of r^2 means we have a good fit, and a low value of r^2 means we have a poor fit. However, it is no accident that we define the coefficient of determination as r^2. This is because it is always numerically equal to the square of the correlation coefficient r in any simple regression problem. We can use this relationship to aid in the interpretation of the correlation coefficient. An r^2 of .80 implies that variable X "explains" 80 percent of the variation in variable Y. This corresponds to a correlation coefficient of approximately $r = .90$. We can thus translate our values of r into values of r^2 and get a better "feel" for the strength of the association. Note that the seemingly high correlation of $r = .64$ corresponds to an $r^2 = .41$ and implies that only about 40 percent of the variation in Y can be explained by variable X. A correlation of $r = .40$ might seem quite weak when we realize that only 16 percent of Y is explained by variations in X.

The coefficient of determination should also be used with caution. Geographers often place undue emphasis on obtaining a high value of r^2. Although we speak in terms of explanation, in many instances we do not have a truly causal relationship, and we should speak in terms of statistical explanation, not causal explanation. The lack of a strong theoretical base in geography means that much empirical work is not based on a sufficient theoretical foundation to be spoken of in terms of causal explanation. Also r^2 alone is not a suitable measure to assess the utility of a regression model. We often obtain a low value of r^2 in cases in which there is a large variation across the individual units of observation. In Chapter 11 we witnessed the dependence of the correlation coefficient on the level of aggregation in the data.

Consider again the example of trip generation. If the data are based on individual households, they would probably exhibit a great deal more variation than when they were aggregated to the level of the traffic zone. We may have aggregated a number of elderly immobile households with a number of highly mobile younger households. The average trip rates and incomes of the traffic zones in that case would give a poor representation of the actual variation in these two variables in the metropolitan area. While we might be able to explain the variations in trips at the level of the traffic zone, we would probably not be able to predict or explain the variation in individual household trip generation rates with the same precision. In this case, the coefficient of determination overestimates the goodness of fit of the relationship between these two variables.

Also, it is very difficult to utilize the coefficient of determination as defined in (12-26) in a fixed-X model. A quick glance at Figure 12-5 shows why. It is impossible to predict each observed Y_i exactly since we have different values of Y_i for one individual value of X_i. However, our equation can predict only one value of $\hat{Y}_i$ for all these observations. So $r^2 = 1$ cannot be achieved. A modified r^2 is used in this instance. In some experimental situations in physical geography, we might require this modified r^2 as a true measure of goodness of fit. Suppose we are interested in the relation between stream velocity and sediment load. We set up an experiment in the laboratory using a flume. The bottom of the flume is packed with materials typical of stream beds. We then subject the flume to a given flow velocity and measure the suspended sediment. The experiment is repeated, and measurements are recorded several times at many different flow velocities. This approach generates data for a fixed-X regression model of suspended sediment versus flow velocity. A modified r^2 [see Draper and Smith (1981, 33–43)] is required to assess goodness of fit.

Finally, it is possible to achieve unrealistically high values of r^2 in many *time series* studies. In time series studies we measure one or more variables at different times. One variable growing over time is quite likely to do a good job at explaining the variation in *any* other variable that also is growing over time—whether or not the two variables are linked in any causal way. In sum, the coefficient of determination represents a useful summary measure of goodness of fit when it is interpreted with caution. A high value of r^2 might, on reflection, be virtually meaningless. A low value

TABLE 12-4
Work Table for Calculation of $s_{Y \cdot X}$ and r^2

Observation No.	X_i	Y_i	$\hat{Y}_i$	$Y_i - \hat{Y}_i$	$(Y_i - \hat{Y}_i)^2$	$(Y_i - \bar{Y})^2$
1	10	3	4.156	−1.156	1.336	9.505
2	12	5	4.826	0.174	0.030	1.173
3	11	5	4.491	0.509	0.259	1.173
4	9	4	3.820	0.180	0.032	4.339
5	14	6	5.497	0.503	0.253	0.007
6	16	7	6.167	0.833	0.694	0.841
7	21	9	7.843	1.157	1.339	8.509
8	18	7	6.838	0.162	0.026	0.841
9	22	8	8.179	−0.179	0.032	3.675
10	24	8	8.849	−0.849	0.721	3.675
11	17	6	6.502	−0.502	0.252	0.007
12	15	5	5.832	−0.832	0.692	1.173

$$\bar{Y} = 6.083$$

$$\sum (Y_i - \hat{Y}_i) = 0.0 \qquad \sum (Y_i - \hat{Y}_i)^2 = 5.666 \qquad \sum (Y_i - \bar{Y})^2 = 34.918$$

$$s_{Y \cdot X} = \sqrt{\frac{5.666}{12 - 2}} \qquad r^2 = \frac{34.918 - 5.666}{34.918}$$

$$= .752 \qquad\qquad = .838$$

of r^2 might be very meaningful if the dependent variable is subject to more random than systematic variation.

Let us complete our discussion by illustrating the calculation of r^2 (see Table 12-4). We might wish to keep this table separate from Table 12-2 or else just add the final four columns to Table 12-2 and make a composite table. The total variation is 34.918, the unexplained (or residual, or error) variation is 5.666, and the explained

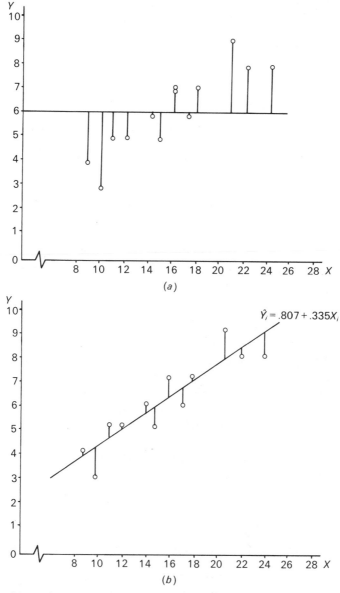

FIGURE 12-12. Geometrical representation of (*a*) total variation, (*b*) residual variation, and (*c*) explained variation.

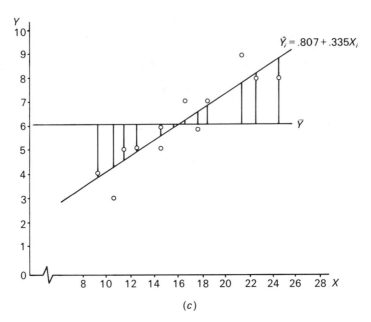

FIGURE 12-12 (continued).

variation is $34.918 - 5.666 = 29.252$. The geometric equivalents to these three components of variation are illustrated in Figure 12-12. To check the calculated value of $r^2 = .838$, we compare it to the square of the correlation coefficient $r = .9153$ from Table 12-3. Notice that $.838 = .9153^2$. This is a useful numerical check for your calculations.

Standard Error of Estimate $s_{Y \cdot X}$

In many applications of regression analysis, the ultimate goal is to come up with a good estimate or prediction for Y by using variable X. Our success is measured by the degree to which we can predict variable Y accurately. In this case, we are primarily interested in the numerical value of the error we are likely to make when utilizing X to predict Y. The best measure of goodness of fit in this instance is the *standard error of the estimate* $s_{Y \cdot X}$. The standard error of the estimate is defined as the standard deviation of the residuals about the regression line:

$$s_{Y \cdot X} = \sqrt{\frac{\Sigma (Y_i - \hat{Y}_i)^2}{n - 2}} \qquad (12\text{-}27)$$

The subscript $Y \cdot X$ is used to denote the fact that we are not calculating the standard deviation of Y but the standard deviation of Y, taking into account that variable Y is regressed on variable X. The *standard error of estimate* is sometimes called the *root mean square error*, or *root MSE*. Notice that it is calculated as the square root of the average or mean squared error. Why do we divide by $n - 2$ and not n? Because, as we shall see in Chapter 13, $s_{Y \cdot X}$ is really an *estimate* of the standard deviation of the error term in our model. We treat our data of n observations as if they were a sample from a population. Recall that when we make an estimate in inferential statistics, we divide by the appropriate degrees of freedom, not the sample size. Most computer programs for regression calculate the standard error of estimate by using $n - 2$, not n, in the denominator. Strictly speaking, we might wish to divide by n if we consider the data a statistical population.

Why are the degrees of freedom equal to $n - 2$? This is because we lose 2 degrees of freedom in the data when we estimate the regression parameters a and b since they require $\bar{X}$ and $\bar{Y}$ in their calculations. We can also use a straightforward geometric argument. Since two points are needed to determine a line, any point in excess of $n = 2$ provides a degree of freedom in defining the location of the line. Thus $n - 2$ degrees of freedom are always associated with a simple regression based on n observations.

Once again let us return to our example of household trip generation. Table 12-4 illustrates the calculations for the standard error of estimate. We simply take the square root of the residual, or ESS, divided by the number of observations minus 2. Note the residual sum of squares is always generated in the computation of r^2. In this case $s_{Y \cdot X} = .752$. Since the standard error of estimate is always measured in units of the dependent variable Y, we can interpret this measure of goodness of fit as corresponding to approximately three-quarters of a trip per household per day. But does this represent a "good" fit? Two previous measures of goodness of fit suggested our relation to be extremely close: $r = .9153$ and $r^2 = .838$. Is a prediction of a household's trip generation rate within three-quarters of a trip per day as good as the values of r and r^2 seem to suggest? *We can make this judgment only within the context of the problem at hand, that is, in relation to some predetermined standard of prediction.*

If the standard error of estimate is .752, then 1 standard deviation on either side of the regression line roughly corresponds to 1.5 trips per household per day. Two standard deviations corresponds to 3 trips. If the errors about the regression line are normally distributed (and in the next chapter we make this very assumption for inferential purposes), then we expect *roughly* 68 percent of our observations to fall within 1 standard deviation about the line and 95 percent to fall within 2 standard deviations. So our predictive ability could be very crudely estimated as a range of about 3 trips per day 95 percent of the time. Unfortunately, for a city with 100,000 households this would mean a potential range of about 300,000 (100,000 × 3) trips in any given day! Put in this context, our equation is not nearly as useful as we might have thought—at least in the context of transportation forecasting. Thus we should never base judgment about goodness of fit on one single measure.

Finally, let us consider a standardized measure of goodness of fit known as the *coefficient of variation*. Recall that in Chapter 2 we defined the coefficient of variation (CV) to be the ratio of the standard deviation to the mean, $s_X / \bar{X}$. We use this statistic to compare the variability of two or more variables with different means and standard deviations. In the context of regression, the appropriate definition is

$$CV = \frac{s_{Y \cdot X}}{\bar{Y}} \times 100 \qquad (12\text{-}28)$$

where we have multiplied by 100 to express the coefficient as a percentage. If CV = 0, this implies that $s_{Y \cdot X} = 0$ and we have a perfect fit. When $b = 0$ and the line of best fit reduces to $\hat{Y} = \bar{Y}$, then CV = 100. In most cases the CV lies between these two extremes. The lower the CV, the better the relative fit of the regression line. However, we must be careful. It is possible to generate an equation with an $s_{Y \cdot X}$ which is acceptable in comparison to some given standard of prediction but which has an associated CV higher than some other equation that may be unacceptable on these grounds.

12.4. Interpreting a Regression Equation

To this point, we have described the situations in which regression analysis can be applied, the method for calculating the line of best fit, and several measures of goodness of fit. What does all this information tell us? In any application of regression analysis, it is often useful to examine our equation within the context of the empirical problem at hand. We have just witnessed the utility of this approach in interpreting the standard error of estimate as a measure of goodness of fit. We now review the output from our regression equation in this light.

Units of Measure

Table 12-5 summarizes the units of measure of the various terms in a regression equation and describes their interpretation in the example of household trip

TABLE 12-5
Dimensions of Regression Model Terms

Term	Notation	Units	Units in example
Dependent variable	Y	—	Trips per household per day
Independent variable	X	—	Thousands of dollars of income
Intercept	a	Y	Trips per household per day
Regression coefficient	b	Units of Y per unit of X	Trips per household per day per \$1000 income
Residuals	$e_i = Y_i - \hat{Y}_i$	Y	Trips per household per day
Standard error of estimate	$s_{Y \cdot X}$	Y	Trips per household per day
Correlation coefficient	r	Dimensionless	
Coefficient of determination	r^2	Dimensionless	
Coefficient of variation	CV	Dimensionless	

generation. First, we must remember that a regression equation provides a formal statistical link between two variables measured in different units. Any regression equation can be expressed as

$$\text{Units of } Y = a + b(\text{units of } X) \tag{12-29}$$

The dimensions on both sides of this equation must be equal. It thus follows the right-hand side must be measured in units of Y. Knowing this, we can say (1) the intercept a must be measured in units of Y and (2) the slope or regression coefficient b is measured in units of Y per unit of X. Making these substitutions, we see

$$\text{Units of } Y = \text{units of } Y + \frac{\text{units of } Y}{\text{units of } X}(\text{units of } X)$$

$$\text{Units of } Y = \text{units of } Y \tag{12-30}$$

and we see that these units of measure ensure dimensional equivalence of both sides of the equation.

The intercept a is thus measured in units of Y and is interpreted as the value of Y when $X = 0$. For the problem in household trip generation, $a = .807$ implies that a household with no income makes on the average .807 trips per day. Is this plausible? The answer depends on whether the range of X over which the equation is estimated includes the value $X = 0$. If it does, we might have some a priori reason to expect a to take on a positive value, a negative value, or zero. In our case we might prefer a zero intercept, since a household with no income should make no trips. However, the range of X in the data is limited to $9 \leq X \leq 24$. So we really have no knowledge of the value of Y in the vicinity of $X = 0$. Were we to have data from a broader range of income levels, we might have different expectations. It is possible that the regression model depicted in Figure 12-13 is applicable to the problem. The true regression is a nonlinear model which has a zero intercept. However, when the data are limited to the region labeled *observed range of X*, a linear equation is a reasonable estimate of the relationship in the range. In this case a nonzero intercept is meaningful. Of course, we could also argue that there is some minimum threshold income below which no observation exists and therefore our intercept is really meaningless.

Now let us turn to the regression coefficient b. It is interpreted to be the rate of change of the dependent variable with respect to the independent variable X. We might also have some a priori grounds for expecting b to be of a certain sign and/or of a certain order of magnitude. If we are modeling a truly causal relationship, then we most certainly should know at least the sign of the regression coefficient. What if our regression calculations yield a value of b with the opposite sign to our expectations, but the goodness-of-fit statistics indicate high values of r and r^2? Suppose that the regression coefficient in the trip generation example turned out to be $b = -.335$, not $b = +.335$? Given our expectations that increasing incomes lead to higher trip rates—especially over the income range in the data—we would have to be skeptical of the regression results. We may have a "poor" set of data with enormous measurement errors of perhaps just a poorly chosen sample. Or the aggregation might have

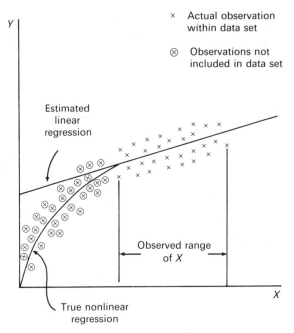

FIGURE 12-13. Regression model.

spoiled the data. If not, we must turn to the theory upon which we base our expectation for the sign of b. Perhaps certain preconditions or assumptions are inherent in the theory that render it inapplicable to our data. Or, we might even be able to improve the theory by suggesting other variables that must be incorporated.

Change of Scale

What if we were to change the units of measurement of either the dependent variable Y or the independent variable X? What happens to the values of the regression equation parameters and the goodness-of-fit statistics? It turns out that the changes are exactly what we would expect and hope for. First, let us consider the independent variable X. Let us change X by measuring household income in tens of thousands of dollars rather than in thousands of dollars. Instead of the sequence 10, 12, ... , 15 for X, we have $X' = 1, 1.2, \ldots , 1.5$. Each X' is one-tenth as large as each X, so $X' = 0.1X$. The new equation is

$$\hat{Y}_i = .807 + 3.35X_i' \tag{12-31}$$

and the new regression coefficient is 10 times as large as the old one, $b' = 10b$. But has the response of trip generation to changes in income been affected? No. Note that a \$10,000 change in income in the old regression equation (12-13) predicts a $10 \times .335 = 3.35$ increase in trips per household per day. This is exactly the same as the estimate given by the regression coefficient in (12-31). In general, if we replace $X' = kX$, we find $b' = (1/k)(b)$. In the previous example $k = .1$ and thus $b' = 10b$.

There is no change in intercept a. All we have done is to stretch or compress the X axis without changing the Y axis. You may wish to prove these two results by substituting $X' = kX$ into (12-8) and (12-12).

We can also determine the effects of changes in the units of measure of variable Y. A similar analysis reveals that a change to $Y' = kY$ leads to $a' = ka$ and $b' = kb$. Whether we change X or Y or both X and Y, we do not change the magnitudes of the dimensionless measures r, r^2, or CV. All are invariant with respect to scale changes in variables X and Y. These results are very important. Regression analysis would be of limited value if we could change the goodness of fit of the equations simply by changing the scale of the numbers involved.

Interpolation and Extrapolation

One of the most common uses of regression analysis is prediction. We seek to predict or estimate the value of Y given any value of X. To illustrate the use of our regression equation, suppose we wish to predict the number of trips per household per day made by a household with an income of $19,000. We simply substitute $X = 19$ (remember, we measure income in thousands of dollars) into the equation and obtain

$$\hat{Y} = .807 + .335(19) = 7.172 \qquad (12\text{-}32)$$

trips per day. Whenever the value of X lies within the range of our data, this process is known as *interpolation*. We can even roughly interpolate by eye, using our scattergram and regression equation. Remember, we assumed that the regression line follows the path of the means within our range of X. So we expect the regression line to pass through the mean value of Y for any value of X. Where the data contain large gaps in the value of X, this prediction may be subject to error.

Sometimes we seek to predict Y for a value of X beyond the range of the data. This is known as *extrapolation*. Generally, we are in a much poorer position because we have no knowledge of the value of Y in this range of X. We encountered the same difficulty when we tried to interpret the intercept value a when the value $X = 0$ was not within the range of the data. The problem is that the function which passes through the means may not continue to follow the linear function postulated for purposes of calibration. In Figure 12-4 notice how many different functions are linear or near linear over part of the range, but then make significant changes in slope over other parts of the range. In Chapter 13 we return to this forecasting problem and illustrate the method used to make *point and interval estimates* for $\hat{Y}$ at any desired level of confidence. Our results clearly illustrate the dangers of extrapolation.

12.5. Technical and Methodological Issues

So far we have focused on the mechanics of fitting a regression line, assessing its goodness of fit, and interpreting the parameters of the model. We close this chapter with a brief exploration of three particular problem areas. All three are relevant to both the computational operations involved in a regression model and questions of interpretation.

Data Aggregation

We have repeatedly stressed that all descriptive and inferential statistics are inextricably bound to the data from which they are derived. Any changes in these data can lead to significant changes in the numerical values of the summary statistics and their interpretations. This issue arises in regression analysis in much the same way as we saw in correlation analysis. Let us examine the consequences of two potential changes to the data. First, suppose we aggregate the data by taking the original observations and substituting an average measure for groups of individual observations. In Chapter 11 we termed the variations in the value of the correlation coefficients with increased aggregation the *scale problem*. Given the close relationship between correlation and regression analysis, it is not surprising that the same problem exists when we estimate the linear regression equation.

A simple example will illustrate. Let us aggregate the original 12 traffic zones into six *traffic districts* according to the scheme summarized in Table 12-6 and illustrated as aggregation *A* in Figure 12-14(*a*). For example, traffic district 1 is formed

TABLE 12-6
Summary of Data for Alternative Spatial Aggregations

Traffic district	Traffic zones included	X	Y
	Aggregation *A*		
1	1, 9	16	5.5
2	2, 8	15	6
3	3, 12	13	5
4	4, 5	11.5	5
5	6, 7	18.5	8
6	10, 11	20.5	7
	Aggregation *B*		
1	1, 2	11	4
2	3, 4	10	4.5
3	5, 6	15	6.5
4	7, 8	19.5	8
5	9, 10	23	8
6	11, 12	16	5.5
	Aggregation *C*		
1	1, 10	17	5.5
2	2, 7	16.5	7
3	3, 11	14	5.5
4	4, 6	12.5	5.5
5	5, 12	14.5	5.5
6	8, 9	20	7.5

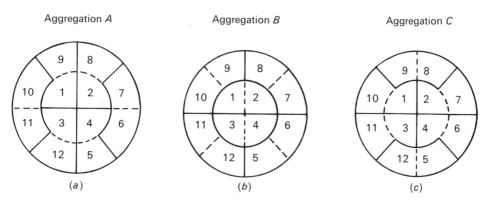

FIGURE 12-14. Alternative spatial aggregations.

from traffic zones 1 and 9. Assuming the zones have an equal number of households, we compute new values of X and Y as the mean values for the two zones:

$$\left(\frac{X_1 + X_9}{2}, \frac{Y_1 + Y_9}{2}\right) = \left(\frac{10 + 22}{2}, \frac{3 + 8}{2}\right) = (16, 5.5)$$

Table 12-7 summarizes the results of the regression analysis for the original data with $n = 12$ observations and the $n = 6$ observations of aggregation A.

First, we notice the changes in the estimates of the regression parameters a and b. The intercept *increases* to 1.299, and the coefficient *decreases* to .304. Second, there are substantial differences in the various measures of goodness of fit. The value of r^2 declines appreciably, suggesting that income explains only 72 percent of the variation in household trip generation rates. In this sense, our equation is worse with this new aggregation. But note the reduction in the standard error of the estimate. Does this mean we can now predict trip rates more accurately by using this equation? Unfortunately, no. The reason for this reduction in $s_{Y \cdot X}$ is related to the significant decline in the overall variation in Y as measured by TSS. Note that the total variation, or sum of squares, falls from 34.918 to 7.208. Put simply, there is simply less variation to explain. A quick examination of the data for the aggregation in Table 12-6 reveals that trip rates for the zones vary only from 5.5 to 8. In the original data in Table 12-1, household trip generation rates vary from 3 to 9 trips per day.

To see why this equation is *not* superior to the original equation, we calculate the standard error of estimate for this new equation, using the original 12 observations. We substitute the values of X corresponding to the 12 traffic zones and compare the estimated trip rates to those given in Table 12-1. We find $s_{Y \cdot X} = .799$. This is clearly above the standard error of estimate from the traffic zone data of .753.

We can also illustrate variations in our equations and summary statistics if we examine different spatial aggregations for traffic districts. Even at the same scale of analysis, the use of different areal units leads to different results. Many geographers call this the *modifiable areal units problem*. Table 12-6 and Figures 12-14(a) to (c) compare three different aggregations of the original 12 traffic zones into six traffic

TABLE 12-7
A Comparison of Regression Results for Different Aggregation Schemes

	Original data
	$\hat{Y}_i = .807 + .335X_i$
$n = 12$	$r = .9153 \qquad s_{Y \cdot X} = .753$
	$r^2 = .838 \qquad \text{TSS} = 34.918$
	Aggregation A
	$\hat{Y}_i = 1.299 + .304X_i$
$n = 6$	$r = .850 \qquad s_{Y \cdot X} = .708$
	$r^2 = .722 \qquad \text{TSS} = 7.208$
	Aggregation B
	$\hat{Y}_i = 1.004 + .322X_i$
$n = 6$	$r = .932 \qquad s_{Y \cdot X} = .695$
	$r^2 = .869 \qquad \text{TSS} = 14.708$
	Aggregation C
	$\hat{Y}_i = 1.798 + .272X_i$
$n = 6$	$r = .789 \qquad s_{Y \cdot X} = .630$
	$r^2 = .622 \qquad \text{TSS} = 4.208$

districts. We label these aggregations *A*, *B*, and *C*, respectively. The regression results for each are compared in Table 12-7. Note that the regression coefficient *b* is always lower than in the original equation, and the intercept is always higher. In fact, it is possible to define spatial aggregations that raise the slope and lower the intercept. We can never be sure of the effects of aggregation. Some aggregations lead to lower estimates of *b* while others lead to higher estimates. Also note the variations in the values of the correlation coefficients and the coefficients of determination. Aggregation *B* leads to an improvement in r^2 while the other two aggregations lead to a reduced r^2.

Finally, compare the three standard errors of estimate. All equations appear to be superior predictors of trip rates compared to the original equation. But this is really illusory. Most often the reduction in the standard error results from a significant decline in total variation TSS. Usually, these two go hand in hand. Note that aggregation *C* reduces the total variation in trip rates by over 83 percent. The sensitivity of regression equations to the size and spatial arrangement of the zones cannot be overemphasized. Our results are valid only for the particular set of zones we use in the analysis. When we must aggregate, it is again useful to follow the rules suggested in Chapter 11.

Specification of a Regression Equation

Although there is not a specific requirement that the dependent variable be causally related to the independent variable, we should nevertheless be extremely careful in the specification of the regression equation. Why? Because, unlike with correlation analysis, regression relations are *not* symmetric. The regression of Y on X is not the same as the regression of X on Y unless the relation is perfect and all points lie on a straight line. If there is no relation between X and Y and $b = 0$, then the two regression lines are perpendicular. The closer the correlation coefficient is to $+1$ or -1, the closer the two regression lines are. To show why the regression of Y on X and that of X and Y are different, we need only consult Equation (12-12), which defines b. For the regression of Y on X we use (12-12) to calculate b. For the regression of X on Y we must interchange X and Y in the formula. This will not change the numerator but will change the denominator substantially since it would be a function of Y and not X.

In our example we find $r = .9153$ and expect two similar regression lines. Figure 12-15 verifies this. The difference between the lines is that the regression of Y on X is generated by minimizing the *vertical* deviations from the points to the line, whereas the regression of X on Y uses the *horizontal* ones. If we express both functions in the form $Y = f(X)$, the regression of X on Y can be reexpressed as $Y = -.220 + .400X$. (To do this yourself, simply solve $X = .549 + 2.499Y$ for Y.) This is extremely close to the regression of Y on X which is $Y = .807 + .335X$.

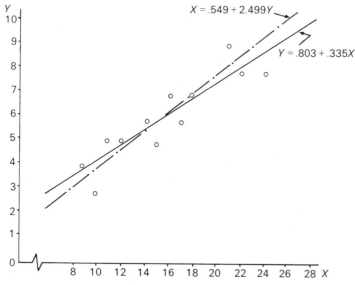

FIGURE 12-15. Comparing the regressions of Y on X and X on Y.

Of course, this lack of symmetry is of little concern if we have a causal relation that can be specified on a priori grounds. In our example, it would be foolish to make a household's income a function of the number of trips made by the household. When we are utilizing our equation for prediction, either we must be capable of controlling the value of X (as in an experimental situation) or we must be able to predict its value independently. An equation to predict Y by using some independent variable X is of little value if we cannot provide the future values of X that we wish to utilize in our prediction! If we plan to use our trip generation equation to predict *future* household trip generation rates, we must have some way of predicting the *future* zonal pattern of household income.

Outliers and Extreme Values

Sometimes when we calculate the least squares regression of a relationship, we notice that one or more of observations have an extremely large residual. Such observations are termed *outliers*—points greater than some arbitrarily defined distance from the regression line. Our attention should be drawn to any observation that is far removed from the systematic relation specified by the regression line. The quickest way to identify outliers is to plot the residuals versus the predicted value of Y. Table 12-8 lists the residuals for the 12 traffic zones, and Figure 12-16(a) illustrates the plot of these residuals versus Y. We notice no outliers, although traffic zones 1 and 7 have larger residuals than most others. One problem in examining residual plots such as Figure 12-16(a) is that the residuals are measured in units of the dependent variable. It is difficult to compare the residuals from two different regressions that are based on different dependent variables.

TABLE 12-8
Residuals for Trip Generation Example

Observation No.	$\hat{Y}_i$	Residual $Y_i - \hat{Y}_i$	Standardized residual $\dfrac{Y_i - \hat{Y}_i}{s_{Y \cdot X}}$
1	4.156	−1.156	−1.537
2	4.826	0.174	0.231
3	4.491	0.509	0.677
4	3.820	0.180	0.239
5	5.497	0.503	0.669
6	6.167	0.833	1.108
7	7.843	1.157	1.539
8	6.838	0.162	0.215
9	8.179	−0.179	−0.238
10	8.849	−0.849	−1.129
11	6.502	−0.502	−0.668
12	5.832	−0.832	−1.106

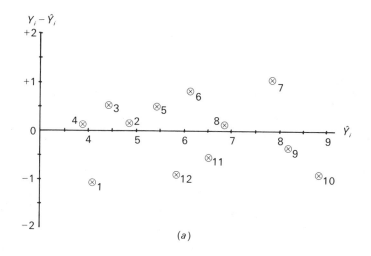

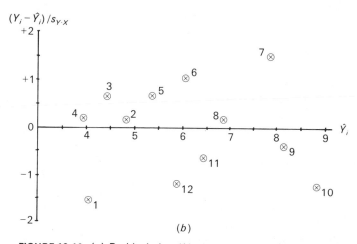

FIGURE 12-16. (*a*) Residual plot; (*b*) plot of standardized residuals.

One way of overcoming this problem is to compute and plot *standardized residuals* from

$$Z_i = \frac{Y_i - \hat{Y}_i}{s_{Y \cdot X}} \qquad (12\text{-}33)$$

This expresses each residual as a standard normal deviate. We can now define an *outlier* as any observation with a standard score Z less than -3.0 or greater than 3.0. If the residuals are normally distributed, then an outlier would have to be far removed from the regression line, at least 3 standard deviations away. We have selected 3 standard deviations quite arbitrarily. If most of the residuals are quite small, then a residual 2.5 standard deviations away from the line might be an outlier. Some

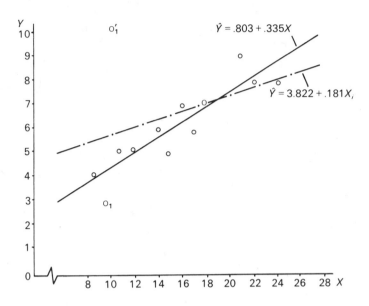

FIGURE 12-17. Effects of an outlier.

judgment is involved here. However, since we have standardized the residuals, we can compare different regression equations, using this criterion. Figure 12-16(*b*) illustrates a plot of the standardized residuals for the trip generation problem. We note that all the residuals are within 2 standard deviations of the regression line, and there are no outliers. Also the standardization of the residuals does not change the *pattern* of the residuals, and Figure 12-16(*a*) and (*b*) reveals quite similar patterns.

One characteristic of the least squares procedure is that the parameter estimates *a* and *b* are extremely sensitive to the presence of outliers, because we *square the deviations*. Large residuals are penalized in the fitting procedure. This effect is similar to one we first noticed in Chapter 2. Recall the sensitivity of the mean of a set of numbers to the presence of an extreme value. For example, suppose we made a mistake in recording the trip generation rate of traffic zone 1. Rather than it being 3 trips per household per day, suppose it is actually 10 trips per household per day. As you can readily see in Figure 12-17, this dramatically changes the scatter of observations. A recalculation of the regression equation with this corrected information yields

$$\hat{Y}_i = 3.822 + .181X_i \qquad (12\text{-}34)$$

Notice the changes in the two parameters. Not surprisingly, we also get much different goodness-of-fit statistics. The new r^2 is .2315, and the new standard error of estimate is 1.679. Figure 12-17 illustrates the two regression lines. Note the large residual of traffic zone 1 from the first regression line, and see how it is significantly reduced in the new equation. The moral is clear: Regression lines are "drawn" to outliers.

In this case we have illustrated the sensitivity of the least squares estimating procedure to outliers by using the example of a measurement error. What if the outlier is a "true" data point and not simply the result of measurement error? Should we just eliminate the point and proceed? It would probably lead to a better regression in that we would get a higher value of r^2. In general, this is unwise. A more appropriate solution is to report the results both with the outlier included in the data set and with it removed. Then we can judge the sensitivity of the results to the nature of the underlying data. Many computer packages for regression analysis have techniques for judging the importance of individual observations on the least squares estimates. In all cases outliers deserve careful examination. They represent important pieces of information about the complexity of the relationship between the two variables. Further study of the outlier might reveal an important variable that has been omitted from the analysis, or it might indicate the existence of a nonlinear relationship.

Another potential difficulty arises when the data contain an *extreme* value of X far removed from the other observations. This situation is depicted in Figure 12-18. A single observation (the problem is the same if there are two or three observations) is located well beyond the range of X which encompasses all other observations. We would not uncover this situation by examining residual plots since the observation is really not an outlier. But the observation is very influential in determining the location of the regression line. Notice the lack of correlation between X and Y exhibited by the remaining observations. This problem illustrates the importance of *basics* in regression analysis. The initial examination of our scattergram can tell us much about the data. We just have to know what to look for.

Once again, we are faced with the same problem—to eliminate the observation or not. Perhaps the best solution is to try generate some new observations in the intervening range of X. Perhaps the extreme value is really "typical" in this range of X. If we eliminate the observation, it is always wise to indicate that the results are limited to a specific range of the independent variable. Suppose we are examining the

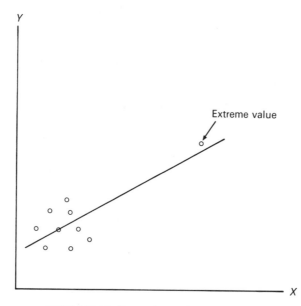

FIGURE 12-18. Regression and extreme values.

relationship between city size and in-migration in a developing country that has a primate (largest) city. We might find that the primate city is 5 to 10 times the size of the second largest city. In terms of regression analysis this presents a difficulty. The alternatives are limited. We cannot simply "invent" new cities that fall in the range of sizes missing from the data. Yet, to drop the primate city would certainly be inadvisable if we really wish to understand the migration pattern of the country. Perhaps we can say just as much without regression analysis as with it. But a picture is worth a thousand words. Careful scrutiny and explanation of the scattergram would nicely complement the regression results. In general, we might say that the appropriate course of action depends on the particular problem at hand and the interests of the geographer or social scientist undertaking the study.

12.6. Summary

This chapter has presented many new ideas and a great deal of information on the mechanics of fitting a linear regression equation. Until you have experience in applying regression analysis to a variety of sets of data in widely different contexts, much of the material appears much more difficult than it really is. Yet we have barely scratched the surface. We have merely introduced regression analysis as a descriptive tool for studying statistical relations among variables. In Chapter 13 we examine the role of statistical inference in regression analysis.

Appendix 12A. Review of the Elementary Geometry of Lines

A straight line is precisely defined by the coordinates of any two points on the line, since it continues forever in the same direction. Thus lines have a constant slope. Consider the points $P_1(X_1, Y_1)$ and $P_2(X_2, Y_2)$ of Figure 12-19. We define the slope of the line passing through these two points to be

$$\frac{\Delta Y}{\Delta X} = \frac{Y_2 - Y_1}{X_2 - X_1} = b \qquad (12\text{-}35)$$

where Δ means change, or difference. When $\Delta X = 1$, $\Delta Y = b$, and clearly we can interpret b as the increase in Y that results from a unit increase in X.

If we know two points on a line or one point and the slope of the line, we can determine the unique equation for that line. Since any two points on the line will suffice, we choose the two points of Figure 12-20. The coordinates are $(0, a)$ for P_0 and (X, Y) for point P. If the line is to have constant slope, then, by (12-35),

$$\frac{Y - a}{X - 0} = b$$

and

$$Y - a = bX$$

or

$$Y = a + bX \qquad (12\text{-}36)$$

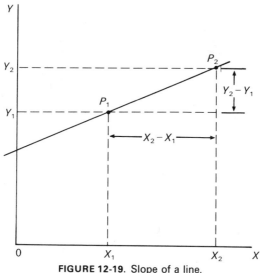

FIGURE 12-19. Slope of a line.

This is the slope-intercept form of the equation for a straight line with the intercept a and slope b. The *parameters a and b* are unrestricted in sign or value.

Figure 12-21 illustrates several examples of straight lines and their equations. The line in Figure 12-21(b) differs from (a) in that is has a steeper slope ($1 > 0.5$). The line depicted in Figure 12-21(c) is parallel to (b) since it has the same slope ($+1$), but differs since it has a lower intercept value ($2 < 5$). In (d) we see the graph of a line

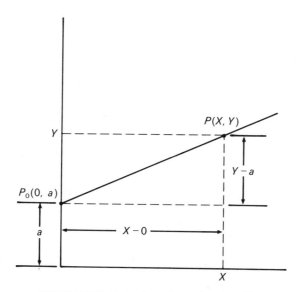

FIGURE 12-20. Derivation of equation for a line.

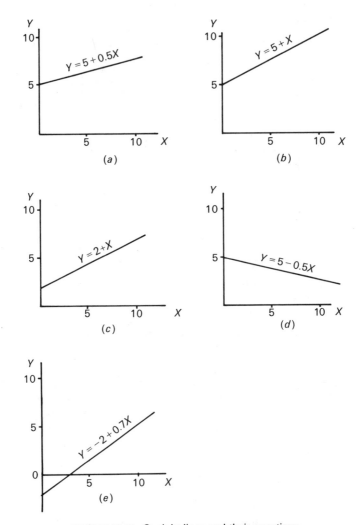

FIGURE 12-21. Straight lines and their equations.

with a negative slope (-0.5), and in (e) we see the graph of a line with a negative intercept (-2). Most of the interest in linear equations in the context of regression analysis is limited to lines passing through the first quadrant since variables X and Y usually take on positive values. However, no restrictions exist on the values to be taken by either the parameters of the line, a and b, or variables X and Y.

Appendix 12B. Least Squares Solution via Elementary Calculus

The normal equations (12-5) and (12-6) which lead directly to the computational formulas for a and b can be easily derived from elementary calculus. Let

$$S(a, b) = \sum_{i=1}^{n} (Y_i - a - bX_i)^2 \tag{12-37}$$

be the total sum of squared deviations from the line to the n points. The values of a and b that minimize S can be derived by differentiating (12-37) with respect to a and b, to obtain

$$\frac{\partial S}{\partial a} = -2 \sum_{i=1}^{n} (Y_i - a - bX_i)$$

and

$$\frac{\partial S}{\partial b} = -2 \sum_{i=1}^{n} X_i(Y_i - a - bX_i)$$

Setting each derivative equal to zero and simplifying yields

$$\sum_{i=1}^{n} (Y_i - a - bX_i) = 0$$

$$\sum_{i=1}^{n} X_i(Y_i - a - bX_i) = 0$$

These equations can be expanded by using simple rules of summations to yield

$$\sum Y_i - na - b \sum X_i = 0$$

$$\sum X_i Y_i - a \sum X_i - b \sum X_i^2 = 0$$

These equations are identical to the normal equations (12-5) and (12-6) with a simple rearrangement of terms.

REFERENCE

N. Draper and H. Smith, *Applied Regression Analysis*, 2nd ed. (New York: Wiley, 1981).

PROBLEMS

Problems in regression analysis are included in Chapter 13.

13

Inferential Aspects of Regression Analysis

So far, our analysis of the regression model has involved only the mechanical procedure whereby a least squares line is fit to a set of sample observations for two variables X and Y. This is but one step in the process of building a simple linear regression model. The following steps are usually components of any regression model building procedure:

1. *Specify the variables in the model and the exact form of the relationship between them.* In Chapter 12, the number of trips made per household per day was selected as the dependent variable. This variable was made a linear function of a single independent variable, household income. In many instances more than one independent variable is related to the dependent variable, or a nonlinear function is more appropriate for the systematic part of the relationship. These two issues are explored in Sections 13.4 and 13.5. Also the model may not be completely specified until the data have been collected. For example, if we were unsure of the nature of the function linking the two variables, we might explore different functions, using the available empirical data. In still other cases the structure of the regression model may be specified a priori from some given theory that links the two variables of concern. Usually, though, the specification of the regression equation is the initial task in regression model building.

2. *Collect data.* The data collection procedure must fulfill the requirements of the underlying assumptions of the simple regression model. These assumptions are detailed in Section 13.1. The exact form of the data depends on whether the fixed-X or random-X model is used as a basis for the regression.

3. *Estimate the parameters of the model.* To estimate the slope and intercept of the regression model, the least squares procedure is used. This has been fully explained in Chapter 12.

4. *Statistically test the utility of the developed model, and check whether the assumptions of the simple linear regression model are satisfied.* This is the inferential side of regression analysis. The purpose of this step is twofold. First, significance tests are undertaken on the slope, the intercept, and the regression model as a whole. Second, hypothesis testing and interval estimation procedures are explained in Section 13.2. For these tests to be valid, several assumptions must be met. Section 13.3 describes methods that can be used to verify the required assumptions.

5. *Use the model for prediction.* Once it is developed, we often wish to use the regression equation to predict values of the dependent variable. In the context of trip generation, it is common to utilize the estimated trip generation equation to predict future trip generation rates based on future values of expected household income. Normally confidence intervals are used to establish a range within which the dependent variable can be predicted at any level of confidence $1 - \alpha$.

Whenever a regression analysis is undertaken, it may not necessarily include all these steps. Sometimes we may not be interested in prediciton at all, and the ultimate goal is only to evaluate the strength of the empirical evidence in support of some prespecified theoretical model. Other times the overriding goal may be to use the model for prediction, and interest in hypothesis testing may be secondary. For example, we might wish to develop a regression equation, using data collected for one period, and then evaluate its predictive capabilities, using data from another period. Quite often both issues are important. The remainder of this chapter explores the computational techniques and issues involved in both hypothesis testing and prediction.

13.1. Assumptions of the Simple Linear Regression Model

In Section 12.2, two different forms of the simple linear regression model were distinguished. In the fixed-X model, it is assumed that an experiment is undertaken, and the values of the independent variable X are under the control of the researcher. Values of X—$X_1, X_2, \ldots, X_M$—are selected, and for each value we randomly sample one or more values of Y. In this representation only Y is a random variable. Alternatively, the data for the regression model can be generated by a survey in which pairs of (X, Y) values are randomly selected from some known population. This random-X model leads to two different random variables, X and Y.

No matter which model is used to derive the sample data for a regression problem, the method of least squares is used to derive the two parameters of the linear equation.

To extend the analysis to include inferential statistics, it is necessary to make several assumptions about the parent populations from which the sample data are drawn. To clarify this notion, it is useful to examine Figure 13-1. The true (population) regression line of Y as a linear function of X is the heavy line. The equation of this line is

$$Y = \alpha + \beta X \tag{13-1}$$

where the convention of utilizing Greek letters for population parameters has been followed. Depending on the particular sample drawn from the population of X and Y values, we may estimate this true regression by any of a possibly infinite number of sample regression lines. Each of these equations is expressed as $Y = a + bX$, where a and b estimate (in the statistical sense) the two population parameters α and β, respectively. If we wish to make inferences about this true population regression line, then additional assumptions are required. These assumptions are identical, with one exception, for both the fixed-X and random-X models. For present purposes, the discussion of these assumptions is couched in the context of the fixed-X model.

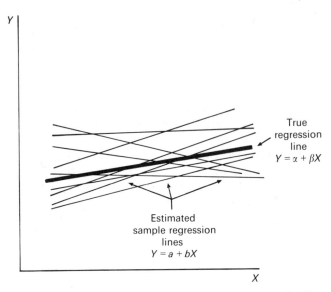

FIGURE 13-1. True regression line and several sample regression lines.

Formally, the simple linear regression model for the fixed-X case is

$$Y_i = \alpha + \beta X_i + \epsilon_i \qquad i = 1, 2, \ldots, M \qquad (13\text{-}2)$$

where Y_i is the value of the dependent random variable, X_i is the value of the ith level of the fixed variable X, and ϵ_i is the random error or disturbance term. In this specification, the subscript i denotes not a particular observation but one of the levels at which the independent variable is set in an experiment. The graphical form of the fixed-X model is seen in Figure 13-2. For each level of X—$X_1, X_2, \ldots, X_M$—there is a probability distribution for the random variable Y. The mean value of the dependent variable Y for $X = X_1$ is given by $Y_1 = \alpha + \beta X_1$. To this systematic component of Y we must add the value of the random component ϵ_i. It is this random error component which makes the dependent variable Y a random variable—even when the independent variable X is fixed.

What is the source of this random error component? The error term may be conceived as being derived from two sources. First, the relationship between X and Y may be imperfect owing to *measurement error*. This source of error must be minimized. The other source of error is termed *stochastic error*. Suppose that an experiment is conducted to predict crop yield Y based on the amount of fertilizer applied X. If several experimental plots are all treated with exactly 300 lb of fertilizer, the recorded crop yields will not be exactly equal. Why? For one thing, the soils of the various plots may not be perfectly homogeneous. If the drainage of the plots is unequal or if the purity of the fertilizer applied to the plots varies, then we might expect some variation in yields. In fact, a large number of other factors may affect the experiment which cannot be *completely* controlled, each having a small influence on yield. The *net* effect of these influences appears to be random. As we shall see in Section 13.5, we can incorporate several independent variables into the regression equation. However, some may be impossible to exactly control and are best treated as random or stochastic error.

Figure 13-2 is a crucial diagram for understanding the inferential aspects of regression analysis. Notice that the vertical axis is labeled *Probability density* since the dependent variable Y is a random variable. Moreover, associated with each level of X is a *conditional probability distribution* of the random variable Y. It is important to distinguish between this *unobserved* population and the observed sample observations upon which the regression equation $Y = \alpha + \beta X$ is estimated. One such sample value is illustrated for each value of X. In fact, there may be more than one sample observation for each value of X. Normally, the experiment would be designed so that a number of different observations of Y are obtained for each value of X. However, even if there are a large number of observations for each value of X and even if the values of X are closely spaced, it would be very fortunate if the estimated regression line $Y = a + bX$ exactly coincided with this true regression equation. By making a few assumptions about the regularity of the distribution of the random error ϵ_i (or equivalently Y_i), it is possible to make statistical inferences about $Y_i = \alpha + \beta X_i$ on the basis of the sample information $Y_i = a + bX_i$. There are four essential assumptions.

ASSUMPTION 1

For the ith level of the independent variable X_i, the expected value of the error component ϵ_i is equal to zero. This is usually expressed as $E(\epsilon_i) = 0$ for $i = 1, 2, \ldots, M$.

This first assumption states that, on the average, the random error component

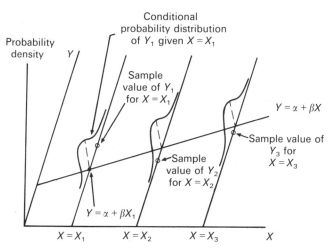

FIGURE 13-2. Fixed-X model.

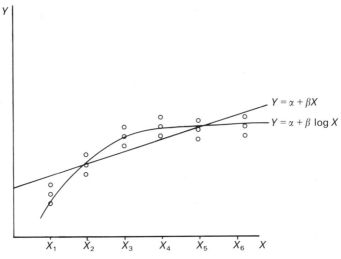

FIGURE 13-3. Violation of Assumption 1.

of Y_i is zero. If there are a series of sample observations of Y_i at some particular value of X, say $X = X_i$, then some of these observations should lie above the regression line and others below it. Recall that the regression line passes through the mean value of Y_i. To show this, note that $E(Y_i) = E(\alpha + \beta X_i + \epsilon_i)$. Since α, β, and X_i are constants, $E(Y_i) = \alpha + \beta X_i + E(\epsilon_i)$. By this first assumption $E(\epsilon_i) = 0$ and thus $E(Y_i) = \alpha + \beta X_i$.

How can this assumption be violated? Suppose we were to fit a linear regression model to two variables that are actually related by a nonlinear function. In Figure 13-3, for example, a linear equation $Y = \alpha + \beta X$ is fit to a set of data. Note that Assumption 1 is not satisfied since the equation does not appear to follow the path of the means. In this instance, if a nonlinear equation of the form $Y = a + b \log X$ is fit to the data, it appears that Assumption 1 could be satisfied. But what if the exact form of the true regression equation that links the two variables is unknown? It may not be immediately apparent that the assumption has been violated. An examination of the residual errors from the estimated equation $Y = a + bX$ may uncover certain violations. In Figure 13-3, for example, the residuals from the linear regression $Y = a + bX$ show a definite pattern and do not appear random at all. On the average, the Y values for $X = X_1$ and X_6 are below the line, but the Y values for X_3 and X_4 lie above the line. This does not appear random. The examination of residuals is discussed in Section 13.3. Nonlinear regression models are briefly introduced in Section 13.4.

For Assumption 1 to hold, the regression model must be correctly specified, that is, it must represent the true underlying process. Once the correct *specification* is formally assumed, estimating the model parameters becomes a relatively straightforward mechanical procedure. In reality, however, we can never be sure that the model is correctly specified. Two types of problems exist. First, we may have omitted relevant variables from the regression equation. If so, the random error term ϵ_i cannot be considered random since it contains a systematic component. Second, the wrong functional form may have been chosen. Applied researchers usually examine more than one possible specification in an attempt to find the specification that best describes the process. A detailed examination of the residuals is one of the best ways to make sound judgments concerning model specification. Relevant material is included in Sections 13.3, 13.4, and 13.5.

ASSUMPTION 2

The variance of the error component ϵ_i is constant for all levels of X_i; that is, $V(\epsilon_i) = \sigma^2$ for $i = 1, 2, \ldots, M$.

The second assumption also concerns the error term ϵ_i. The variance of the probability distribution of ϵ_i is equal to σ^2. Also, this is constant for all values of X. This variance is unknown, but it can be estimated from the sample residuals. As introduced in Chapter 10, this is known as the assumption of *homoscedasticity*, or simply *equal variance*. Since $E(\epsilon_i) = 0$ and $V(\epsilon_i) = \sigma^2$, it is possible to specify the variance of the conditional probability distribution of Y_i, or $V(Y_i)$. Note that $V(Y_i) = V(\alpha + \beta X_i + \epsilon_i)$. Since $\alpha + \beta X_i$ is a constant, it has no variance and can be omitted. Simplifying, we see that $V(Y_i) = V(\epsilon_i) = \sigma^2$. The probability distributions of Y_i and ϵ_i differ only by their means. Violations of the assumption of homoscedasticity are also usually uncovered through the analysis of residuals (see Section 13.3).

ASSUMPTION 3

The values of the error component for any two ϵ_i and ϵ_j are pairwise uncorrelated.

By this assumption, the outcome Y_i for any given value of X, say $X = X_i$, neither affects nor is affected by the outcome Y_j for any other value of X, say $X = X_j$. In terms of the covariance operator described in Chapter 12, this assumption implies cov $(\epsilon_i, \epsilon_j) = 0$ for all i and j. If this assumption is not satisfied, then *autocorrelation* is present in the error term. Autocorrelation can exist in many forms. A detailed discussion of this topic is offered in Section 13.3.

ASSUMPTION 4

The error components ϵ_i are normally distributed.

This last assumption completes the specification of the error term in the fixed-X regression model. It *is not required* for calculating point estimates of α, β, and σ (a, b, and $s_{Y \cdot X}$, respectively), *nor is it required* to make point estimates of the mean value of Y for any given X, $\hat{Y}_i$. It *is required* for the construction of confidence intervals and/or tests of hypotheses concerning these parameters. This assumption should not be surprising since virtually all the confidence interval and hypothesis testing methods described in Chapters 7 and 8 are based on the assumption of the normality of the population distribution. The verification of this assumption is also based on an examination of the residuals from the sample regression line $Y = a + bX$. The specific technique used is explained in Section 13.3.

For a random-X model this assumption must be augmented with an additional constraint governing the probability distribution of the random variable X. One of the most important members of the class of random-X models assumes that the joint distribution of X and Y is bivariate normal. Data conforming to the random-X model might be realized by a sampling procedure in which individuals, households, industrial firms, zones, regions, or cities are randomly selected as observations from a statistical population. If the underlying probability distribution of the two variables is bivariate normal (see Figure 5-20), then the standard inferential tests described in Section 13-2 are still valid. Often, however, the data used by geographers in a regression model do not conform to either the fixed-X or the random-X model. The data for the trip generation example of Chapter 12 illustrate this point. The validity of the use of the standard inferential tests in this situation is unclear. Sometimes other justifications for their use are given.

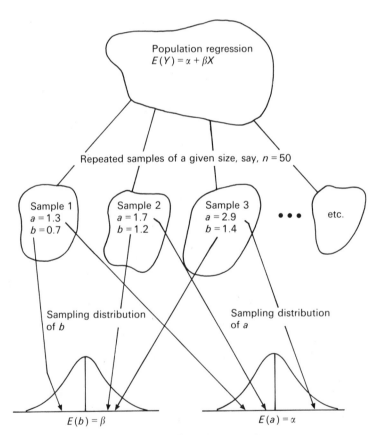

FIGURE 13-4. Sampling procedure.

Gauss-Markov Theorem

The principal analytical results of regression theory derive from the *Gauss-Markov theorem*. This theorem justifies the use of the least squares procedure for estimating the two parameters of the linear regression equation. Also, it establishes the principal properties of the least squares estimators a and b.

GAUSS-MARKOV THEOREM

Given the four assumptions of the simple linear regression model, the least squares estimators a and b are unbiased and have the minimum variance among all the linear unbiased estimators of α and β.

To examine the implications of this theorem, consider the sampling procedure depicted in Figure 13-4. From some statistical population, repeated random samples of some given size, say $n = 50$, are drawn and least squares estimators a and b calculated from (12-7) and (12-12). If the four assumptions are satisfied, then several characteristics of the sampling distributions of a and b follow from the Gauss-Markov theorem. First, estimators a and b are unbiased estimators of α and β. In Figure 13-4 the sampling distributions of a and b are centered on $E(a) = \alpha$ and $E(b) = \beta$. These sampling distributions are themselves normal. Finally, and most important, a and b are the most *efficient* estimators in the sense that the sampling distributions of a and b have less variability (smaller variances) than any other linear and unbiased estimators of α and β. The least squares estimators a and b are thus often called *BLUE*, an acronym for *b*est *l*inear *u*nbiased *e*stimators. Statistical tests on various aspects of the fitted regression equation $Y = a + bX$ follow directly from the Gauss-Markov theorem.

13.2. Inferences in Regression Analysis

Point Estimates

In many applications of regression analysis, it is important to estimate the value of some parameter in the regression equation. Statistical estimation in the context of regression is based on the process of statistical estimation described in Chapter 7. In Chapter 7 two types of statistical estimation were distinguished: point estimation and interval estimation. In point estimation, a single number is computed from the sample information and is used as an estimate of some population parameter. Note that all the point estimates in regression analysis require only that the *first three* assumptions explained in Section 13.1 be satisfied. *No assumption of normality for the error term is required.* However, the assumption of normality is essential for both hypothesis testing and the construction of confidence intervals. Table 13-1 summarizes the principal parameters and their estimators in regression analysis.

From the Gauss-Markov theorem, we know that a and b are the BLUE estimators of α and β. Similarly, $s_{Y \cdot X}$, r, and r^2 provide point estimates of σ, ρ, and ρ^2,

TABLE 13-1
Estimators in Regression Analysis

Regression model parameter	Point estimator	Equation
α	a	(12-8)
β	b	(12-12)
$\sigma^2[\sigma]$	$s_{Y \cdot X}^2[s_{Y \cdot X}]$	(12-27)
ρ	r	(12-14)
ρ^2	r^2	(12-26)
$\mu_{Y \cdot X}[E(Y)]$	$a + bX_i$	(12-1)
$\hat{Y}_{\text{new}}$	$a + X_{\text{new}}$	(12-1)

respectively. If we wish to predict the mean value of Y for any level of X within the range of the sample data [or $E(Y)$], the best point estimate is clearly $E(Y) = a + bX_i$. This should not be surprising since the regression equation is defined to follow the path of the means if the first three assumptions outlined in Section 13.1 hold. The expressions $E(Y)$ and $\mu_{Y \cdot X}$ are interchangeable. Here $E(Y)$ is used since Y_i is a random variable, and $\mu_{Y \cdot X}$ is used to reflect the fact that it is a population parameter which clearly depends on the value of variable X. Finally, suppose we have a single new value of X, denoted X_{new}, and wish to predict the most likely value of Y for X_{new}, or Y_{new}. The best estimator is again $Y_{\text{new}} = a + bX_{\text{new}}$. This is the same as $\mu_{Y \cdot X}$. However, as we shall see, the interval estimates of these two parameters are much different. This is expected. It is easier to make a prediction of the mean of several numbers than a single number. In the same sense we found the sampling distribution of $\bar{X}$ to be much more compact than the variable X itself. The standard error of the mean is equal to the standard deviation of variable X only when the sample size is 1!

Inferences concerning the Slope of the Regression Line β

The slope of the regression line is particularly important in regression analysis. Sometimes the interest in β derives from the fact that it measures the sensitivity of variable Y to changes in variable X. For example, the slope of the trip generation equation tells a transportation planner the *magnitude* of the increase in the number of trips expected from increasing incomes. Estimating β is thus extremely important. For this purpose a confidence interval for β can be constructed by using the least squares estimator b. There is a second reason for examining β. If it is possible to infer from a sample regression line that $\beta = 0$, then it can be concluded that the regression equation is of *no use* as a predictor since $E(Y) = \alpha + (0)X = \bar{Y}$. For every value of X, the best estimate of Y is $\bar{Y}$. Whether it is possible to infer that the slope of the population regression is nonzero therefore has important ramifications for the use of the regression equation for predictive purposes. To establish confidence interval and hypothesis testing formulas for β, it is necessary to describe the sampling distribution of the least squares estimator b.

The sampling distribution of b is fully characterized in the Gauss-Markov theorem (see Section 13.1 and especially Figure 13-4). It has the following features:

1. It is normal.
2. It has a mean value centered on β; that is, it is unbiased and $E(b) = \beta$.
3. It has a standard error, denoted σ_b,

$$\sigma_b = \frac{\sigma}{\sqrt{\sum_{i=1}^{n} X_i^2 - (\sum X_i)^2/n}}$$

Note that the formula for the standard error of b contains one unknown, σ. However, σ can be estimated from the sample data by $s_{Y \cdot X}$, and an estimate of σ_b, denoted s_b, is obtained:

$$s_b = \frac{s_{Y \cdot X}}{\sqrt{\sum_{i=1}^{n} X_i^2 - (\sum_{i=1}^{n} X_i)^2/n}} \qquad (13\text{-}3)$$

It turns out that the sampling distribution of b is no longer normal once $s_{Y \cdot X}$ has been substituted for σ. It is t-distributed with $n - 2$ degrees of freedom. There are only $n - 2$ degrees of freedom since 2 degrees of freedom are lost in the estimation of $s_{Y \cdot X}$.

The value of the standard error s_b depends on two separate factors. First, s_b will increase with increases in the standard error of estimate $s_{Y \cdot X}$. This is expected. If the standard deviation of the conditional probability distribution of ϵ_i is large, one will be less confident about estimating the slope of the regression line. However, the greater the sum of squared deviations of X (that is, the greater the variability in X), the lower s_b. To see why, compare Figure 13-5(a) with (b). Suppose all the values of X used to fit the regression line are concentrated at a few values of X quite close to $\bar{X}$. This is the situation in Figure 13-5(a). The estimated slopes of various sample regression lines are likely to be quite variable. By comparison, the estimated slopes of possible regression lines in situations in which there is a considerable variability in the values of X are all quite similar. In general, then, we are more confident about predicting the true value for β when the prediction is based on observations over a wide range of the independent variable X.

The classical test of hypothesis for β is straightforward. The test statistic is

$$t = \frac{b - \beta}{s_b} \qquad (13\text{-}4)$$

It is most common to test the null hypothesis $\beta = 0$, but this test statistic can be used to test a null hypothesis concerning any desired value of β. When the null hypothesis is $\beta = 0$, Equation (13-4) reduces to $t = b/s_b$.

As an alternative to a formal test of hypothesis, it is possible to construct a confidence interval for β from the point estimate b:

$$b - t_{\alpha/2, n-2} s_b \leq \beta \leq b + t_{\alpha/2, n-2} s_b \qquad (13\text{-}5)$$

If this interval contains zero, then we will accept the null hypothesis that $\beta = 0$ and conclude that the regression line is of no use in prediction.

Again, since σ is unknown, it is estimated by $s_{Y \cdot X}$, and the estimated standard error of a becomes

$$s_a = s_{Y \cdot X} \sqrt{\frac{1}{n} + \frac{\bar{X}^2}{\sum_{i=1}^{n} X_i^2 - (\sum_{i=1}^{n} X_i)^2/n}} \qquad (13\text{-}6)$$

The appropriate test statistic

$$t = \frac{a - \alpha}{s_a} \qquad (13\text{-}7)$$

is t-distributed with $n - 2$ degrees of freedom. It is most common to test the null hypothesis that $\alpha = 0$, but this test statistic can be used to test a null hypothesis concerning any value of α. By using the standard t statistic, the $100(1 - \alpha)$ percent confidence interval for α from a given least squares estimate a is

$$a - t_{\alpha/2, n-2} s_a \leq \alpha \leq a + t_{\alpha/2, n-2} s_a \qquad (13\text{-}8)$$

As we have noted, the intercept of the household trip generation equation is not particularly meaningful. Nevertheless, a test of this intercept is useful to illustrate the necessary calculations of this test. Table 13-3 includes a complete test of hypothesis

TABLE 13-3
Hypothesis Test for Regression Intercept in Trip Generation Example

Hypotheses
H_0: $\alpha = 0$
H_A: $\alpha \neq 0$

Test statistic and test to be employed
Test statistic is $(a - \alpha)/s_a$, Equation (13-7).

Level of significance
Set $\alpha = .05$.

Decision rule
If $t < -2.228$ or $t > 2.228$, reject H_0;
otherwise, accept H_0.

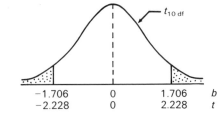

−1.706	0	1.706	b
−2.228	0	2.228	t

Computation of test statistic

$$t = \frac{.807}{.752\sqrt{\frac{1}{12} + (15.75)^2/[3237 - (189)^2/12]}} = \frac{.807}{.7656} = 1.054$$

Decision
Since $-2.228 < 1.054 < 2.228$, we accept the null hypothesis.

on the null hypothesis that $\alpha = 0$. The calculated t value is well below the critical value necessary for rejection at a significance level of .05, and we accept the null hypothesis that $\alpha = 0$. This seems to be a reasonable conclusion. Without any income, households in a traffic zone might be expected to make zero trips. Since the intercept is well out of the range of the available data, it is of minor concern in any case. If the data were to suggest a rejection of the null hypothesis, this would also be of no real significance. Given the value $a = .807$, the 95 percent confidence limits are

$$.807 - 2.228(.7656) \leq \alpha \leq .807 + 2.228(.7656)$$

$$-.899 \leq \alpha \leq 2.513$$

The fact that this interval contains $\alpha = 0$ verifies the acceptance of the null hypothesis, which was concluded in Table 13-3.

Analysis of Variance for Regression

It is also possible to test the null hypothesis $\beta = 0$ by using the F distribution. The analysis of variance for regression always produces results equivalent to those obtained from the t test for the null hypothesis $\beta = 0$. In fact, the value of the F statistic computed in the analysis of variance is the exact square of the value of the sample t statistic given by (13-4). This will be demonstrated for the houshold trip generation equation. Although these two tests are exactly equivalent for simple linear regression, they are not equivalent for cases in which the regression equation includes more than one independent variable. In multiple regression the t test is used to test the significance of a single independent variable given the null hypothesis $\beta_j = 0$. The analysis of variance is used to test the joint hypothesis that all the independent variables have an insignificant effect on the dependent variable Y, that is, $\beta_1 = \beta_2 = \cdots = \beta_m = 0$.

    The general form of the analysis of variance table for simple linear regression is given in Table 13-4. The total sum of squares (TSS) is equal to $\Sigma_{i=1}^{n} (Y_i - \bar{Y})^2$. Associated with this total variation are $n - 1$ degrees of freedom, 1 less than the number of observations. The two mean squares are calculated by apportioning these

TABLE 13-4
Analysis of Variance Table in Simple Regression

Source of variation	Sum of squares	Degrees of freedom	Mean square
Regression (model)	RSS	1	$\dfrac{\text{RSS}}{1}$
Error (residual)	ESS	$n - 2$	$\dfrac{\text{ESS}}{n - 2}$
	TSS	$n - 1$	
	$F = \dfrac{\text{RSS}/1}{\text{ESS}/(n - 2)}$		

TABLE 13-5
Analysis of Variance for Trip Generation Equation

Sources of variation	Sum of squares	Degrees of freedom	Mean square
Regression	29.252	1	29.252
Error	5.666	10	.5666
	34.918	11	

$$F = \frac{29.252}{.5666} = 51.6$$

Note. Students may get slightly different results depending on the number of places kept in the calculations.

$n - 1$ degrees of freedom to two sources: the regression model and the error, or residual, variation. The degrees of freedom for the residual error (ESS) are always equal to the number of observations minus the number of independent variables minus 1. In simple regression this is always equal to $n - 2$. The degrees of freedom attributable to the model are always equal to 1. For a test of hypothesis with a significance level of α, the critical value may be found in the F table with degrees of freedom $(1, n - 2)$. If the calculated F value is less than $F(1, n - 2)$, then the null hypothesis H_0: $\beta = 0$ is not rejected, and we conclude that variable X is not useful in predicting the value of the dependent variable Y. As usual, if the calculated value of F exceeds the value of F in the table, the null hypothesis is rejected. The analysis of variance is given in Table 13-5 for the household trip generation data. From the F table in the Appendix, we see $F(1, 10) = 4.96$ at $\alpha = .05$, and the calculated value of $F = 51.68$ clearly exceeds this critical value. We conclude that household income is useful in predicting houshold trip generation rates. Note that the calculated value of $F = 51.68$ is almost the exact square of the t value calculated in Table 13-2 for the t test on the same null hypothesis: $(7.189)^2 \simeq 51.628$. The difference between these two figures is due solely to the number of places used in the calculation of each.

Not only are the F test for the analysis of variance and the t test identical in simple linear regression, but also they are both equivalent to the test of significance for the simple correlation coefficient r. This should not be surprising, given the relationship between correlation and regression expressed by Equation (12-19): $b = r \, s_Y/s_X$. To see this equivalency, let us rewrite the F statistic computed for the analysis of variance in terms of the coefficient of determination r^2. Since TSS = ESS + RSS and r^2 = RSS/TSS, we can rewrite the mean square due to regression, RSS/1, as r^2(TSS)/1. Similarly, we rewrite the residual or error mean square, ESS/$(n - 2)$, as $(1 - r^2)$(TSS)/$(n - 2)$. The F statistic can then be rewritten as

$$F = \frac{r^2(\text{TSS})/1}{(1 - r^2)(\text{TSS})/(n - 2)} = \frac{r^2(n - 2)}{1 - r^2} \tag{13-9}$$

Now, let us compare the F statistic expressed by (13-9) to the t statistic used to test the significance of the simple correlation coefficient (11-5): $t = r\sqrt{n - 2}/\sqrt{(1 - r^2)}$. Equation (13-9) is the square of (11-5).

Confidence Interval for $\mu_{Y \cdot X}$ of Given X

One of the most important applications of regression analysis is to use the derived equation $Y = a + bX$ to estimate a value for the conditional mean of Y, denoted $\mu_{Y \cdot X}$, for a given value of X. We have already noted that the point estimate of $\mu_{Y \cdot X}$ is determined by substituting the appropriate value of X, say X_0, into the equation so that $Y_0 = \mu_{Y \cdot X_0} = a + bX_0$. Of course, there is some error involved in such an estimate, and it is usual to construct a confidence interval for $\mu_{Y \cdot X_0}$. To develop this confidence interval, it is necessary to specify the sampling distribution of $\hat{Y}_0$. The characteristics of this sampling distribution are as follows:

1. It is normal.
2. It has a mean value of $\mu_{Y \cdot X_0}$, and it is unbiased.
3. It has a standard error $\sigma_{\hat{Y}_0}$ of

$$\sigma_{\hat{Y}_0} = \sigma \sqrt{\frac{1}{n} + \frac{(X_0 - \bar{X})^2}{\sum_{i=1}^n X_i^2 - (\sum_{i=1}^n X_i)^2/n}}$$

Estimating σ by $s_{Y \cdot X}$ leads to

$$s_{\hat{Y}_0} = s_{Y \cdot X} \sqrt{\frac{1}{n} + \frac{(X_0 - \bar{X})^2}{\sum_{i=1}^n X_i^2 - (\sum_{i=1}^n X_i)^2/n}} \qquad (13\text{-}10)$$

The standard error of Y_0 increases with (1) the standard error of estimate $s_{Y \cdot X}$, (2) the reciprocal of the sum of squared deviations, (3) the sample size n, and (4) the difference between the value X_0 and the mean $\bar{X}$. The standard error of estimate and

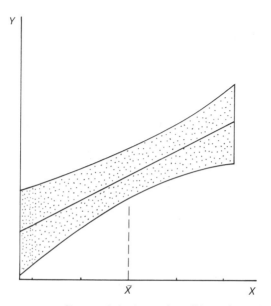

FIGURE 13-6. Characteristic shape of confidence intervals.

TABLE 13-6
Work Table

Confidence limit for $\hat{Y}$, at $X_0 = \bar{X}$

$\bar{X} = 15.75$ $\bar{Y} = \hat{Y} = 6.083$ $s_{Y \cdot X} = .752$

$n = 12$ $\sum X_i^2 = 3237 \left(\sum X_i \right) = 189$ $t_{.025, 10 \text{ df}} = 2.228$

1. Determine $s_{\hat{Y}}$:

$$s_{\hat{Y}} = s_{Y \cdot X} \sqrt{\frac{1}{n} + \frac{(X_0 - \bar{X})^2}{\sum X_i^2 - (\sum X_i)^2 / n}}$$

$$= .752 \sqrt{\frac{1}{12} + \frac{0^2}{3237 - (189)^2 / 12}}$$

$$= .752(.288) = .217$$

2. Determine confidence interval:

$$\hat{Y}_0 - t_{\alpha/2}(s_{\hat{Y}_i}) \le \mu_{\hat{Y}_0} \le \hat{Y}_0 + t_{\alpha/2}(s_{\hat{Y}_i})$$

$$6.083 - 2.228(.217) \le \mu_{\hat{Y}_0} \le 6.083 + 2.228(.217)$$

$$5.600 \le \mu_{\hat{Y}_0} \le 5.567$$

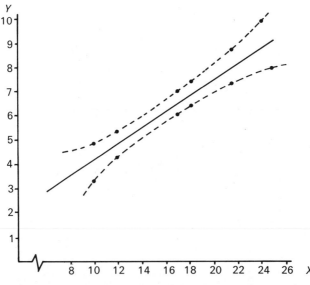

FIGURE 13-7. Confidence intervals for trip generation equation.

TABLE 13-7
Confidence Intervals for $\hat{Y}$

Annual household income, $000	Estimated mean trips per day	95 percent confidence interval		Upper – lower estimate
		Lower estimate	Upper estimate	
10	4.156	3.386	4.925	1.539
12	4.826	4.205	5.448	1.243
14	5.491	4.979	6.014	1.035
15.75	6.083	5.600	6.567	0.967
18	6.838	6.300	7.375	1.075
20	7.507	6.852	8.162	1.310
22	8.179	7.368	8.989	1.621

the sum of squared deviations of X are important for the same reasons outlined for s_b. A larger sample size is always better for estimating a mean such as $\hat{Y}_0$.

The implications of the final factor, the difference between X_0 and $\bar{X}$, are illustrated in Figure 13-6. The confidence interval is narrowest at $X_0 = \bar{X}$. The farther X_0 lies from $\bar{X}$, the wider the interval. There is a simple explanation for the peculiar shape of these confidence intervals. Suppose that the bands were parallel to the regression line and not shaped as the interval in Figure 13-6. This would imply that the only error in predicting Y derives from mistakes in estimating α. Once it is recognized that both a and b can be in error, then the shape of these limits is easier to understand. The farther the point is from $\bar{X}$, the greater the impact of an error in estimating β on the estimated value $\hat{Y}_0$. The dangers to extrapolation are clear.

Because of the importance of household trip generation equations in making urban area trip forecasts, the confidence interval about $Y = .807 + .335X$ holds special interest. Consider first the confidence interval about the line at the point $X_0 = \bar{X}$. For the average income household of 15.75 (thousands of dollars), the estimated number of trips per day is $\hat{Y}_0 = \bar{Y} = 6.083$ [recall the regression line always passes through $(\bar{X}, \bar{Y})$]. In Table 13-6 the calculations necessary to specify the 95 percent confidence interval at $X_0 = \bar{X}$ are illustrated. Table 13-7 summarizes the estimates for $\hat{Y}_0$ at several other values of X within the observed range of the independent variable X. The interval is narrowest at $X = \bar{X}$ and wider for values of X greater than or less than $\bar{X} = 15.75$. This pattern is depicted most clearly in Figure 13-7.

Prediction Interval for Y_{new}

Often we are not interested in making inferences about the mean of the conditional distribution of Y_0 at any given level of X, $X = X_0$. Instead, we wish to make inferences about the value of a single new observation of X, say X_{new}. Suppose we must predict the number of trips per household per day made by households in a traffic zone with a forecasted income of $X = \$20,000$. We cannot use the confidence interval for this purpose since it is based on the expected mean number of trips made by households

TABLE 13-8
Prediction Interval for $\hat{Y}$ at $X_{new} = \bar{X}$

$\bar{X} = 15.75$ $\bar{Y} = \hat{Y}_{new} = 6.083$ $s_{Y \cdot X} = .752$

$n = 12$ $\sum X_i^2 = 3237 \sum X_i = 189$ $t_{.025, \, 10 \, df} = 2.228$

1. Determine $s_{\hat{Y}_{new}}$:

$$s_{\hat{Y}_{new}} = s_{Y \cdot X} \sqrt{1 + \frac{1}{n} + \frac{(X_{new} - \bar{X})^2}{\sum X_i^2 - (\sum X_i)^2/n}}$$

$$= .752 \sqrt{1 + \frac{1}{12} + \frac{0^2}{3237 - (189)^2/12}}$$

$$= .752(1.04)$$

$$= .782$$

2. Determine the prediction interval:

$$\hat{Y}_{new} - t_{\alpha/2, 10 \, df} \, s_{\hat{Y}_{new}} \le \mu_{Y_{new}} \le Y_{new} + t_{\alpha/2, 10 \, df} \, s_{\hat{Y}_{new}}$$
$$6.083 - 2.228(.782) \le \mu_{Y_{new}} \le 6.083 + 2.228(.782)$$
$$4.341 \le \mu_{Y_{new}} \le 7.826$$

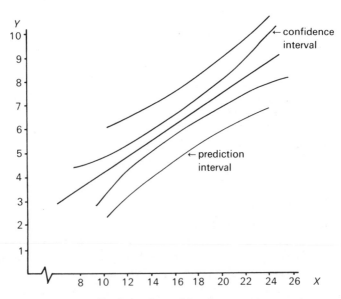

FIGURE 13-8. Prediction interval for trip generation equation.

TABLE 13-9
Prediction Intervals for $\hat{Y}$

Annual household income, $000	Estimated mean trips per day	95 percent prediction interval		Upper − lower estimate
		Lower estimate	Upper estimate	
10	4.156	2.310	6.001	3.691
12	4.826	3.037	6.615	3.578
14	5.491	3.742	7.252	3.510
15.75	6.083	4.341	7.826	3.485
18	6.838	5.076	8.599	3.523
20	7.507	5.709	9.305	3.596
22	8.179	6.316	10.041	3.725

in several traffic zones with an income of $20,000. It is much more difficult to estimate a single value than a mean, so we expect that this prediction interval will be wider than the corresponding confidence interval at any value of X.

The construction of a prediction interval is virtually identical to the procedure followed for a confidence interval. The only difference is the standard error of $\hat{Y}_{new}$ is

$$s_{\hat{Y}_{new}} = s_{Y \cdot X} \sqrt{1 + \frac{1}{n} + \frac{(X_{new} - \bar{X})^2}{\Sigma_{i=1}^{n} X_i^2 - (\Sigma_{i=1}^{n} X_i)^2/n}} \qquad (13\text{-}11)$$

This standard error is almost identical to $s_{\hat{Y}_0}$. In fact, it is larger by an additional $s_{Y \cdot X}$. To distinguish (13-10) from (13-11), the quantity $s_{\hat{Y}_{new}}$ is often called the *standard error of the forecast*. The $100(1 - \alpha)$ percent prediction or forecast interval is

$$\hat{Y}_{new} - t_{\alpha/2, n-2} s_{\hat{Y}_{new}} \leq Y_{new} \leq \hat{Y}_{new} + t_{\alpha/2, n-2} s_{\hat{Y}_{new}} \qquad (13\text{-}12)$$

Consider now the specification of prediction intervals for the trip generation equation. The prediction interval for $X_{new} = \bar{X}$ is calculated in Table 13-8, and prediction intervals for several values of X are summarized in Table 13-9. Figure 13-8 compares the confidence and prediction intervals for the household trip generation equation. Note the greater width of the prediction interval.

13.3. Violation of the Assumptions of the Linear Regression Model

Specific statistical tests are available for checking whether the assumptions underlying the use of the linear regression model are satisfied by the data utilized in any empirical study. These tests employ the residuals of the sample regression equation $Y_i = a + bX_i + e_i$ as estimates of the error term of the true population regression model $Y_i = \alpha + \beta X_i + \epsilon_i$. Before presenting these tests in details, we examine a widely used technique that can provide a quick, but not foolproof, check of these assumptions. Often, a *residual plot* can reveal clear or possible violations of these assumptions. Two forms of these residual plots are commonly used. First, an *overall plot* of

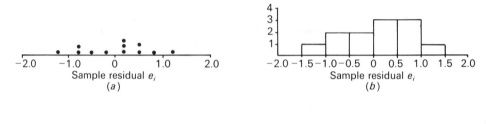

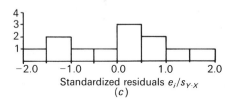

FIGURE 13-9. Residual plots for trip generation equation. (*a*) Dot diagram; (*b*) histogram; (*c*) histogram of standardized residuals.

the residuals can be made as a dot diagram or histogram, as in Figure 13-9. Consider the 12 residuals to the household trip generation problem. To one decimal place these residuals are -1.2, 0.2, 0.5, 0.2, 0.5, 0.8, 1.2, 0.2, -0.2, -0.8, -0.5, and -0.8. In Figure 13-9(*a*) these observations are depicted as dots along an axis measuring the value of the residual e_i. This procedure can be awkward if there are a large number of residuals and it is usual to graph the residuals from larger data sets by using a histogram. The histogram format is illustrated in Figure 13-9(*b*).

Overall plots can also be *standardized* so that comparisons among different residual plots are possible. This is done by standardizing the sample residuals by dividing by the standard error of the estimate. Now under Assumption 4 of the linear regression model, the residuals from the sample regression line should be normally distributed with mean 0 and variance σ^2. The standardized residuals are thus also normally distributed with zero mean, but have a unit variance and standard deviation. What should we look for in a standardized residual plot? Roughly 95 percent of the observations should fall in the range $[-2, +2]$, or more exactly $[-1.96, +1.96]$. In Figure 13-9(*c*), the standardized residuals are plotted for the household trip generation problem. Notice that 100 percent of observations lie within the range $[-2, +2]$. Given the small sample size of $n = 12$, this is not unexpected. Checking for outliers is also convenient when the residuals are expressed in this form.

Examination of Residual Plots

A more common type of residual plot graphs the sample residuals e_i (or standardized residuals) versus the predicted value of Y, denoted $\hat{Y}_i$. How should such a plot appear? If the first three assumptions of the linear regression model hold, then the residual plot should be rectangular, as in Figure 13-10(*a*). Any significant departure from this pattern is usually evidence of some violation in one of these three assumptions. There are several common patterns indicative of specific problems.

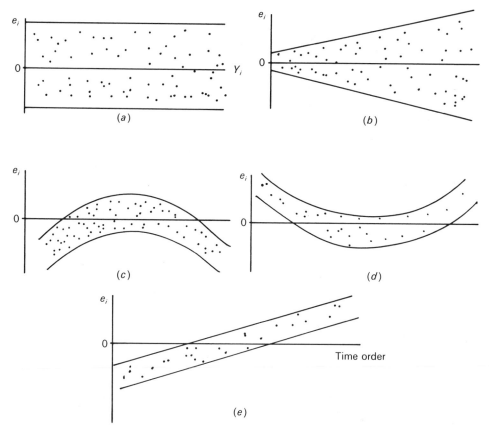

FIGURE 13-10. Residual plots for a regression equation. (*a*) No violation of assumptions; (*b*) heteroscedastic; (*c*) and (*d*) incorrect functional forms; (*e*) residuals linearly related to time.

1. A plot in which the width of the residual band increases with Y indicates that the assumption of homoscedasticity has been violated. In Figure 13-10(*b*) the residuals appear to have a greater variance for large values of Y than for smaller values. Corrective action for this condition must be undertaken, and the least squares estimators a and b do not have the properties specified by the Gauss-Markov theorem [see Draper and Smith (1981) or Pindyck and Rubinfeld (1981)]. Although it is less common, the variance of the residuals may decrease with Y or even increase at both extremes.

2. Unsatisfactory residual plots can also be obtained when the wrong functional form is used for the systematic component of the relationship. In Figure 13-10(*c*), for example, we notice that positive residuals occur at moderate values of Y and negative residuals for high and low values of Y. This can be corrected by fitting an appropriate nonlinear model. Figure 13-10(*d*) illustrates another violation of this type. Nonlinear models are briefly introduced in Section 13.4.

3. If the data have been generated by an experiment that took place over time,

TABLE 13-10
Residuals from Trip Generation Equation

Traffic zone	e_i	Average household size
1	−1.156	2.2
2	0.174	3.2
3	0.509	3.3
4	0.180	3.4
5	0.503	3.8
6	0.833	3.4
7	1.157	4.0
8	0.162	2.8
9	−0.179	2.9
10	−0.849	2.3
11	−0.502	2.6
12	−0.832	2.5

it is often instructive to plot the residuals in time order. In Figure 13-10(e) the residuals appear to increase with time. The appropriate corrective action is to introduce a linear function of time into the regression equation by developing a multiple regression model. A multiple regression equation incorporating a time variable would solve the technical violation illustrated in Figure 13-10(e), but the real task of the researcher is to find some assignable cause for this trend.

The residuals e_i and the fitted values $\hat{Y}_i$ of the household trip generation are listed in Table 12-4 and plotted in Figure 12-16(a) and (b). There are only 12 residuals, and it is difficult to find any serious deficiency of the simple linear regression model. Since other variables besides income can affect household trip generation rates, the residuals can hardly be thought of as random. The final type of residual plot to consider is the plot from the sample regression equation versus other possible

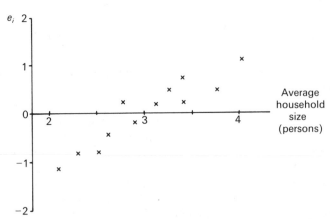

FIGURE 13-11. Residuals and household sizes for 12 traffic zones.

independent variables. For example, a plot of the residuals from the household trip generation equation versus the variable of household size reveals the essential non-randomness of the error component in the simple linear regression model. The residuals and the household sizes for the 12 traffic zones are listed in Table 13-10 and plotted in Figure 13-11. The residuals are strongly linearly related to average household size. Traffic zones with small average household sizes are overpredicted by the regression equation, and those traffic zones with large average household sizes are underpredicted. There is a clear indication that a systematic relationship exists between the error component and household size. It is not unreasonable to expect larger households to make more trips simply because more trips are required to maintain larger households. Or, there may be more than one or even two working members. In Section 13.5 we see the improvement that can be made in the household trip generation equation by introducing the household size variable.

Testing for Homoscedasticity

When a residual plot gives the impression that the variance increases or decreases in a systematic way, a number of statistical tests can be undertaken to confirm the existence of *heteroscedasticity*. If the data are of the fixed-X variety, a simple, direct test can be undertaken. At each level of X for which data are available, an estimate of σ^2 is made by using

$$s_j^2 = \frac{\sum_{i=1}^{n_j} (\hat{Y}_{ij} - \hat{Y}_j)^2}{n_j - 2} \tag{13-13}$$

where $\hat{Y}_j$ is the predicted value of Y at the level $X = X_j$ and $\hat{Y}_{1j}, \hat{Y}_{2j}, \ldots, \hat{Y}_{n_j j}$, are the n_j residuals from the regression line at $X = X_j$. Note that Equation (13-13) is simply the variance of the residuals at $X = X_j$. It is equivalent to $s_{Y \cdot X}^2$. Similar estimates of σ^2 can be made at each level of X_j for which sufficient replications exist. The constancy of the error variance can then be tested by using an F test on each pair of levels. If the data conform to the random-X model, then a slightly different test is used. First, divide the observations into two equal parts based on the value of X. The lowest-valued observations are put in the first group, and the observations for the highest values of X are put in the second group. Then calculate separate regressions for each data set. The estimates of σ^2 from each data set can be compared by using the F test. Many software packages routinely perform a variant of this test. The sample is divided into two equal parts based on the value of X, and an estimate of the variance about the regression is calculated from the *single* regression line generated with all the observations. An F test can then be performed on the two estimates. An even simpler but less powerful alternative exists. The rank correlation between the absolute value of the residuals and the value of the independent variable for each observation is determined. This is tested for significance by using the procedure outlined in Chapter 11.

 If any of these tests indicates the assumption of homoscedasticity cannot be sustained, then appropriate remedial action must be undertaken. A number of different procedures can be followed. These techniques are fully described in Draper and Smith (1981), Pindyck and Rubinfeld (1981), and Neter, Wasserman, and Kutner (1983).

TABLE 13-11
Kolmogorov-Smirnov Test for Normality of Residuals

Z	Cumulative number of residuals	$S(Z)$, observed cumulative relative frequency	$F(Z)$, normal cumulative probability	Absolute value of difference
-3	0	.0000	.0013	.0013
-2	0	.0000	.0228	.0228
-1	3	.2500	.1587	.0913
0	5	.3750	.5000	.1250
1	11	.8333	.8413	.0080
2	12	1.0000	.9772	.0228
3	12	1.0000	.9987	.0013

$$\text{Test statistic } D = \max_{Z} |F(Z) - S(Z)| = .1250$$

Testing the Normality of the Residuals

The normality of the residuals can be checked by using either of the two goodness-of-fit tests described in Chapter 10: the χ^2 and Kolmogorov-Smirnov tests. The Kolmogorov-Smirnov test is preferred since it is the more powerful. To apply the Kolmogorov-Smirnov test, we simply compare the S-shaped cumulative distribution of the standard normal to the sample cumulative distribution function of the standardized residuals. Although it is possible to determine the sample cumulative distribution function for all residuals, it is common to calculate the cumulative function for the seven integer values of Z starting at -3.0. The required calculations for the trip generation equation are summarized in Table 13-11. Of the 12 residuals, 3 have standardized values below $Z = -1$, and therefore $S(-1) = \frac{3}{12} = .2500$. The two cumulative functions are illustrated in Figure 13-12. The test statistic $D = .1250$ is well below the maximum allowable difference of .483 for $n = 7$ at $\alpha = .05$. There is insufficient information to reject the assumption of normality.

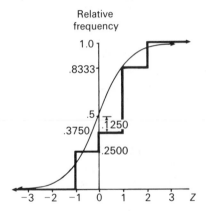

FIGURE 13-12. Two cumulative functions.

TABLE 13-12
Join Structure for Residuals from Trip Generation Example

Traffic zone	Color	Joins L_i	$L_i(L_i - 1)$	Joins BB	Joins WW	Joins BW
1	W	4	12	0	2	2
2	B	4	12	3	0	1
3	B	4	12	1	0	3
4	B	4	12	4	0	0
5	B	3	6	2	0	1
6	B	3	6	3	0	0
7	B	3	6	3	0	0
8	B	3	6	2	0	1
9	W	3	6	0	2	1
10	W	3	6	0	3	0
11	W	3	6	0	2	1
12	W	3	6	0	1	2
		$\sum\limits_{i=1}^{i=12} L_i = 40$	$\sum\limits_{i=1}^{i=12} L_i(L_i - 1) = 96$	18	10	12

$$L = \sum_{i=1}^{i=12} \frac{L_i}{2} = 20 \qquad K = \tfrac{1}{2}\sum L_i(L_i - 1) = 48$$

$$BB = \frac{18}{2} = 9 \qquad WW = \frac{10}{2} = 5 \qquad BW = \frac{12}{2} = 6$$

Testing for Spatial Autocorrelation

Where the observations used in a regression problem represent contiguous areas of a map, the residuals can be plotted on a map and examined for randomness. Any of the autocorrelation measures described in Chapter 11 can be used for this purpose. The simple two-color test for contiguity, though by no means the most powerful test, is the easiest to apply. Since this test utilizes a nominal variable, the level of the measurement of the residuals is reduced by dividing them into two nominal categories: positive and negative residuals. The map is developed by coloring the areas corresponding to positive residuals black and the areas corresponding to negative residuals white. Referring once more to the household trip generation equation, we see that this procedure generates the two-color map of Figure 13-13. Note that all seven positive residuals are contiguous as are the five negative residuals. Apparently there is strong *positive* autocorrelation since adjacent traffic zones tend to have identically signed residuals. The join structure is summarized in Table 13-12. There are $L = 20$ total joins, 6 of these BW (black-white), 9 are BB, and 5 are WW.

The required calculations to generate the sampling distributions of the three join types are summarized in Table 13-13. From the visual impression of Figure 13-13, we expect too many BB and WW joins and too few BW joins. Although this is the case, none of the results are statistically significant. This somewhat surprising result can be partially explained by the rather small sample size. Usually, at least 40

TABLE 13-13
Tests of Spatial Autocorrelation for Trip Generation Residuals

$$p = \tfrac{7}{12} = .5833 \qquad q = \tfrac{5}{12} = .4167 \qquad L = 20 \qquad K = 48$$

$$\mu(BB) = p^2 L = (.5833)^2(20) = 6.805$$

$$\sigma(BB) = \sqrt{p^2 L + p^3 K - p^4(L + K)}$$

$$= \sqrt{\begin{array}{c} 6.805 + (.5833)^3(48) \\ -(.5833)^4(20 + 48) \end{array}}$$

$$= 2.908$$

$$Z(BB) = \frac{9 - 6.805}{2.908}$$

$$= .755$$

$$\mu(BW) = 2pqL = 2(.5833)(.4167)(20) = 9.722$$

$$\sigma(BW) = \sqrt{2pqL + pqK - 4p^2q^2(L + K)}$$

$$= \sqrt{\begin{array}{c} 9.722 + (.5833)(.4167)(48) \\ -4(.5833)^2(.4167)^2(20 + 48) \end{array}}$$

$$= 2.306$$

$$Z(BW) = \frac{6 - 9.722}{2.306}$$

$$= -1.614$$

$$\mu(WW) = q^2 L = (.4167)^2(20) = 3.473$$

$$\sigma(WW) = \sqrt{q^2 L + q^3 K - q^4(L + K)}$$

$$= \sqrt{\begin{array}{c} 3.473 + (.4167)^3(48) \\ -(.4167)^4(20 + 48) \end{array}}$$

$$= 2.213$$

$$Z(WW) = \frac{5 - 3.473}{2.213}$$

$$= .690$$

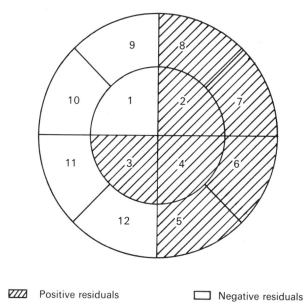

ZZZ Positive residuals   ☐ Negative residuals

FIGURE 13-13. Two-color map.

observations are required before the assumption of normality of the sampling distributions can be accepted. The loss of information resulting from the reduction in level of measurement of the residuals from interval to nominal is also a problem. More sophisticated tests are described in Chapter 11 and in detail by Cliff and Ord (1973) and Lewis (1977).

13.4. Nonlinear Regression Models

If regression analysis were to be applicable only to purely linear functions, it would be of limited value. Fortunately, this is not the case. Nonlinear models can also be utilized in regression analysis. Nonlinear models can be divided into two types: intrinsically linear models and intrinsically nonlinear models. A nonlinear model is said to be *intrinsically linear* if it can be expressed in the standard linear form given by Equation (13-2) by using a suitable transformation of one or both variables in the model. If such a transformation cannot be found, then the model is *intrinsically nonlinear*. Intrinsically linear models are of particular interest because the normal least squares estimating procedures can be used to determine the two parameters in the equation. Linearity can be produced by transforming the independent variable X, the dependent variable Y, or both X and Y. The variety of possibilities is illustrated by using models of each type.

Models Obtained by Transforming the Independent Variable

Consider the following three potential functional forms describing the regression relationship:

$$Y_i = \alpha + \frac{\beta}{X_i} + \epsilon_i \tag{13-14}$$

$$Y_i = \alpha + \beta X_i^r + \epsilon_i \tag{13-15}$$

$$Y_i = \alpha + \beta \log X_i + \epsilon_i \tag{13-16}$$

Each of these functions can be made linear by an obvious transformation. In Equation (13-14) if $1/X_i$ is replaced by Z_i, the equation becomes $Y_i = \alpha + \beta Z_i + \epsilon_i$. This is the reciprocal transformation. For (13-15), the appropriate substitution is $Z_i = X_i^r$. This leads to a family of regression models depending on the value of r. If $r = 1$, Equation (13-15) reduces to the simple linear regression model, for $r = \frac{1}{2}$ it specifies a square root transformation, and for positive integer powers of r it specifies a series of power functions. Equation (13-16) specifies a logarithmic transformation of the independent variable. These three equations are all intrinsically linear since the proper transformation produces a linear equation of the form (13-2). These models are termed *intrinsically linear* since they are linear in the regression model parameters α and β.

The effects of these transformations are best illustrated by a simple numerical example. Suppose the entire data set consists of the following five observations:

X	0.50	0.75	1.50	4	10
Y	25	20	15	10	5

When they are graphed in Figure 13-14(a), the data clearly suggest that the relationship between the two variables is nonlinear. If the transformation $Z = 1/X$ is applied to the data, the input to the regression model is

Z	2	1.33	0.67	0.25	0.10
Y	25	20	15	10	5

This transformation yields the scatter diagram of Figure 13-14(b). There appears to be reasonable cause to fit a linear function to these data. The intercept and regression coefficient obtained from this regression equation $Y = \alpha + \beta Z$ can be substituted directly into the reciprocal equation $Y = \alpha + \beta(1/X)$.

Although the reciprocal transformation seems to have solved the problem, caution is in order. The goal is not necessarily to find a transformation of the variables that creates a scattergram suggesting a linear relation. Many different transformations can generate virtually identically shaped functional relations. For example, if the same observations are transformed by substitution $Z = \log X$, the following data are obtained:

Z	-0.30	-0.12	0.18	0.6	1.0
Y	25	20	15	10	5

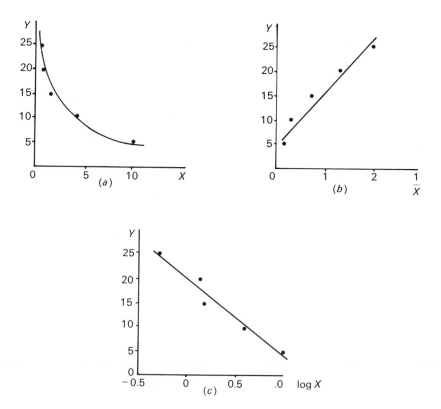

FIGURE 13-14. (*a*) Nonlinear function; (*b*) reciprocal transformation of (*a*); (*c*) logarithmic transformation of (*a*).

These data are graphed in Figure 13-14(*c*). Clearly, the logarithmic transformation might also be used to produce a linear relation. In many instances the data might be subjected to several different transformations with near identical results. The choice among these transformations is best made on the basis of a priori knowledge of the nature of relationship linking the two variables.

Models Obtained by Transforming the Dependent Variable

It is also possible to transform the dependent variable. As an example, consider the function

$$Y_i = \frac{1}{1 + e^{\alpha + \beta X_i + \epsilon_i}} \tag{13-17}$$

Taking the reciprocals of both sides and then subtracting 1 yields

$$\frac{1}{Y_i} - 1 = e^{\alpha + \beta X_i + \epsilon_i} \tag{13-18}$$

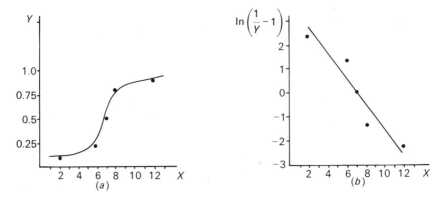

FIGURE 13-15. (*a*) Logistic function; (*b*) linear transformation of (*a*).

Taking the natural logarithm, denoted ln, of each side produces

$$\ln\left(\frac{1}{Y_i} - 1\right) = \alpha + \beta X_i + \epsilon_i \tag{13-19}$$

which is linear in the two parameters α and β. The effect of this transformation is best illustrated by using another set of hypothetical data:

X	Y	$\ln(1/Y - 1)$
3	0.1	2.2
6	0.2	1.4
7	0.5	0.0
8	0.8	−1.4
10	0.9	−2.2

In Figure 13-15 we see that the logistic S-shaped function suggested in the scattergram of the raw (X, Y) data is changed to a near-linear relation by this transformation. Logistic functions are typically found in the examination of the temporal pace of the diffusion of an innovation. Variable Y is the percentage of some population adopting some innovation, and X is the time or year of adoption.

Models Obtained by the Transformation of Both X and Y

In many situations in geography, a *multiplicative* model such as a simple power function proves to be an adequate representation of the relationship between two variables. Consider the regression model

$$Y_i = \alpha X_i^\beta \epsilon_i \tag{13-20}$$

where α and β are unknown parameters and ϵ_i is the usual population error term. By taking the logarithms of both sides of (13-20), the model is converted to:

$$\log Y_i = \log \alpha + \beta \log X_i + \log \epsilon_i \tag{13-21}$$

This can be estimated by normal least squares procedures. There is one important difference between the multiplicative model (13-20) and (13-2). For the requirements for hypothesis testing and confidence interval construction to be met, the log of the error term must be normally distributed. An alternate specification of $Y_i = \alpha X_i^\beta + \epsilon_i$ has an additive error term. This model is intrinsically nonlinear since there are no substitutions for X and Y that can reduce it to the form of (13-2).

Nonlinear Regression Models in Geography

Many regression models utilized by geographers require some sort of variable transformation. Consider the following two examples, one chosen from human geography and the other from physical geography.

DECLINE OF POPULATION DENSITY WITH DISTANCE FROM CENTRAL BUSINESS DISTRICT

Considerable empirical evidence has been collected to confirm that the pattern of population density with distance from the central business district (CBD) of most cities follows a negative exponential function of the form

$$D(X) = D(0)e^{-\beta X} \tag{13-22}$$

where $D(X)$ is the population density in the city at a distance X mi from the CBD. There are two parameters in this model, $D(0)$ and β. Parameter $D(0)$ can be interpreted as the estimated population density at the center of the city. Note that at a distance of $X = 0$, $e^{-\beta(0)} = 1$ and therefore $D(X) = D(0)$. Now $D(0)$ is not an observed data point, but an extrapolation from the observed data. The parameter β is termed the *population density gradient* since it controls the rate at which population densities decline with distance from the CBD. The role of the two parameters is clearly illustrated in Figure 13-16.

To test the applicability of the negative exponential model to a given city, it is necessary to collect data for small subareas of the city, usually census tracts. Let X_i be the distance of the centroid of census tract i to the center of the CBD, and let the

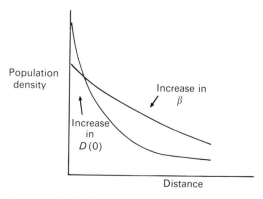

FIGURE 13-16. Effect of changes in $D(0)$ and β on population density-distance relation, Equation (13–22).

population density of the census tract be $D_i(X)$. Assuming a multiplicative error term, we can express the statistical model as

$$D_i(X) = D(0)e^{-\beta X_i}\epsilon_i \qquad (13\text{-}23)$$

which can be transformed to the form of Equation (13-2) simply by taking the natural logarithm of both sides. This leads to

$$\ln D_i(X) = \ln D(0) - \beta X_i + \ln \epsilon_i \qquad (13\text{-}24)$$

which can be fit by using conventional least squares techniques. The regression coefficient β is an estimate of the density gradient, and the antilogarithm of the intercept is the estimated value of $D(0)$. For example, the estimated equation $D_i(X) = 10.309 - 0.73X_i$ implies a density gradient of $\beta = 0.73$ and an estimated central density of 30,000 per square mile. Further reading on this topic is suggested at the end of the chapter.

RIVER CHANNEL MORPHOLOGY

All existing evidence indicated that the greater the quantity of water that moves through a river channel, the larger the cross section of the channel. For example, many different studies have demonstrated that a multiplicative model can be used to relate water surface width b of most rivers to the increase in the mean annual discharge Q_m in a downstream direction:

$$b = kQ_m^\beta \qquad (13\text{-}25)$$

Usually the coefficients k and β are different for each river, but several studies have utilized field data collected from a large number of different rivers in roughly the same geographic area. Research has demonstrated that for most rivers an equation of the form $b = kQ^{0.5}$ could be derived, with each river having a different value of coefficient k. In a study using data aggregated for several different streams, the relation $b = 2.38Q_2^{0.527}$ was shown to provide an excellent statistical fit. In this case the independent variable was not mean annual discharge but the discharge at the river reach for a flood with a 2-yr recurrence interval. Differences in the values of the parameters for these hydraulic relations depend on the exact specification of the dependent and independent variables as well as the characteristics of the sediment load of the rivers involved. In turn, characteristics of the sediment load depend on the river bed material. Coarse-load gravel-fed streams tend to have channels with a high width/depth ratio, whereas fine sediment beds tend to produce narrow and deep cross sections.

Many different hydraulic relations tend to be described best by using power functions requiring a log-log transform of the variables in the equation. For example, research into stream meandering has shown that many meander dimensions are related to river channel discharge. Meander wavelength appears to vary with discharge according to a simple multiplicative law similar to (13-25). All such relations appear as straight lines when they are graphed on log-log paper.

13.5. Introduction to Multiple Regression Analysis

Seldom in the social or physical sciences is it possible to satisfactorily explain the variations in a dependent variable by using a single independent variable. Fortunately, it is possible to extend regression analysis to situations in which several independent variables are used to account for the variability of a single response variable. This is termed *multiple regression analysis*. From the point of view of prediction, it is an obvious advantage to be able to use the information in two or more variables in an explanatory equation. We should be able to provide better predictions by using multiple regression analysis. How can appropriate variables be found? At times, a theory linking the two variables might suggest the proper specification of a multiple linear equation. In other instances previous empirical research might suggest logical variables to be included. For example, household trip generation equations generally include other variables besides household income. The number of cars available to the household and the size of the household are two frequently used predictor variables for household trip generation rates. Larger households with two or more cars tend to make more trips than smaller households with one car or no cars available.

There are two ways of identifying or verifying the usefulness of potential variables. First, any residual plot that indicates a linear relation between a new variable and the residuals of a simple regression equation *usually* points to the need for a new variable to be included in the model. Figure 13-11 clearly illustrates the need to introduce household size as an explanatory variable in the trip generation equation. Second, it is possible to calculate the simple correlation coefficient between the residuals and any new variable. Whenever a significant correlation exists, a multiple regression equation is *usually* necessary for proper specification.

Although a complete discussion of multiple linear regression is beyond the scope of this text, a few of the concepts are introduced in the context of household trip generation. Relabel the independent variable of average household income as X_1, and let X_2 be the average household size in the traffic zone. For these variables, it is now possible to fit the following multiple linear regression equation:

$$Y_i = a + b_1 X_{1i} + b_2 X_{2i} + c_i \qquad (13\text{-}26)$$

The subscript i refers to the observation number or index of the traffic zone. Least squares can still be applied to the estimation of the three parameters a, b_1, and b_2. Consider Figure 13-17. The dependent variable Y (household trip generation rate) is illustrated as the vertical axis in the three-dimensional space formed by including both household income X_1 and household size X_2 as predictors. Each observation can be located as a single point in this three-dimensional space. The regression equation $Y = a + b_1 X_1 + b_2 X_2$ defines a plane in this space. The exact orientation of this plane is determined by the values of the three parameters a, b_1, and b_2. Parameter a is the intercept on the Y axis, where $X_1 = X_2 = 0$. The regression coefficient b_1 defines the slope of the plane with respect to the X_1 axis, and the regression coefficient b_2 controls the orientation with respect to the X_2 axis. The values of these three parameters are chosen so that the sum of squared deviations $\Sigma_{i=1}^{n} (Y_i - a - b_1 X_{1i} - b_2 X_{2i})^2$ is minimized. These deviations represent the vertical distances from the observations to the

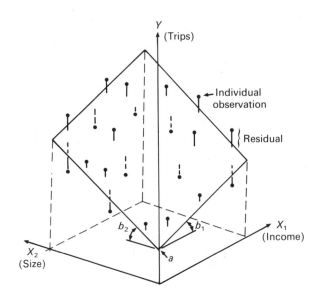

FIGURE 13-17. Least squares estimation of a multiple regression equation with two independent variables.

regression plane. As usual, observations above the plane appear as positive residuals and those below the plane as negative residuals.

The least squares regression equation for the household trip generation model

$$Y = -3.01 + 0.350X_1 + 1.18X_2 \qquad (13\text{-}27)$$

indicates that both household size and household income are positive influences on trip generation rates. The intercept of -3.01 is clearly out of the range of the input data and can be ignored. It is meaningless to suggest that households with no members and no income make -3.01 trips per day! How has the multiple regression equation improved the prediction of trip generation rates? First, note in Table 13-14 that the residuals are generally smaller once the household size variable has been included. Note particularly the significant decline in the sum of the squared residuals ESS. It has fallen from 5.667 to 0.6285. Second, we notice a similar improvement in the standard error of estimate.

The superiority of the new equation is best illustrated with the use of the coefficient of determination. No matter how many variables are added to the regression equation, the total variation in Y, or TSS, remains constant at $\sum_{i=1}^{n}(Y - \bar{Y})^2$. As we have just witnessed, the addition of a new variable reduces the ESS and increases the RSS. At worst, the introduction of a new variable can offer no change in ESS and RSS. Except in the most unlikely cases, the new variable will cause some reduction in ESS. Define the *coefficient of multiple determination* as

$$R^2 = \frac{\text{TSS} - \text{ESS}}{\text{TSS}} = \frac{\text{RSS}}{\text{TSS}}$$

TABLE 13-14
Goodness-of-Fit Measures for Multiple Regression Trip Generation Example

Traffic zone	Y_i	$\hat{Y}_i$	$Y_i - \hat{Y}_i$	$(Y_i - \hat{Y}_i)^2$	$(Y_i - \bar{Y})^2$
1	3	3.09	−.09	.0081	9.505
2	5	4.97	.03	.0009	1.173
3	5	4.73	.27	.0729	1.173
4	4	4.15	−.15	.0225	4.339
5	6	6.37	−.37	.1369	.007
6	7	6.60	.40	.1600	.841
7	9	9.06	−.06	.0036	8.509
8	7	6.60	.40	.1600	.841
9	8	8.12	−.12	.0144	3.675
10	8	8.11	−.11	.0121	3.675
11	6	6.01	−.01	.0001	.007
12	5	5.19	−.19	.0361	1.173
			0.0	ESS = .6276	TSS = 34.918

$$s_{Y \cdot X_1 X_2} = \sqrt{\frac{.6276}{n-3}} = .264$$

$$R^2 = \frac{34.918 - .6276}{34.918}$$

$$= .982$$

We have used uppercase for this statistic to distinguish it from the simple coefficient of determination r^2. The formula for R^2 is exactly the same as for r^2. The only difference is that the ESS in the multiple regression equation is based on the residuals from the multiple linear regression equation. For this case, the reduction in ESS from 5.667 to 0.6285 leads to an R^2 value of .982. The obvious statistical questions are:

1. Has the new variable significantly improved the prediction of household trip generation rates?
2. Is the new equation superior to the simple linear regression?
3. Does this new equation satisfy the assumptions of regression analysis?

Multiple regression analysis is probably the single most widely used statistical technique in the social sciences. The generalization from simple to multiple regression introduces a number of problems beyond the scope of this text. An especially readable, yet technically excellent, presentation is contained in Neter, Wasserman, and Kutner (1983).

13.6. Summary

In this chapter we have extended the role of regression analysis beyond the simple descriptive function introduced in Chapter 12. First, the descriptive role of regression analysis is identified as but one function undertaken in a more comprehensive five-step model building procedure. Testing the statistical significance of the equation and using the derived model for prediction purposes require four specific assumptions. With these assumptions it is possible to test the hypothesis that the slope of the regression line equals zero. If this hypothesis can be rejected at a given level of significance, then the model is necessarily useful for purposes of prediction. For this test to be valid, certain other assumptions must be met. Statistical tests of these assumptions—normality, homoscedasticity, randomness of the error term—have also been discussed. We have briefly introduced two other modifications of regression analysis: First, it is possible to use conventional least squares regression analysis to fit some nonlinear relationships by using variable transformations. Second, a brief introduction to multiple regression analysis suggests the wide applicability of this statistical technique.

Appendix 13A. SAS Program and Output for Trip Generation Example

```
DATA EX1;
INPUT INCOME TRIPS;
CARDS;
10 10
12  5
11  5
 9  4
14  6
16  7
21  9
18  7
22  8
24  8
17  6
15  5
;
PROC PRINT DATA=EX1;
TITLE 'RAW DATA INPUT';
PROC CORR DATA=EX1;
TITLE 'CORRELATION OF INPUT VARIABLES';
PROC REG DATA=EX1;
TITLE 'SIMPLE LINEAR REGRESSION TRIPS GENERATION ON INCOME';
MODEL TRIPS=INCOME/P R CLI CLM;
OUTPUT OUT = PLOTDATA
           P=PRED
           L95=L95
           U95=U95
           R=RESID;
PROC PLOT DATA=PLOTDATA;
PLOT TRIPS*INCOME='*'
     PRED*INCOME='P'
     U95*INCOME='U'
     L95*INCOME='L'/OVERLAY;
TITLE 'PLOT OF INCOME VS TRIPS WITH CONFIDENCE INTERVALS';
LABEL X = 'INCOME TRIPS IN THOUSANDS'
      Y = 'TRIPS GENERATED' ;
```

RAW DATA INPUT

OBS INCOME TRIPS

1	10	10
2	12	5
3	11	5
4	9	4
5	14	6
6	16	7
7	21	9
8	18	7
9	22	8
10	24	8
11	17	6
12	15	5

CORRELATION OF INPUT VARIABLES

VARIABLE	N	MEAN	STD DEV	SUM	MINIMUM	MAXIMU
INCOME	12	15.75000	4.864061	189.0000	9.000000	24.000000
TRIPS	12	6.66667	1.825742	80.0000	4.000000	10.000000

PEARSON CORRELATION COEFFICIENTS / PROB > |R| UNDER H0:RHO=0 / N = 12

	INCOME	TRIPS
INCOME	1.00000	0.48113
	0.0000	0.1133
TRIPS	0.48113	1.00000
	0.1133	0.0000

SIMPLE LINEAR REGRESSION TRIPS GENERATION ON INCOME

DEP VARIABLE: TRIPS
ANALYSIS OF VARIANCE

SOURCE	DF	SUM OF SQUARES	MEAN SQUARE	F VALUE	PROB>F
MODEL	1	8.48799232	8.48799232	3.012	0.1133
ERROR	10	28.17867435	2.81786744		
C TOTAL	11	36.66666667			

ROOT MSE	1.67865	R-SQUARE	0.2315
DEP MEAN	6.666667	ADJ R-SQ	0.1546
C.V.	25.17976		

PARAMETER ESTIMATES

VARIABLE	DF	PARAMETER ESTIMATE	STANDARD ERROR	T FOR H0: PARAMETER=0	PROB > |T|
INTERCEP	1	3.82228626	1.70901406	2.237	0.0493
INCOME	1	0.18059558	0.10405547	1.736	0.1133

OBS	ACTUAL	PREDICT VALUE	STD ERR PREDICT	LOWER95% MEAN	UPPER95% MEAN	LOWER95% PREDICT
1	10.0000	5.6282	0.7699	3.9127	7.3438	1.5133
2	5.0000	5.9894	0.6222	4.6032	7.3757	2.0005
3	5.0000	5.8088	0.6922	4.2665	7.3511	1.7630
4	4.0000	5.4476	0.8533	3.5463	7.3490	1.2518
5	6.0000	6.3506	0.5177	5.1972	7.5041	2.4365
6	7.0000	6.7118	0.4853	5.6305	7.7931	2.8184
7	9.0000	7.6148	0.7302	5.9877	9.2419	3.5359
8	7.0000	7.0730	0.5382	5.8739	8.2722	3.1452
9	8.0000	7.7954	0.8110	5.9883	9.6025	3.6414
10	8.0000	8.1566	0.9858	5.9601	10.3531	3.8190
11	6.0000	6.8924	0.5017	5.7745	8.0104	2.9886
12	5.0000	6.5312	0.4908	5.4376	7.6249	2.6343

OBS	UPPER95% PREDICT	RESIDUAL	STD ERR RESIDUAL	STUDENT RESIDUAL	-2-1-0 1 2
1	9.7432	4.3718	1.4917	2.9308	| |***** |
2	9.9784	-0.9894	1.5591	-0.6346	| *| |
3	9.8546	-0.8088	1.5293	-0.5289	| *| |
4	9.6435	-1.4476	1.4456	-1.0014	| **| |
5	10.2647	-0.3506	1.5968	-0.2196	| | |
6	10.6053	0.2882	1.6070	0.1793	| | |
7	11.6937	1.3852	1.5115	0.9164	| |* |
8	11.0008	-0.0730	1.5900	-0.0459	| | |
9	11.9493	0.2046	1.4697	0.1392	| | |
10	12.4941	-0.1566	1.3587	-0.1152	| | |
11	10.7962	-0.8924	1.6019	-0.5571	| *| |
12	10.4281	-1.5312	1.6053	-0.9539	| *| |

OBS	COOK'S D
1	1.144
2	0.032
3	0.029
4	0.175
5	0.003
6	0.001
7	0.098
8	0.000
9	0.003
10	0.003
11	0.015
12	0.043

```
SUM OF RESIDUALS                1.55431E-15
SUM OF SQUARED RESIDUALS          28.17867
PREDICTED RESID SS (PRESS)        43.79562
```

PLOT OF INCOME VS TRIPS WITH CONFIDENCE INTERVALS

```
PLOT OF TRIPS*INCOME     SYMBOL USED IS *
PLOT OF PRED*INCOME      SYMBOL USED IS P
PLOT OF U95*INCOME       SYMBOL USED IS U
PLOT OF L95*INCOME       SYMBOL USED IS L
```

NOTE: 2 OBS HIDDEN

REFERENCES

A. D. Cliff and J. K. Ord, *Spatial Autocorrelation* (London: Pion, 1973).

N. Draper and H. Smith, *Applied Regression Analysis*, 2d ed. (New York: Wiley, 1981).

P. Lewis, *Maps and Statistics* (London: Methuen, 1977).

J. Neter, W. Wasserman, and M. Kutner, *Applied Linear Regression Models* (Homewood, Ill.: Irwin, 1983).

R. Pindyck and D. L. Rubinfeld, *Econometric Models and Economic Forecasts*, 2d ed. (New York: McGraw-Hill, 1981).

FURTHER READING

Examples of the use of regression analysis in geographical research can be found in several sources. Recent issues of many journals in the field contain summaries of research findings, many of which are based on the use of regression analysis. Introductory textbooks also describe several examples of the use of regression analysis. Useful references are listed in Chapter 2. The more advanced textbooks cited in the References section of this chapter also contain relevant examples. Finally, the problems at the end of this chapter, as well as the examples contained in Chapters 12 and 13, may be consulted. Note that most of the applications in the technical journals utilize quite advanced forms of regression analysis. Many of these techniques have been only mentioned in passing in this introductory text.

For the specific topics of this chapter, useful references include Cliff and Ord (1973) for spatial autocorrelation; Clark (1971), Latham and Yeates (1970), McDonald (1976), and Newling (1966) for population density gradients; and Leopold, Wolman, and Miller (1984) for hydraulic relations of streams.

C. Clark, "Urban Population Densities," *Journal of the Royal Statistical Society, Series A* 114 (1951) 490–494.

R. Latham and M. Yeates, "Population Density Growth in Metropolitan Toronto," *Geographical Analysis* 2 (1970) 177–185.

L. Leopold, G. Wolman, and J. Miller, *Fluvial Processes in Geomorphology* (San Francisco: Freeman, 1984).

J. F. McDonald, "Some Tests of Alternative Urban Population Density Functions," *Journal of Urban Economics* 3 (1976) 242–252.

B. Newling, "Urban Growth and Spatial Structure: Mathematical Models and Empirical Evidence," *Geographical Review* 56 (1966) 213–225.

PROBLEMS

1. Explain the meaning of the following terms or concepts:
   a. Dependent (or response or endogenous) variable
   b. Independent (or predictor or exogenous) variable
   c. Intercept
   d. Regression coefficient
   e. Coefficient of determination
   f. Correlation coefficient
   g. Residual
   h. Coefficient of variation
   i. Outlier
   j. Standardized residual
   k. Time series
   l. Standard error of estimate [or root mean square error (root MSE)]

2. Differentiate between
   a. A functional relationship and a statistical one
   b. The fixed-X model and a random-X model
   c. The regression sum of squares and the error sum of squares
   d. Interpolation and extrapolation
   e. Measurement error and stochastic error
   f. Intrinsically linear and intrinsically nonlinear regression models

3. Briefly explain the assumptions of linear regression analysis and the implications of the Gauss-Markov theorem.

4. How are correlation and regression analysis linked?

5. Why should a regression equation not be used to extrapolate a value of Y for a value of X beyond the range of the input data?

6. Measurements of the L_{10} sound levels and distance of the sound recorder to the centerline of an urban expressway are as follows:

Observation No.	Distance, ft	Sound level, dB
1	45	74
2	63	81
3	160	66
4	225	68
5	305	69
6	390	66
7	461	58
8	515	67
9	605	55
10	625	64

   a. Draw a scattergram of these data with distance as the independent variable on the X axis.
   b. Estimate the regression equation, using least squares. Draw the regression equation within the scattergram. Use the format of Table 12-2 for your calculations.
   c. Interpret the meaning of the regression coefficient and intercept. In what units are they measured?
   d. What would you estimate the sound level (in decibels) to be at a distance of 100 ft?
   e. Using a work table of the form of Table 12-4, calculate $s_{Y \cdot X}$ and r^2 for this problem.
   f. Check your answer to part (e) by calculating the correlation coefficient for these data. Verify that the square of the correlation coefficient equals the coefficient of determination.
   g. Use the data of this problem to verify Equation (12-19).
   h. Suppose that the distance variable is measured in hundreds of feet and not feet. What is the impact on the regression equation?

7. The following table contains the population of a set of 10 cities and the total number of serious crimes which occurred in that city in 1980. Serious crimes are defined as murder, rape, assault, robbery, and car theft.

Observation No.	Population (000)	Total serious crimes
1	7017	393,162
2	4370	294,466
3	1832	106,482
4	1306	99,293
5	970	54,854
6	794	30,771
7	610	32,146
8	435	25,650
9	352	11,273
10	288	18,173

a. Convert the variable called total serious crimes to a rate by dividing by the total population.
b. Draw a scattergram with city size as the independent variable on the X axis.
c. Estimate the regression equation.
d. Interpret the meaning of the regression coefficient and intercept.
e. Using appropriate work table, calculate $s_{Y \cdot X}$ and r^2.
f. Generate a table of residuals and standardized residuals.
g. Are there any outliers?
h. Which observation is most closely predicted by the equation? Least closely?

8. Consider the sound data of Problem 6.
a. Construct a residual plot in the form of Figure 12-16(a). Do the assumptions of the simple linear regression model appear to be met?
b. Test the null hypothesis that $\beta_0 = 0$ at the $\alpha = .05$ level of significance.
c. Test the null hypothesis that $\beta_1 = 0$ at the $\alpha = .05$ level of significance.
d. Construct an ANOVA table for this problem in the form of Table 13-5, and evaluate the appropriate hypothesis.
9. Complete parts (a) to (d) of Problem 8, using the crime data of Problem 7.

10. Test the residuals from the multiple regression model for household trip generation (Table 13-14) for
a. Autocorrelation, using the contiguity test for two-color maps described in Chapter 11
b. Normality, using the χ^3 test described in Chapter 10
c. Randomness, using the runs test described in Chapter 10

11. Measurements are taken of the residential density of various neighborhoods in a city, Y, and the distance from the middle of each neighborhood to the central business district, X. The measurements are as follows:

Observation No.	Distance to CBD, km	Residential density, persons/km^2
1	1.8	1000
2	2.0	800
3	3.0	300
4	4.3	200
5	5.4	100
6	6.8	90
7	7.3	80
8	8.6	60
9	9.4	40
10	10.9	30
11	13.7	20
12	15.5	15

a. Draw a scattergram of these data.
b. Fit a regression equation of the form $Y = a + bX$ to these data. Draw the least squares line within the scattergram.
c. Fit a regression equation of the form of (13-22) by transforming the variable X to $\ln X$. Draw the least squares regression line within a scattergram of the data.
d. Which equation appears to best describe the data? Examine the lines within the scattergram, the values of r^2 and $s_{Y \cdot X}$, and the residuals from each equation. It may be useful to draw the log-transformed equation within the same scattergram as the untransformed model. This will require some knowledge of logarithms.

12. Using the data of Problem 7,
    a. Construct a 95 percent confidence interval for the mean crime rate for a city of 5,000,000.
    b. Construct a 95 percent prediction interval for the crime rate for a city of 5,000,000 persons.

13. For the data of Problem 11,
    a. Construct a 95 percent confidence interval for the mean residential density of a neighborhood located 10 km from the CBD.
    b. Construct a 95 percent prediction interval for the residential density of a single neighborhood 6 km from the CBD.
    c. Which interval is more useful for inferential purposes?

APPENDIX: STATISTICAL TABLES

TABLE A-1
Binomial Distribution

Entry is probability $P(X = x) = \binom{n}{x} p^x (1-p)^{n-x}$

n x	.01	.02	.03	.04	.05	.06	.07	.08	.09	
2 0	0.9801	0.9604	0.9409	0.9216	0.9025	0.8836	0.8649	0.8464	0.8281	2
1	0.0198	0.0392	0.0582	0.0768	0.0950	0.1128	0.1302	0.1472	0.1638	1
2	0.0001	0.0004	0.0009	0.0016	0.0025	0.0036	0.0049	0.0064	0.0081	0 2
3 0	0.9703	0.9412	0.9127	0.8847	0.8574	0.8306	0.8044	0.7787	0.7536	3
1	0.0294	0.0576	0.0847	0.1106	0.1354	0.1590	0.1816	0.2031	0.2236	2
2	0.0003	0.0012	0.0026	0.0046	0.0071	0.0102	0.0137	0.0177	0.0221	1
3	0.0000	0.0000	0.0000	0.0001	0.0001	0.0002	0.0003	0.0005	0.0007	0 3
4 0	0.9606	0.9224	0.8853	0.8493	0.8145	0.7807	0.7481	0.7164	0.6857	4
1	0.0388	0.0753	0.1095	0.1416	0.1715	0.1993	0.2252	0.2492	0.2713	3
2	0.0006	0.0023	0.0051	0.0088	0.0135	0.0191	0.0254	0.0325	0.0402	2
3	0.0000	0.0000	0.0001	0.0002	0.0005	0.0008	0.0013	0.0019	0.0027	1
4	0.0000	0.0000	0.0000	0.0000	0.0000	0.0000	0.0000	0.0000	0.0001	0 1
5 0	0.9510	0.9039	0.8587	0.8154	0.7738	0.7339	0.6957	0.6591	0.6240	5
1	0.0480	0.0922	0.1328	0.1699	0.2036	0.2342	0.2618	0.2866	0.3086	4
2	0.0010	0.0038	0.0082	0.0142	0.0214	0.0299	0.0394	0.0498	0.0610	3
3	0.0000	0.0001	0.0003	0.0006	0.0011	0.0019	0.0030	0.0043	0.0060	2
4	0.0000	0.0000	0.0000	0.0000	0.0000	0.0001	0.0001	0.0002	0.0003	1
5	0.0000	0.0000	0.0000	0.0000	0.0000	0.0000	0.0000	0.0000	0.0000	0 5
6 0	0.9415	0.8858	0.8330	0.7828	0.7351	0.6899	0.6470	0.6064	0.5679	6
1	0.0571	0.1085	0.1546	0.1957	0.2321	0.2642	0.2922	0.3164	0.3370	5
2	0.0014	0.0055	0.0120	0.0204	0.0305	0.0422	0.0550	0.0688	0.0833	4
3	0.0000	0.0002	0.0005	0.0011	0.0021	0.0036	0.0055	0.0080	0.0110	3
4	0.0000	0.0000	0.0000	0.0000	0.0001	0.0002	0.0003	0.0005	0.0008	2
5	0.0000	0.0000	0.0000	0.0000	0.0000	0.0000	0.0000	0.0000	0.0000	1
6	0.0000	0.0000	0.0000	0.0000	0.0000	0.0000	0.0000	0.0000	0.0000	0 6
7 0	0.9321	0.8681	0.8080	0.7514	0.6983	0.6485	0.6017	0.5578	0.5168	7
1	0.0659	0.1240	0.1749	0.2192	0.2573	0.2897	0.3170	0.3396	0.3578	6
2	0.0020	0.0076	0.0162	0.0274	0.0406	0.0555	0.0716	0.0886	0.1061	5
3	0.0000	0.0003	0.0008	0.0019	0.0036	0.0059	0.0090	0.0128	0.0175	4
4	0.0000	0.0000	0.0000	0.0001	0.0002	0.0004	0.0007	0.0011	0.0017	3
5	0.0000	0.0000	0.0000	0.0000	0.0000	0.0000	0.0000	0.0001	0.0001	2
6	0.0000	0.0000	0.0000	0.0000	0.0000	0.0000	0.0000	0.0000	0.0000	1
7	0.0000	0.0000	0.0000	0.0000	0.0000	0.0000	0.0000	0.0000	0.0000	0 7
8 0	0.9227	0.8508	0.7837	0.7214	0.6634	0.6096	0.5596	0.5132	0.4703	8
1	0.0746	0.1389	0.1939	0.2405	0.2793	0.3113	0.3370	0.3570	0.3721	7
2	0.0026	0.0099	0.0210	0.0351	0.0515	0.0695	0.0888	0.1087	0.1288	6
3	0.0001	0.0004	0.0013	0.0029	0.0054	0.0089	0.0134	0.0189	0.0255	5
4	0.0000	0.0000	0.0001	0.0002	0.0004	0.0007	0.0013	0.0021	0.0031	4
5	0.0000	0.0000	0.0000	0.0000	0.0000	0.0000	0.0001	0.0001	0.0002	3
6	0.0000	0.0000	0.0000	0.0000	0.0000	0.0000	0.0000	0.0000	0.0000	2
7	0.0000	0.0000	0.0000	0.0000	0.0000	0.0000	0.0000	0.0000	0.0000	1
8	0.0000	0.0000	0.0000	0.0000	0.0000	0.0000	0.0000	0.0000	0.0000	0 8
9 0	0.9135	0.8337	0.7602	0.6925	0.6302	0.5730	0.5204	0.4722	0.4279	9
1	0.0830	0.1531	0.2116	0.2597	0.2985	0.3292	0.3525	0.3695	0.3809	8
2	0.0034	0.0125	0.1262	0.0433	0.0629	0.0840	0.1061	0.1285	0.1507	7
3	0.0001	0.0006	0.0019	0.0042	0.0077	0.0125	0.0186	0.0261	0.0348	6
4	0.0000	0.0000	0.0001	0.0003	0.0006	0.0012	0.0021	0.0034	0.0052	5
5	0.0000	0.0000	0.0000	0.0000	0.0000	0.0001	0.0002	0.0003	0.0005	4
6	0.0000	0.0000	0.0000	0.0000	0.0000	0.0000	0.0000	0.0000	0.0000	3
7	0.0000	0.0000	0.0000	0.0000	0.0000	0.0000	0.0000	0.0000	0.0000	2
8	0.0000	0.0000	0.0000	0.0000	0.0000	0.0000	0.0000	0.0000	0.0000	1
9	0.0000	0.0000	0.0000	0.0000	0.0000	0.0000	0.0000	0.0000	0.0000	0 9
	.99	.98	.97	.96	.95	.94	.93	.92	.91	x n

p

(continued)

TABLE A-1 (continued)

					p						
n x	.10	.15	.20	.25	.30	.35	.40	.45	.50		
2 0	0.8100	0.7225	0.6400	0.5625	0.4900	0.4225	0.3600	0.3025	0.2500	2	
1	0.1800	0.2550	0.3200	0.3750	0.4200	0.4550	0.4800	0.4950	0.5000	1	
2	0.0100	0.0225	0.0400	0.0625	0.0900	0.1225	0.1600	0.2025	0.2500	0	2
3 0	0.7290	0.6141	0.5120	0.4219	0.3430	0.2746	0.2160	0.1664	0.1250	3	
1	0.2430	0.3251	0.3840	0.4219	0.4410	0.4436	0.4320	0.4084	0.3750	2	
2	0.0270	0.0574	0.0960	0.1406	0.1890	0.2389	0.2880	0.3341	0.3750	1	
3	0.0010	0.0034	0.0080	0.0156	0.0270	0.0429	0.0640	0.0911	0.1250	0	3
4 0	0.6561	0.5220	0.4096	0.3164	0.2401	0.1785	0.1296	0.0915	0.0625	4	
1	0.2916	0.3685	0.4096	0.4219	0.4116	0.3845	0.3456	0.2995	0.2500	3	
2	0.0486	0.0975	0.1536	0.2109	0.2646	0.3105	0.3456	0.3675	0.3750	2	
3	0.0036	0.0115	0.0256	0.0469	0.0756	0.1115	0.1536	0.2005	0.2500	1	
4	0.0001	0.0005	0.0016	0.0039	0.0081	0.0150	0.0256	0.0410	0.0625	0	4
5 0	0.5905	0.4437	0.3277	0.2373	0.1681	0.1160	0.0778	0.0503	0.0312	5	
1	0.3280	0.3915	0.4096	0.3955	0.3601	0.3124	0.2592	0.2059	0.1562	4	
2	0.0729	0.1382	0.2048	0.2637	0.3087	0.3364	0.3456	0.3369	0.3125	3	
3	0.0081	0.0244	0.0512	0.0879	0.1323	0.1811	0.2304	0.2757	0.3125	2	
4	0.0004	0.0022	0.0064	0.0146	0.0283	0.0488	0.0768	0.1128	0.1562	1	
5	0.0000	0.0001	0.0003	0.0010	0.0024	0.0053	0.0102	0.0185	0.0312	0	5
6 0	0.5314	0.3771	0.2621	0.1780	0.1176	0.0754	0.0467	0.0277	0.0156	6	
1	0.3543	0.3993	0.3932	0.3560	0.3025	0.2437	0.1866	0.1359	0.0938	5	
2	0.0984	0.1762	0.2458	0.2966	0.3241	0.3280	0.3110	0.2780	0.2344	4	
3	0.0146	0.0415	0.0819	0.1318	0.1852	0.2355	0.2765	0.3032	0.3125	3	
4	0.0012	0.0055	0.0154	0.0330	0.0595	0.0951	0.1382	0.1861	0.2344	2	
5	0.0001	0.0004	0.0015	0.0044	0.0102	0.0205	0.0369	0.0609	0.0938	1	
6	0.0000	0.0000	0.0001	0.0002	0.0007	0.0018	0.0041	0.0083	0.0156	0	6
7 0	0.4783	0.3206	0.2097	0.1335	0.0824	0.0490	0.0280	0.0152	0.0078	7	
1	0.3720	0.3960	0.3670	0.3115	0.2471	0.1848	0.1306	0.0872	0.0547	6	
2	0.1240	0.2097	0.2753	0.3115	0.3177	0.2985	0.2613	0.2140	0.1641	5	
3	0.0230	0.0617	0.1147	0.1730	0.2269	0.2679	0.2903	0.2918	0.2734	4	
4	0.0026	0.0109	0.0287	0.0577	0.0972	0.1442	0.1935	0.2388	0.2734	3	
5	0.0002	0.0012	0.0043	0.0115	0.0250	0.0466	0.0774	0.1172	0.1641	2	
6	0.0000	0.0001	0.0004	0.0013	0.0036	0.0084	0.0172	0.0320	0.0547	1	
7	0.0000	0.0000	0.0000	0.0001	0.0002	0.0006	0.0016	0.0037	0.0078	0	7
8 0	0.4305	0.2725	0.1678	0.1001	0.0576	0.0319	0.0168	0.0084	0.0039	8	
1	0.3826	0.3847	0.3355	0.2670	0.1977	0.1373	0.0896	0.0548	0.0312	7	
2	0.1488	0.2376	0.2936	0.3115	0.2965	0.2587	0.2090	0.1569	0.1094	6	
3	0.0331	0.0839	0.1468	0.2076	0.2541	0.2786	0.2787	0.2568	0.2188	5	
4	0.0046	0.0185	0.0459	0.0865	0.1361	0.1875	0.2322	0.2627	0.2734	4	
5	0.0004	0.0026	0.0092	0.0231	0.0467	0.0808	0.1239	0.1719	0.2188	3	
6	0.0000	0.0002	0.0011	0.0038	0.0100	0.0217	0.0413	0.0703	0.1094	2	
7	0.0000	0.0000	0.0001	0.0004	0.0012	0.0033	0.0079	0.0164	0.0312	1	
8	0.0000	0.0000	0.0000	0.0000	0.0001	0.0002	0.0007	0.0017	0.0039	0	8
9 0	0.3874	0.2316	0.1342	0.0751	0.0404	0.0207	0.0101	0.0046	0.0020	9	
1	0.3874	0.3679	0.3020	0.2253	0.1556	0.1004	0.0605	0.0339	0.0176	8	
2	0.1722	0.2597	0.3020	0.3003	0.2668	0.2162	0.1612	0.1110	0.0703	7	
3	0.0446	0.1069	0.1762	0.2336	0.2668	0.2716	0.2508	0.2119	0.1641	6	
4	0.0074	0.0283	0.0661	0.1168	0.1715	0.2194	0.2508	0.2600	0.2461	5	
5	0.0008	0.0050	0.0165	0.0389	0.0735	0.1181	0.1672	0.2128	0.2461	4	
6	0.0001	0.0006	0.0028	0.0087	0.0210	0.0424	0.0743	0.1160	0.1641	3	
7	0.0000	0.0000	0.0003	0.0012	0.0039	0.0098	0.0212	0.0407	0.0703	2	
8	0.0000	0.0000	0.0000	0.0001	0.0004	0.0013	0.0035	0.0083	0.0176	1	
9	0.0000	0.0000	0.0000	0.0000	0.0000	0.0001	0.0003	0.0008	0.0020	0	9
	.90	.85	.80	.75	.70	.65	.60	.55	.50	x n	
					p						

TABLE A-1 (continued)

						p					
n x	.01	.02	.03	.04	.05	.06	.07	.08	.09		
10 0	0.9044	0.8171	0.7374	0.6648	0.5987	0.5386	0.4840	0.4344	0.3894	10	
1	0.0914	0.1667	0.2281	0.2770	0.3151	0.3438	0.3643	0.3777	0.3851	9	
2	0.0042	0.0153	0.0317	0.0519	0.0746	0.0988	0.1234	0.1478	0.1714	8	
3	0.0001	0.0008	0.0026	0.0058	0.0105	0.0168	0.0248	0.0343	0.0452	7	
4	0.0000	0.0000	0.0001	0.0004	0.0010	0.0019	0.0033	0.0052	0.0078	6	
5	0.0000	0.0000	0.0000	0.0000	0.0001	0.0001	0.0003	0.0005	0.0009	5	
6	0.0000	0.0000	0.0000	0.0000	0.0000	0.0000	0.0000	0.0000	0.0001	4	
7	0.0000	0.0000	0.0000	0.0000	0.0000	0.0000	0.0000	0.0000	0.0000	3	
8	0.0000	0.0000	0.0000	0.0000	0.0000	0.0000	0.0000	0.0000	0.0000	2	
9	0.0000	0.0000	0.0000	0.0000	0.0000	0.0000	0.0000	0.0000	0.0000	1	
10	0.0000	0.0000	0.0000	0.0000	0.0000	0.0000	0.0000	0.0000	0.0000	0	10
12 0	0.8864	0.7847	0.6938	0.6127	0.5404	0.4759	0.4186	0.3677	0.3225	12	
1	0.1074	0.1922	0.2575	0.3064	0.3413	0.3645	0.3781	0.3837	0.3827	11	
2	0.0060	0.0216	0.0438	0.0702	0.0988	0.1280	0.1565	0.1835	0.2082	10	
3	0.0002	0.0015	0.0045	0.0098	0.0173	0.0272	0.0393	0.0532	0.0686	9	
4	0.0000	0.0001	0.0003	0.0009	0.0021	0.0039	0.0067	0.0104	0.0153	8	
5	0.0000	0.0000	0.0000	0.0001	0.0002	0.0004	0.0008	0.0014	0.0024	7	
6	0.0000	0.0000	0.0000	0.0000	0.0000	0.0000	0.0001	0.0001	0.0003	6	
7	0.0000	0.0000	0.0000	0.0000	0.0000	0.0000	0.0000	0.0000	0.0000	5	
8	0.0000	0.0000	0.0000	0.0000	0.0000	0.0000	0.0000	0.0000	0.0000	4	
9	0.0000	0.0000	0.0000	0.0000	0.0000	0.0000	0.0000	0.0000	0.0000	3	
10	0.0000	0.0000	0.0000	0.0000	0.0000	0.0000	0.0000	0.0000	0.0000	2	
11	0.0000	0.0000	0.0000	0.0000	0.0000	0.0000	0.0000	0.0000	0.0000	1	
12	0.0000	0.0000	0.0000	0.0000	0.0000	0.0000	0.0000	0.0000	0.0000	0	12
15 0	0.8601	0.7386	0.6333	0.5421	0.4633	0.3953	0.3367	0.2863	0.2430	15	
1	0.1303	0.2261	0.2938	0.3388	0.3658	0.3785	0.3801	0.3734	0.3605	14	
2	0.0092	0.0323	0.0636	0.0988	0.1348	0.1691	0.2003	0.2273	0.2496	13	
3	0.0004	0.0029	0.0085	0.0178	0.0307	0.0468	0.0653	0.0857	0.1070	12	
4	0.0000	0.0002	0.0008	0.0022	0.0049	0.0090	0.0148	0.0223	0.0317	11	
5	0.0000	0.0000	0.0001	0.0002	0.0006	0.0013	0.0024	0.0043	0.0069	10	
6	0.0000	0.0000	0.0000	0.0000	0.0000	0.0001	0.0003	0.0006	0.0011	9	
7	0.0000	0.0000	0.0000	0.0000	0.0000	0.0000	0.0000	0.0001	0.0001	8	
8	0.0000	0.0000	0.0000	0.0000	0.0000	0.0000	0.0000	0.0000	0.0000	7	
9	0.0000	0.0000	0.0000	0.0000	0.0000	0.0000	0.0000	0.0000	0.0000	6	
10	0.0000	0.0000	0.0000	0.0000	0.0000	0.0000	0.0000	0.0000	0.0000	5	
11	0.0000	0.0000	0.0000	0.0000	0.0000	0.0000	0.0000	0.0000	0.0000	4	
12	0.0000	0.0000	0.0000	0.0000	0.0000	0.0000	0.0000	0.0000	0.0000	3	
13	0.0000	0.0000	0.0000	0.0000	0.0000	0.0000	0.0000	0.0000	0.0000	2	
14	0.0000	0.0000	0.0000	0.0000	0.0000	0.0000	0.0000	0.0000	0.0000	1	
15	0.0000	0.0000	0.0000	0.0000	0.0000	0.0000	0.0000	0.0000	0.0000	0	15
20 0	0.8179	0.6676	0.5438	0.4420	0.3585	0.2901	0.2342	0.1887	0.1516	20	
1	0.1652	0.2725	0.3364	0.3683	0.3774	0.3703	0.3526	0.3282	0.3000	19	
2	0.0159	0.0528	0.0988	0.1458	0.1887	0.2246	0.2521	0.2711	0.2818	18	
3	0.0010	0.0065	0.0183	0.0364	0.0596	0.0860	0.1139	0.1414	0.1672	17	
4	0.0000	0.0006	0.0024	0.0065	0.0133	0.0233	0.0364	0.0523	0.0703	16	
5	0.0000	0.0000	0.0002	0.0009	0.0022	0.0048	0.0088	0.0145	0.0222	15	
6	0.0000	0.0000	0.0000	0.0001	0.0003	0.0008	0.0017	0.0032	0.0055	14	
7	0.0000	0.0000	0.0000	0.0000	0.0000	0.0001	0.0002	0.0005	0.0011	13	
8	0.0000	0.0000	0.0000	0.0000	0.0000	0.0000	0.0000	0.0001	0.0002	12	
9	0.0000	0.0000	0.0000	0.0000	0.0000	0.0000	0.0000	0.0000	0.0000	11	
10	0.0000	0.0000	0.0000	0.0000	0.0000	0.0000	0.0000	0.0000	0.0000	10	
11	0.0000	0.0000	0.0000	0.0000	0.0000	0.0000	0.0000	0.0000	0.0000	9	
12	0.0000	0.0000	0.0000	0.0000	0.0000	0.0000	0.0000	0.0000	0.0000	8	
13	0.0000	0.0000	0.0000	0.0000	0.0000	0.0000	0.0000	0.0000	0.0000	7	
14	0.0000	0.0000	0.0000	0.0000	0.0000	0.0000	0.0000	0.0000	0.0000	6	
15	0.0000	0.0000	0.0000	0.0000	0.0000	0.0000	0.0000	0.0000	0.0000	5	
16	0.0000	0.0000	0.0000	0.0000	0.0000	0.0000	0.0000	0.0000	0.0000	4	
17	0.0000	0.0000	0.0000	0.0000	0.0000	0.0000	0.0000	0.0000	0.0000	3	
18	0.0000	0.0000	0.0000	0.0000	0.0000	0.0000	0.0000	0.0000	0.0000	2	
19	0.0000	0.0000	0.0000	0.0000	0.0000	0.0000	0.0000	0.0000	0.0000	1	
20	0.0000	0.0000	0.0000	0.0000	0.0000	0.0000	0.0000	0.0000	0.0000	0	20
	.99	.98	.97	.96	.95	.94	.93	.92	.91	x n	
						p					

(continued)

TABLE A-1 (continued)

						p						
n x		.10	.15	.20	.25	.30	.35	.40	.45	.50		
10	0	0.3487	0.1969	0.1074	0.0563	0.0282	0.0135	0.0060	0.0025	0.0010	10	
	1	0.3874	0.3474	0.2684	0.1877	0.1211	0.0725	0.0403	0.0207	0.0098	9	
	2	0.1937	0.2759	0.3020	0.2816	0.2335	0.1757	0.1209	0.0763	0.0439	8	
	3	0.0574	0.1298	0.2013	0.2503	0.2668	0.2522	0.2150	0.1665	0.1172	7	
	4	0.0112	0.0401	0.0881	0.1460	0.2001	0.2377	0.2508	0.2384	0.2051	6	
	5	0.0015	0.0085	0.0264	0.0584	0.1029	0.1536	0.2007	0.2340	0.2461	5	
	6	0.0001	0.0012	0.0055	0.0162	0.0368	0.0689	0.1115	0.1596	0.2051	4	
	7	0.0000	0.0001	0.0008	0.0031	0.0090	0.0212	0.0425	0.0746	0.1172	3	
	8	0.0000	0.0000	0.0001	0.0004	0.0014	0.0043	0.0106	0.0229	0.0439	2	
	9	0.0000	0.0000	0.0000	0.0000	0.0001	0.0005	0.0016	0.0042	0.0098	1	
	10	0.0000	0.0000	0.0000	0.0000	0.0000	0.0000	0.0001	0.0003	0.0010	0	10
12	0	0.2824	0.1422	0.0687	0.0317	0.0138	0.0057	0.0022	0.0008	0.0002	12	
	1	0.3766	0.3012	0.2062	0.1267	0.0712	0.0368	0.0174	0.0075	0.0029	11	
	2	0.2301	0.2924	0.2835	0.2323	0.1678	0.1088	0.0639	0.0339	0.0161	10	
	3	0.0852	0.1720	0.2362	0.2581	0.2397	0.1954	0.1419	0.0923	0.0537	9	
	4	0.0213	0.0683	0.1329	0.1936	0.2311	0.2367	0.2128	0.1700	0.1208	8	
	5	0.0038	0.0193	0.0532	0.1032	0.1585	0.2039	0.2270	0.2225	0.1934	7	
	6	0.0005	0.0040	0.0155	0.0401	0.0792	0.1281	0.1766	0.2124	0.2256	6	
	7	0.0000	0.0006	0.0033	0.0115	0.0291	0.0591	0.1009	0.1489	0.1934	5	
	8	0.0000	0.0001	0.0005	0.0024	0.0078	0.0199	0.0420	0.0762	0.1208	4	
	9	0.0000	0.0000	0.0001	0.0004	0.0015	0.0048	0.0125	0.0277	0.0537	3	
	10	0.0000	0.0000	0.0000	0.0000	0.0002	0.0008	0.0025	0.0068	0.0161	2	
	11	0.0000	0.0000	0.0000	0.0000	0.0000	0.0001	0.0003	0.0010	0.0029	1	
	12	0.0000	0.0000	0.0000	0.0000	0.0000	0.0000	0.0000	0.0001	0.0002	0	12
15	0	0.2059	0.0874	0.0352	0.0134	0.0047	0.0016	0.0005	0.0001	0.0000	15	
	1	0.3432	0.2312	0.1319	0.0668	0.0305	0.0126	0.0047	0.0016	0.0005	14	
	2	0.2669	0.2856	0.2309	0.1559	0.0916	0.0476	0.0219	0.0090	0.0032	13	
	3	0.1285	0.2184	0.2501	0.2252	0.1700	0.1110	0.0634	0.0318	0.0139	12	
	4	0.0428	0.1156	0.1876	0.2252	0.2186	0.1792	0.1268	0.0780	0.0417	11	
	5	0.0105	0.0449	0.1032	0.1651	0.2061	0.2123	0.1859	0.1404	0.0916	10	
	6	0.0019	0.0132	0.0430	0.0917	0.1472	0.1906	0.2066	0.1914	0.1527	9	
	7	0.0003	0.0030	0.0138	0.0393	0.0811	0.1319	0.1771	0.2013	0.1964	8	
	8	0.0000	0.0005	0.0035	0.0131	0.0348	0.0710	0.1181	0.1647	0.1964	7	
	9	0.0000	0.0001	0.0007	0.0034	0.0116	0.0298	0.0612	0.1048	0.1527	6	
	10	0.0000	0.0000	0.0001	0.0007	0.0030	0.0096	0.0245	0.0515	0.0916	5	
	11	0.0000	0.0000	0.0000	0.0001	0.0006	0.0024	0.0074	0.0191	0.0417	4	
	12	0.0000	0.0000	0.0000	0.0000	0.0001	0.0004	0.0016	0.0052	0.0139	3	
	13	0.0000	0.0000	0.0000	0.0000	0.0000	0.0001	0.0003	0.0010	0.0032	2	
	14	0.0000	0.0000	0.0000	0.0000	0.0000	0.0000	0.0000	0.0001	0.0005	1	
	15	0.0000	0.0000	0.0000	0.0000	0.0000	0.0000	0.0000	0.0000	0.0000	0	15
20	0	0.1216	0.0388	0.0115	0.0032	0.0008	0.0002	0.0000	0.0000	0.0000	20	
	1	0.2702	0.1368	0.0576	0.0211	0.0068	0.0020	0.0005	0.0001	0.0000	19	
	2	0.2852	0.2293	0.1369	0.0669	0.0278	0.0100	0.0031	0.0008	0.0002	18	
	3	0.1901	0.2428	0.2054	0.1339	0.0716	0.0323	0.0123	0.0040	0.0011	17	
	4	0.0898	0.1821	0.2182	0.1897	0.1304	0.0738	0.0350	0.0139	0.0046	16	
	5	0.0319	0.1028	0.1746	0.2023	0.1789	0.1272	0.0746	0.0365	0.0148	15	
	6	0.0089	0.0454	0.1091	0.1686	0.1916	0.1712	0.1244	0.0746	0.0370	14	
	7	0.0020	0.0160	0.0545	0.1124	0.1643	0.1844	0.1659	0.1221	0.0739	13	
	8	0.0004	0.0046	0.0222	0.0609	0.1144	0.1614	0.1797	0.1623	0.1201	12	
	9	0.0001	0.0011	0.0074	0.0271	0.0654	0.1158	0.1597	0.1771	0.1602	11	
	10	0.0000	0.0002	0.0020	0.0099	0.0308	0.0686	0.1171	0.1593	0.1762	10	
	11	0.0000	0.0000	0.0005	0.0030	0.0120	0.0336	0.0710	0.1185	0.1602	9	
	12	0.0000	0.0000	0.0001	0.0008	0.0039	0.0136	0.0355	0.0727	0.1201	8	
	13	0.0000	0.0000	0.0000	0.0002	0.0010	0.0045	0.0146	0.0366	0.0739	7	
	14	0.0000	0.0000	0.0000	0.0000	0.0002	0.0012	0.0049	0.0150	0.0370	6	
	15	0.0000	0.0000	0.0000	0.0000	0.0000	0.0003	0.0013	0.0049	0.0148	5	
	16	0.0000	0.0000	0.0000	0.0000	0.0000	0.0000	0.0003	0.0013	0.0046	4	
	17	0.0000	0.0000	0.0000	0.0000	0.0000	0.0000	0.0000	0.0002	0.0011	3	
	18	0.0000	0.0000	0.0000	0.0000	0.0000	0.0000	0.0000	0.0000	0.0002	2	
	19	0.0000	0.0000	0.0000	0.0000	0.0000	0.0000	0.0000	0.0000	0.0000	1	
	20	0.0000	0.0000	0.0000	0.0000	0.0000	0.0000	0.0000	0.0000	0.0000	0	20
		.90	.85	.80	.75	.70	.65	.60	.55	.50	x n	
						p						

TABLE A-2
Poisson Distribution

Poisson probabilities

λ

X	0.1	0.2	0.3	0.4	0.5	0.6	0.7	0.8	0.9	1.0
0	.9048	.8187	.7408	.6703	.6065	.5488	.4966	.4493	.4066	.3679
1	.0905	.1637	.2222	.2681	.3033	.3293	.3476	.3595	.3659	.3679
2	.0045	.0164	.0333	.0536	.0758	.0988	.1217	.1438	.1647	.1839
3	.0002	.0011	.0033	.0072	.0126	.0198	.0284	.0383	.0494	.0613
4		.0001	.0002	.0007	.0016	.0030	.0050	.0077	.0111	.0153
5				.0001	.0002	.0004	.0007	.0012	.0020	.0031
6							.0001	.0002	.0003	.0005
7										.0001

λ

X	1.5	2.0	2.5	3.0	3.5	4.0	4.5	5.0	6.0	7.0
0	.2231	.1353	.0821	.0498	.0302	.0183	.0111	.0067	.0025	.0009
1	.3347	.2707	.2052	.1494	.1057	.0733	.0500	.0337	.0149	.0064
2	.2510	.2707	.2565	.2240	.1850	.1465	.1125	.0842	.0446	.0223
3	.1255	.1804	.2138	.2240	.2158	.1954	.1687	.1404	.0892	.0521
4	.0471	.0902	.1336	.1680	.1888	.1954	.1898	.1755	.1339	.0912
5	.0141	.0361	.0668	.1008	.1322	.1563	.1708	.1755	.1606	.1277
6	.0035	.0120	.0278	.0504	.0771	.1042	.1281	.1462	.1606	.1490
7	.0008	.0034	.0099	.0216	.0385	.0595	.0824	.1044	.1377	.1490
8	.0001	.0009	.0031	.0081	.0169	.0298	.0463	.0653	.1033	.1304
9		.0002	.0009	.0027	.0066	.0132	.0232	.0363	.0688	.1014
10			.0002	.0008	.0023	.0053	.0104	.0181	.0413	.0710
11				.0002	.0007	.0019	.0043	.0082	.0225	.0452
12				.0001	.0002	.0006	.0016	.0034	.0113	.0264
13					.0001	.0002	.0006	.0013	.0052	.0142
14						.0001	.0002	.0005	.0022	.0071
15							.0001	.0002	.0009	.0033
16									.0003	.0014
17									.0001	.0006
18										.0002
19										.0001

Table entries are $P(X = x/\lambda)$.

TABLE A-3
Standard Normal Distribution

Standard normal distribution areas

Mean   z

z	.00	.01	.02	.03	.04	.05	.06	.07	.08	.09
0.0	.0000	.0040	.0080	.0120	.0160	.0199	.0239	.0279	.0319	.0359
0.1	.0398	.0438	.0478	.0517	.0557	.0596	.0636	.0675	.0714	.0753
0.2	.0793	.0832	.0871	.0910	.0948	.0987	.1026	.1064	.1103	.1141
0.3	.1179	.1217	.1255	.1293	.1331	.1368	.1406	.1443	.1480	.1517
0.4	.1554	.1591	.1628	.1664	.1700	.1736	.1772	.1808	.1844	.1879
0.5	.1915	.1950	.1985	.2019	.2054	.2088	.2123	.2157	.2190	.2224
0.6	.2257	.2291	.2324	.2357	.2389	.2422	.2454	.2486	.2518	.2549
0.7	.2580	.2612	.2642	.2673	.2704	.2734	.2764	.2794	.2823	.2852
0.8	.2881	.2910	.2939	.2967	.2995	.3023	.3051	.3078	.3106	.3133
0.9	.3159	.3186	.3212	.3238	.3264	.3289	.3315	.3340	.3365	.3389
1.0	.3413	.3438	.3461	.3485	.3508	.3531	.3554	.3577	.3599	.3621
1.1	.3643	.3665	.3686	.3708	.3729	.3749	.3770	.3790	.3810	.3830
1.2	.3849	.3869	.3888	.3907	.3925	.3944	.3962	.3980	.3997	.4015
1.3	.4032	.4049	.4066	.4082	.4099	.4115	.4131	.4147	.4162	.4177
1.4	.4192	.4207	.4222	.4236	.4251	.4265	.4279	.4292	.4306	.4319
1.5	.4332	.4345	.4357	.4370	.4382	.4394	.4406	.4418	.4429	.4441
1.6	.4452	.4463	.4474	.4484	.4495	.4505	.4515	.4525	.4535	.4545
1.7	.4554	.4564	.4573	.4582	.4591	.4599	.4608	.4616	.4625	.4633
1.8	.4641	.4649	.4656	.4664	.4671	.4678	.4686	.4693	.4699	.4706
1.9	.4713	.4719	.4726	.4732	.4738	.4744	.4750	.4756	.4761	.4767
2.0	.4772	.4778	.4783	.4788	.4793	.4798	.4803	.4808	.4812	.4817
2.1	.4821	.4826	.4830	.4834	.4838	.4842	.4846	.4850	.4854	.4857
2.2	.4861	.4864	.4868	.4871	.4875	.4878	.4881	.4884	.4887	.4890
2.3	.4893	.4896	.4898	.4901	.4904	.4906	.4909	.4911	.4913	.4916
2.4	.4918	.4920	.4922	.4925	.4927	.4929	.4931	.4932	.4934	.4936
2.5	.4938	.4940	.4941	.4943	.4945	.4946	.4948	.4949	.4951	.4952
2.6	.4953	.4955	.4956	.4957	.4959	.4960	.4961	.4962	.4963	.4964
2.7	.4965	.4966	.4967	.4968	.4969	.4970	.4971	.4972	.4973	.4974
2.8	.4974	.4975	.4976	.4977	.4977	.4978	.4979	.4979	.4980	.4981
2.9	.4981	.4982	.4982	.4983	.4984	.4984	.4985	.4985	.4986	.4986
3.0	.49865	.4987	.4987	.4988	.4988	.4989	.4989	.4989	.4990	.4990
4.0	.4999683									

TABLE A-4
t Distribution

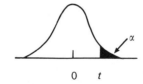

α df	.10	.05	.025	.01	.005
1	3.078	6.314	12.706	31.821	63.657
2	1.886	2.920	4.303	6.965	9.925
3	1.638	2.353	3.182	4.541	5.841
4	1.533	2.132	2.776	3.747	4.604
5	1.476	2.015	2.571	3.365	4.032
6	1.440	1.943	2.447	3.143	3.707
7	1.415	1.895	2.365	2.998	3.499
8	1.397	1.860	2.306	2.896	3.355
9	1.383	1.833	2.262	2.821	3.250
10	1.372	1.812	2.228	2.764	3.169
11	1.363	1.796	2.201	2.718	3.106
12	1.356	1.782	2.179	2.681	3.055
13	1.350	1.771	2.160	2.650	3.012
14	1.345	1.761	2.145	2.624	2.977
15	1.341	1.753	2.131	2.602	2.947
16	1.337	1.746	2.120	2.583	2.921
17	1.333	1.740	2.110	2.567	2.898
18	1.330	1.734	2.101	2.552	2.878
19	1.328	1.729	2.093	2.539	2.861
20	1.325	1.725	2.086	2.528	2.845
21	1.323	1.721	2.080	2.518	2.831
22	1.321	1.717	2.074	2.508	2.819
23	1.319	1.714	2.069	2.500	2.807
24	1.318	1.711	2.064	2.492	2.797
25	1.316	1.708	2.060	2.485	2.787
26	1.315	1.706	2.056	2.479	2.779
27	1.314	1.703	2.052	2.473	2.771
28	1.313	1.701	2.048	2.467	2.763
29	1.311	1.699	2.045	2.462	2.756
30	1.310	1.697	2.042	2.457	2.750
40	1.303	1.684	2.021	2.423	2.704
60	1.296	1.671	2.000	2.390	2.660
120	1.289	1.658	1.980	2.358	2.617
∞	1.282	1.645	1.960	2.326	2.576

TABLE A-5
χ^2 Distribution

Entry is $\chi^2(a; v)$ where $P[\chi^2(v) \le \chi^2(a; v)] = a$

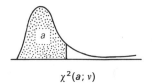

$\chi^2(a; v)$

df v	.005	.010	.025	.050	.100	.900	.950	.975	.990	.995
1	0.0^4393	0.0^3157	0.0^3982	0.0^2393	0.0158	2.71	3.84	5.02	6.63	7.88
2	0.0100	0.0201	0.0506	0.103	0.211	4.61	5.99	7.38	9.21	10.60
3	0.072	0.115	0.216	0.352	0.584	6.25	7.81	9.35	11.34	12.84
4	0.207	0.297	0.484	0.711	1.064	7.78	9.49	11.14	13.28	14.86
5	0.412	0.554	0.831	1.145	1.61	9.24	11.07	12.83	15.09	16.75
6	0.676	0.872	1.24	1.64	2.20	10.64	12.59	14.45	16.81	18.55
7	0.989	1.24	1.69	2.17	2.83	12.02	14.07	16.01	18.48	20.28
8	1.34	1.65	2.18	2.73	3.49	13.36	15.51	17.53	20.09	21.96
9	1.73	2.09	2.70	3.33	4.17	14.68	16.92	19.02	21.67	23.59
10	2.16	2.56	3.25	3.94	4.87	15.99	18.31	20.84	23.21	25.19
11	2.60	3.05	3.82	4.57	5.58	17.28	19.68	21.92	24.73	26.76
12	3.07	3.57	4.40	5.23	6.30	18.55	21.03	23.34	26.22	28.30
13	3.57	4.11	5.01	5.89	7.04	19.81	22.36	24.74	27.69	29.82
14	4.07	4.66	5.63	6.57	7.79	21.06	23.68	26.12	29.14	31.32
15	4.60	5.23	6.26	7.26	8.55	22.31	25.00	27.49	30.58	32.80
16	5.14	5.81	6.91	7.96	9.31	23.54	26.30	28.85	32.00	34.27
17	5.70	6.41	7.56	8.67	10.09	24.77	27.59	30.19	33.41	35.72
18	6.26	7.01	8.23	9.39	10.86	25.99	28.87	31.53	34.81	37.16
19	6.84	7.63	8.91	10.12	11.65	27.20	30.14	32.85	36.19	38.58
20	7.43	8.26	9.59	10.85	12.44	28.41	31.41	34.17	37.57	40.00
21	8.03	8.90	10.28	11.59	13.24	29.62	32.67	35.48	38.93	41.40
22	8.64	9.54	10.98	12.34	14.04	30.81	33.92	36.78	40.29	42.80
23	9.26	10.20	11.69	13.09	14.85	32.01	35.17	38.08	41.64	44.18
24	9.89	10.86	12.40	13.85	15.66	33.20	36.42	39.36	42.98	45.56
25	10.52	11.52	13.12	14.61	16.47	34.38	37.65	40.65	44.31	46.93
26	11.16	12.20	13.84	15.38	17.29	35.56	38.89	41.92	45.64	48.29
27	11.81	12.88	14.57	16.15	18.11	36.74	40.11	43.19	46.96	49.64
28	12.46	13.56	15.31	16.93	18.94	37.92	41.34	44.46	48.28	50.99
29	13.12	14.26	16.05	17.71	19.77	39.09	42.56	45.72	49.59	52.34
30	13.79	14.95	16.79	18.49	20.60	40.26	43.77	46.98	50.89	53.67
40	20.71	22.16	24.43	26.51	29.05	51.81	55.76	59.34	63.69	66.77
50	27.99	29.71	32.36	34.76	37.69	63.17	67.50	71.42	76.15	79.49
60	35.53	37.48	40.48	43.19	46.46	74.40	79.08	83.30	88.38	91.95
70	43.28	45.44	48.76	51.74	55.33	85.53	90.53	95.02	100.4	104.2
80	51.17	53.54	57.15	60.39	64.28	96.58	101.9	106.6	112.3	116.3
90	59.20	61.75	65.65	69.13	73.29	107.6	113.1	118.1	124.1	128.3
100	67.33	70.06	74.22	77.93	82.36	118.5	124.3	129.6	135.8	140.2

TABLE A-6
F Distribution

Percentiles of the *F* distribution: Entry is $F(1-\alpha; r_1, r_2)$ where $P[F \le F(1-\alpha; r_1, r_2)] = 1-\alpha$.

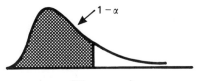

$$F(1-\alpha; r_1, r_2)$$

						r_1				
r_2	$1-\alpha$	1	2	3	4	5	6	7	8	9
1	.50	1.00	1.50	1.71	1.82	1.89	1.94	1.98	2.00	2.03
	.90	39.9	49.5	53.6	55.8	57.2	58.2	58.9	59.4	59.9
	.95	161	200	216	225	230	234	237	239	241
	.975	648	800	864	900	922	937	948	957	963
	.99	4,052	5,000	5,403	5,625	5,764	5,859	5,928	5,981	6,022
	.995	16,211	20,000	21,615	22,500	23,056	23,437	23,715	23,925	24,091
	.999	405,280	500,000	540,380	562,500	567,400	585,940	592,870	598,140	602,280
2	.50	0.667	1.00	1.13	1.21	1.25	1.28	1.30	1.32	1.33
	.90	8.53	9.00	9.16	9.24	9.29	9.33	9.35	9.37	9.38
	.95	18.5	19.0	19.2	19.2	19.3	19.3	19.4	19.4	19.4
	.975	38.5	39.0	39.2	39.2	39.3	39.3	39.4	39.4	39.4
	.99	98.5	99.0	99.2	99.2	99.3	99.3	99.4	99.4	99.4
	.995	199	199	199	199	199	199	199	199	199
	.999	998.5	999.0	999.2	999.2	999.3	999.3	999.4	999.4	999.4
3	.50	0.585	0.881	1.00	1.06	1.10	1.13	1.15	1.16	1.17
	.90	5.54	5.46	5.39	5.34	5.31	5.28	5.27	5.25	5.24
	.95	10.1	9.55	9.28	9.12	9.01	8.94	8.89	8.85	8.81
	.975	17.4	16.0	15.4	15.1	14.9	14.7	14.6	14.5	14.5
	.99	34.1	30.8	29.5	28.7	28.2	27.9	27.7	27.5	27.3
	.995	55.6	49.8	47.5	46.2	45.4	44.8	44.4	44.1	43.9
	.999	167.0	148.5	141.1	137.1	134.6	132.8	131.6	130.6	129.9
4	.50	0.549	0.828	0.941	1.00	1.04	1.06	1.08	1.09	1.10
	.90	4.54	4.32	4.19	4.11	4.05	4.01	3.98	3.95	3.94
	.95	7.71	6.94	6.59	6.39	6.26	6.16	6.09	6.04	6.00
	.975	12.2	10.6	9.98	9.60	9.36	9.20	9.07	8.98	8.90
	.99	21.2	18.0	16.7	16.0	15.5	15.2	15.0	14.8	14.7
	.995	31.3	26.3	24.3	23.2	22.5	22.0	21.6	21.4	21.1
	.999	74.1	61.2	56.2	53.4	51.7	50.5	49.7	49.0	48.5
5	.50	0.528	0.799	0.907	0.965	1.00	1.02	1.04	1.05	1.06
	.90	4.06	3.78	3.62	3.52	3.45	3.40	3.37	3.34	3.32
	.95	6.61	5.79	5.41	5.19	5.05	4.95	4.88	4.82	4.77
	.975	10.0	8.43	7.76	7.39	7.15	6.98	6.85	6.76	6.68
	.99	16.3	13.3	12.1	11.4	11.0	10.7	10.5	10.3	10.2
	.995	22.8	18.3	16.5	15.6	14.9	14.5	14.2	14.0	13.8
	.999	47.2	37.1	33.2	31.1	29.8	28.8	28.2	27.6	27.2
6	.50	0.515	0.780	0.886	0.942	0.977	1.00	1.02	1.03	1.04
	.90	3.78	3.46	3.29	3.18	3.11	3.05	3.01	2.98	2.96
	.95	5.99	5.14	4.76	4.53	4.39	4.28	4.21	4.15	4.10
	.975	8.81	7.26	6.60	6.23	5.99	5.82	5.70	5.60	5.52
	.99	13.7	10.9	9.78	9.15	8.75	8.47	8.26	8.10	7.98
	.995	18.6	14.5	12.9	12.0	11.5	11.1	10.8	10.6	10.4
	.999	35.5	27.0	23.7	21.9	20.8	20.0	19.5	19.0	18.7
7	.50	0.506	0.767	0.871	0.926	0.960	0.983	1.00	1.01	1.02
	.90	3.59	3.26	3.07	2.96	2.88	2.83	2.78	2.75	2.72
	.95	5.59	4.74	4.35	4.12	3.97	3.87	3.79	3.73	3.68
	.975	8.07	6.54	5.89	5.52	5.29	5.12	4.99	4.90	4.82
	.99	12.2	9.55	8.45	7.85	7.46	7.19	6.99	6.84	6.72
	.995	16.2	12.4	10.9	10.1	9.52	9.16	8.89	8.68	8.51
	.999	29.2	21.7	18.8	17.2	16.2	15.5	15.0	14.6	14.3

(*continued*)

TABLE A-6 (continued)

r_2	$1-\alpha$	10	12	15	20	24	30	60	120	∞
1	.50	2.04	2.07	2.09	2.12	3.13	2.15	2.17	2.18	2.20
	.90	60.2	60.7	61.2	61.7	62.0	62.3	62.8	63.1	63.3
	.95	242	244	246	248	249	250	252	253	254
	.975	969	977	985	993	997	1,001	1,010	1,014	1,018
	.99	6,056	6,106	6,157	6,209	6,235	6,261	6,313	6,339	6,366
	.995	24,224	24,426	24,630	24,836	24,940	25,044	25,253	25,359	25,464
	.999	605,620	610,670	615,760	620,910	623,500	626,100	631,340	633,970	636,620
2	.50	1.34	1.36	1.38	1.39	1.40	1.41	1.43	1.43	1.44
	.90	9.39	9.41	9.42	9.44	9.45	9.46	9.47	9.48	9.49
	.95	19.4	19.4	19.4	19.4	19.5	19.5	19.5	19.5	19.5
	.975	39.4	39.4	39.4	39.4	39.5	39.5	39.5	39.5	39.5
	.99	99.4	99.4	99.4	99.4	99.5	99.5	99.5	99.5	99.5
	.995	199	199	199	199	199	199	199	199	200
	.999	999.4	999.4	999.4	999.4	999.5	999.5	999.5	999.5	999.5
3	.50	1.18	1.20	1.21	1.23	1.23	1.24	1.25	1.26	1.27
	.90	5.23	5.22	5.20	5.18	5.18	5.17	5.15	5.14	5.13
	.95	8.79	8.74	8.70	8.66	8.64	8.62	8.57	8.55	8.53
	.975	14.4	14.3	14.3	14.2	14.1	14.1	14.0	13.9	13.9
	.99	27.2	27.1	26.9	26.7	26.6	26.5	26.3	26.2	26.1
	.995	43.7	43.4	43.1	42.8	42.6	42.5	42.1	42.0	41.8
	.999	129.2	128.3	127.4	126.4	125.9	125.4	124.5	124.0	123.5
4	.50	1.11	1.13	1.14	1.15	1.16	1.16	1.18	1.18	1.19
	.90	3.92	3.90	3.87	3.84	3.83	3.82	3.79	3.78	3.76
	.95	5.96	5.91	5.86	5.80	5.77	5.75	5.69	5.66	5.63
	.975	8.84	8.75	8.66	8.56	8.51	8.46	8.36	8.31	8.26
	.99	14.5	14.4	14.2	14.0	13.9	13.8	13.7	13.6	13.5
	.995	21.0	20.7	20.4	20.2	20.0	19.9	19.6	19.5	19.3
	.999	48.1	47.4	46.8	46.1	45.8	45.4	44.7	44.4	44.1
5	.50	1.07	1.09	1.10	1.11	1.12	1.12	1.14	1.14	1.15
	.90	3.30	3.27	3.24	3.21	3.19	3.17	3.14	3.12	3.11
	.95	4.74	4.68	4.62	4.56	4.53	4.50	4.43	4.40	4.37
	.975	6.62	6.52	6.43	6.33	6.28	6.23	6.12	6.07	6.02
	.99	10.1	9.89	9.72	9.55	9.47	9.38	9.20	9.11	9.02
	.995	13.6	13.4	13.1	12.9	12.8	12.7	12.4	12.3	12.1
	.999	26.9	26.4	25.9	25.4	25.1	24.9	24.3	24.1	23.8
6	.50	1.05	1.06	1.07	1.08	1.09	1.10	1.11	1.12	1.12
	.90	2.94	2.90	2.87	2.84	2.82	2.80	2.76	2.74	2.72
	.95	4.06	4.00	3.94	3.87	3.84	3.81	3.74	3.70	3.67
	.975	5.46	5.37	5.27	5.17	5.12	5.07	4.96	4.90	4.85
	.99	7.87	7.72	7.56	7.40	7.31	7.23	7.06	6.97	6.88
	.995	10.2	10.0	9.81	9.59	9.47	9.36	9.12	9.00	8.88
	.999	18.4	18.0	17.6	17.1	16.9	16.7	16.2	16.0	15.7
7	.50	1.03	1.04	1.05	1.07	1.07	1.08	1.09	1.10	1.10
	.90	2.70	2.67	2.63	2.59	2.58	2.56	2.51	2.49	2.47
	.95	3.64	3.57	3.51	3.44	3.41	3.38	3.30	3.27	3.23
	.975	4.76	4.67	4.57	4.47	4.42	4.36	4.25	4.20	4.14
	.99	6.62	6.47	6.31	6.16	6.07	5.99	5.82	5.74	5.65
	.995	8.38	8.18	7.97	7.75	7.65	7.53	7.31	7.19	7.08
	.999	14.1	13.7	13.3	12.9	12.7	12.5	12.1	11.9	11.7

TABLE A-6 (continued)

r_2	$1-\alpha$	1	2	3	4	5	6	7	8	9
						r_1				
8	.50	0.499	0.757	0.860	0.915	0.948	0.971	0.988	1.00	1.01
	.90	3.46	3.11	2.92	2.81	2.73	2.67	2.62	2.59	2.56
	.95	5.32	4.46	4.07	3.84	3.69	3.58	3.50	3.44	3.39
	.975	7.57	6.06	5.42	5.05	4.82	4.65	4.53	4.43	4.36
	.99	11.3	8.65	7.59	7.01	6.63	6.37	6.18	6.03	5.91
	.995	14.7	11.0	9.60	8.81	8.30	7.95	7.69	7.50	7.34
	.999	25.4	18.5	15.8	14.4	13.5	12.9	12.4	12.0	11.8
9	.50	0.494	0.749	0.852	0.906	0.939	0.962	0.978	0.909	1.00
	.90	3.36	3.01	2.81	2.69	2.61	2.55	2.51	2.47	2.44
	.95	5.12	4.26	3.86	3.63	3.48	3.37	3.29	3.23	3.18
	.975	7.21	5.71	5.08	4.72	4.48	4.32	4.20	4.10	4.03
	.99	10.6	8.02	6.99	6.42	6.06	5.80	5.61	5.47	5.35
	.995	13.6	10.1	8.72	7.96	7.47	7.13	6.88	6.69	6.54
	.999	22.9	16.4	13.9	12.6	11.7	11.1	10.7	10.4	10.1
10	.50	0.490	0.743	0.845	0.899	0.932	0.954	0.971	0.983	0.992
	.90	3.29	2.92	2.73	2.61	2.52	2.46	2.41	2.38	2.35
	.95	4.96	4.10	3.71	3.48	3.33	3.22	3.14	3.07	3.02
	.975	6.94	5.46	4.83	4.47	4.24	4.07	3.95	3.85	3.78
	.99	10.0	7.56	6.55	5.99	5.64	5.39	5.20	5.06	4.94
	.995	12.8	9.43	8.08	7.34	6.87	6.54	6.30	6.12	5.97
	.999	21.0	14.9	12.6	11.3	10.5	9.93	9.52	9.20	8.96
12	.50	0.484	0.735	0.835	0.888	0.921	0.943	0.959	0.972	0.981
	.90	3.18	2.81	2.61	2.48	2.39	2.33	2.28	2.24	2.21
	.95	4.75	3.89	3.49	3.26	3.11	3.00	2.91	2.85	2.80
	.975	6.55	5.10	4.47	4.12	3.89	3.73	3.61	3.51	3.44
	.99	9.33	6.93	5.95	5.41	5.06	4.82	4.64	4.50	4.39
	.995	11.8	8.51	7.23	6.52	6.07	5.76	5.52	5.35	5.20
	.999	18.6	13.0	10.8	9.63	8.89	8.38	8.00	7.71	7.48
15	.50	0.478	0.726	0.826	0.878	0.911	0.933	0.949	0.960	0.970
	.90	3.07	2.70	2.49	2.36	2.27	2.21	2.16	2.12	2.09
	.95	4.54	3.68	3.29	3.06	2.90	2.79	2.71	2.64	2.59
	.975	6.20	4.77	4.15	3.80	3.58	3.41	3.29	3.20	3.12
	.99	8.68	6.36	5.42	4.89	4.56	4.32	4.14	4.00	3.89
	.995	10.8	7.70	6.48	5.80	5.37	5.07	4.85	4.67	4.54
	.999	16.6	11.3	9.34	8.25	7.57	7.09	6.74	6.47	6.26
20	.50	0.472	0.718	0.816	0.868	0.900	0.922	0.938	0.950	0.959
	.90	2.97	2.59	2.38	2.25	2.16	2.09	2.04	2.00	1.96
	.95	4.35	3.49	3.10	2.87	2.71	2.60	2.51	2.45	2.39
	.975	5.87	4.46	3.86	3.51	3.29	3.13	3.01	2.91	2.84
	.99	8.10	5.85	4.94	4.43	4.10	3.87	3.70	3.56	3.46
	.995	9.94	6.99	5.82	5.17	4.76	4.47	4.26	4.09	3.96
	.999	14.8	9.95	8.10	7.10	6.46	6.02	5.69	5.44	5.24
24	.50	0.469	0.714	0.812	0.863	0.895	0.917	0.932	0.944	0.953
	.90	2.93	2.54	2.33	2.19	2.10	2.04	1.98	1.94	1.91
	.95	4.26	3.40	3.01	2.78	2.62	2.51	2.42	2.36	2.30
	.975	5.72	4.32	3.72	3.38	3.15	2.99	2.87	2.78	2.70
	.99	7.82	5.61	4.72	4.22	3.90	3.67	3.50	3.36	3.26
	.995	9.55	6.66	5.52	4.89	4.49	4.20	3.99	3.83	3.69
	.999	14.0	9.34	7.55	6.59	5.98	5.55	5.23	4.99	4.80

(continued)

TABLE A-6 (continued)

		r_1								
r_2	$1-\alpha$	10	12	15	20	24	30	60	120	∞
8	.50	1.02	1.03	1.04	1.05	1.06	1.07	1.08	1.08	1.09
	.90	2.54	2.50	2.46	2.42	2.40	2.38	2.34	2.32	2.29
	.95	3.35	3.28	3.22	3.15	3.12	3.08	3.01	2.97	2.93
	.975	4.30	4.20	4.10	4.00	3.95	3.89	3.78	3.73	3.67
	.99	5.81	5.67	5.52	5.36	5.28	5.20	5.03	4.95	4.86
	.995	7.21	7.01	6.81	6.61	6.50	6.40	6.18	6.06	5.95
	.999	11.5	11.2	10.8	10.5	10.3	10.1	9.73	9.53	9.33
9	.50	1.01	1.02	1.03	1.04	1.05	1.05	1.07	1.07	1.08
	.90	2.42	2.38	2.34	2.30	2.28	2.25	2.21	2.18	2.16
	.95	3.14	3.07	3.01	2.94	2.90	2.86	2.79	2.75	2.71
	.975	3.96	3.87	3.77	3.67	3.61	3.56	3.45	3.39	3.33
	.99	5.26	5.11	4.96	4.81	4.73	4.65	4.48	4.40	4.31
	.995	6.42	6.23	6.03	5.83	5.73	5.62	5.41	5.30	5.19
	.999	9.89	9.57	9.24	8.90	8.72	8.55	8.19	8.00	7.81
10	.50	1.00	1.01	1.02	1.03	1.04	1.05	1.06	1.06	1.07
	.90	2.32	2.28	2.24	2.20	2.18	2.16	2.11	2.08	2.06
	.95	2.98	2.91	2.84	2.77	2.74	2.70	2.62	2.58	2.54
	.975	3.72	3.62	3.52	3.42	3.37	3.31	3.20	3.14	3.08
	.99	4.85	4.71	4.56	4.41	4.33	4.25	4.08	4.00	3.91
	.995	5.85	5.66	5.47	5.27	5.17	5.07	4.86	4.75	4.64
	.999	8.75	8.45	8.13	7.80	7.64	7.47	7.12	6.94	6.76
12	.50	0.989	1.00	1.01	1.02	1.03	1.03	1.05	1.05	1.06
	.90	2.19	2.15	2.10	2.06	2.04	2.01	1.96	1.93	1.90
	.95	2.75	2.69	2.62	2.54	2.51	2.47	2.38	2.34	2.30
	.975	3.37	3.28	3.18	3.07	3.02	2.96	2.85	2.79	2.72
	.99	4.30	4.16	4.01	3.86	3.78	3.70	3.54	3.45	3.36
	.995	5.09	4.91	4.72	4.53	4.43	4.33	4.12	4.01	3.90
	.999	7.29	7.00	6.71	6.40	6.25	6.09	5.76	5.59	5.42
15	.50	0.977	0.989	1.00	1.01	1.02	1.02	1.03	1.04	1.05
	.90	2.06	2.02	1.97	1.92	1.90	1.87	1.82	1.79	1.76
	.95	2.54	2.48	2.40	2.33	2.29	2.25	2.16	2.11	2.07
	.975	3.06	2.96	2.86	2.76	2.70	2.64	2.52	2.46	2.40
	.99	3.80	3.67	3.52	3.37	3.29	3.21	3.05	2.96	2.87
	.995	4.42	4.25	4.07	3.88	3.79	3.69	3.48	3.37	3.26
	.999	6.08	5.81	5.54	5.25	5.10	4.95	4.64	4.48	4.31
20	.50	0.966	0.977	0.989	1.00	1.01	1.01	1.02	1.03	1.03
	.90	1.94	1.89	1.84	1.79	1.77	1.74	1.68	1.64	1.61
	.95	2.35	2.28	2.20	2.12	2.08	2.04	1.95	1.90	1.84
	.975	2.77	2.68	2.57	2.46	2.41	2.35	2.22	2.16	2.09
	.99	3.37	3.23	3.09	2.94	2.86	2.78	2.61	2.52	2.42
	.995	3.85	3.68	3.50	3.32	3.22	3.12	2.92	2.81	2.69
	.999	5.08	4.82	4.56	4.29	4.15	4.00	3.70	3.54	3.38
24	.50	0.961	0.972	0.983	0.994	1.00	1.01	1.02	1.02	1.03
	.90	1.88	1.83	1.78	1.73	1.70	1.67	1.61	1.57	1.53
	.95	2.25	2.18	2.11	2.03	1.98	1.94	1.84	1.79	1.73
	.975	2.64	2.54	2.44	2.33	2.27	2.21	2.08	2.01	1.94
	.99	3.17	3.03	2.89	2.74	2.66	2.58	2.40	2.31	2.21
	.995	3.59	3.42	3.25	3.06	2.97	2.87	2.66	2.55	2.43
	.999	4.64	4.39	4.14	3.87	3.74	3.59	3.29	3.14	2.97

TABLE A-6 (continued)

r_2	$1-\alpha$		1	2	3	4	5	6	7	8	9
							r_1				
30	.50		0.466	0.709	0.807	0.858	0.890	0.912	0.927	0.939	0.948
	.90		2.88	2.49	2.28	2.14	2.05	1.98	1.93	1.88	1.85
	.95		4.17	3.32	2.92	2.69	2.53	2.42	2.33	2.27	2.21
	.975		5.57	4.18	3.59	3.25	3.03	2.87	2.75	2.65	2.57
	.99		7.56	5.39	4.51	4.02	3.70	3.47	3.30	3.17	3.07
	.995		9.18	6.35	5.24	4.62	4.23	3.95	3.74	3.58	3.45
	.999		13.3	8.77	7.05	6.12	5.53	5.12	4.82	4.58	4.39
60	.50		0.461	0.701	0.798	0.849	0.880	0.901	0.917	0.928	0.937
	.90		2.79	2.39	2.18	2.04	1.95	1.87	1.82	1.77	1.74
	.95		4.00	3.15	2.76	2.53	2.37	2.25	2.17	2.10	2.04
	.975		5.29	3.93	3.34	3.01	2.79	2.63	2.51	2.41	2.33
	.99		7.08	4.98	4.13	3.65	3.34	3.12	2.95	2.82	2.72
	.995		8.49	5.80	4.73	4.14	3.76	3.49	3.29	3.13	3.01
	.999		12.0	7.77	6.17	5.31	4.76	4.37	4.09	3.86	3.69
120	.50		0.458	0.697	0.793	0.844	0.875	0.896	0.912	0.923	0.932
	.90		2.75	2.35	2.13	1.99	1.90	1.82	1.77	1.72	1.68
	.95		3.92	3.07	2.68	2.45	2.29	2.18	2.09	2.02	1.96
	.975		5.15	3.80	3.23	2.89	2.67	2.52	2.39	2.30	2.22
	.99		6.85	4.79	3.95	3.48	3.17	2.96	2.79	2.66	2.56
	.995		8.18	5.54	4.50	3.92	3.55	3.28	3.09	2.93	2.81
	.999		11.4	7.32	5.78	4.95	4.42	4.04	3.77	3.55	3.38
∞	.50		0.455	0.693	0.789	0.839	0.870	0.891	0.907	0.918	0.927
	.90		2.71	2.30	2.08	1.94	1.85	1.77	1.72	1.67	1.63
	.95		3.84	3.00	2.60	2.37	2.21	2.10	2.01	1.94	1.88
	.975		5.02	3.69	3.12	2.79	2.57	2.41	2.29	2.19	2.11
	.99		6.63	4.61	3.78	3.32	3.02	2.80	2.64	2.51	2.41
	.995		7.88	5.30	4.28	3.72	3.35	3.09	2.90	2.74	2.62
	.999		10.8	6.91	5.42	4.62	4.10	3.74	3.47	3.27	3.10

TABLE A-7
Critical Values of Kolmogorov-Smirnov Test Statistic D

		α						α			
n	.20	.10	.05	.02	.01	n	.20	.10	.05	.02	.01
1	.900	.950	.975	.990	.995	21	.226	.259	.287	.321	.344
2	.684	.776	.842	.900	.929	22	.221	.253	.281	.314	.337
3	.565	.636	.708	.785	.829	23	.216	.247	.275	.307	.330
4	.493	.565	.624	.689	.734	24	.212	.242	.269	.301	.323
5	.447	.509	.563	.627	.669	25	.208	.238	.264	.295	.317
6	.410	.468	.519	.577	.617	26	.204	.233	.259	.290	.311
7	.381	.436	.483	.538	.576	27	.200	.229	.254	.284	.305
8	.358	.410	.454	.507	.542	28	.197	.225	.250	.279	.300
9	.339	.387	.430	.480	.513	29	.193	.221	.246	.275	.295
10	.323	.369	.409	.457	.489	30	.190	.218	.242	.270	.290
11	.308	.352	.391	.437	.468	31	.187	.214	.238	.266	.285
12	.296	.338	.375	.419	.449	32	.184	.211	.234	.262	.281
13	.285	.325	.361	.404	.432	33	.182	.208	.231	.258	.277
14	.275	.314	.349	.390	.418	34	.179	.205	.227	.254	.273
15	.266	.304	.338	.377	.404	35	.177	.202	.224	.251	.269
16	.258	.295	.327	.366	.392	36	.174	.199	.221	.247	.265
17	.250	.286	.318	.355	.381	37	.172	.196	.218	.244	.262
18	.244	.279	.309	.346	.371	38	.170	.194	.215	.241	.258
19	.237	.271	.301	.337	.361	39	.168	.191	.213	.238	.255
20	.232	.265	.294	.329	.352	40	.165	.189	.210	.235	.252
				Approximation for $n > 40$		$\dfrac{1.07}{\sqrt{n}}$	$\dfrac{1.22}{\sqrt{n}}$	$\dfrac{1.36}{\sqrt{n}}$	$\dfrac{1.52}{\sqrt{n}}$	$\dfrac{1.63}{\sqrt{n}}$

Index